Physical Principles of
Astronomical Instrumentation

Series in Astronomy and Astrophysics

The *Series in Astronomy and Astrophysics* includes books on all aspects of theoretical and experimental astronomy and astrophysics. Books in the series range in level from textbooks and handbooks to more advanced expositions of current research.

Series Editors:
M Birkinshaw, *University of Bristol, UK*
J Silk, *University of Oxford, UK*
G Fuller, *University of Manchester, UK*

RECENT BOOKS IN THE SERIES

The Physics of Interstellar Dust
E Krügel

Very High Energy Gamma-Ray Astronomy
T C Weekes

Numerical Methods in Astrophysics: An Introduction
P Bodenheimer, G P Laughlin, M Różyczka, and H W Yorke

An Introduction to the Physics of Interstellar Dust
Endrik Krugel

Astrobiology: An Introduction
Alan Longstaff

Fundamentals of Radio Astronomy: Observational Methods
Jonathan M Marr, Ronald L Snell, and Stanley E Kurtz

Stellar Explosions: Hydrodynamics and Nucleosynthesis
Jordi José

Cosmology for Physicists
David Lyth

Cosmology
Nicola Vittorio

Cosmology and the Early Universe
Pasquale Di Bari

Fundamentals of Radio Astronomy: Astrophysics
Ronald L. Snell, Stanley E. Kurtz, and Jonathan M. Marr

Introduction to Cosmic Inflation and Dark Energy
Konstantinos Dimopoulos

Physical Principles of Astronomical Instrumentation
Peter A. R. Ade, Matthew J. Griffin, and Carole E. Tucker

Physical Principles of Astronomical Instrumentation

Peter A. R. Ade
Matthew J. Griffin
Carole E. Tucker

CRC Press

Taylor & Francis Group
Boca Raton London New York

CRC Press is an imprint of the
Taylor & Francis Group, an **Informa** business

First edition published 2022
by CRC Press
6000 Broken Sound Parkway NW, Suite 300, Boca Raton, FL 33487-2742

and by CRC Press
2 Park Square, Milton Park, Abingdon, Oxon, OX14 4RN

Library of Congress Cataloging-in-Publication Data
Names: Ade, Peter (Peter A. R.),
author. | Griffin, M. J. (Matt J.), author. | Tucker, Carole, author.
Title: Physical principles of astronomical instrumentation / Peter Ade, Matthew
Griffin, Carole Tucker.
Description: First edition. | Boca Raton : CRC Press, 2021. | Series:
Series in astronomy and astrophysics | Includes bibliographical
references and index.
Identifiers: LCCN 2021009251 (print) | LCCN 2021009252 (ebook) | ISBN
9781439871898 (hardback) | ISBN 9781032040035 (paperback) | ISBN
9781315374659 (ebook)
Subjects: LCSH: Astronomical instruments. | Astrophysics.
Classification: LCC QB86 .G75 2021 (print) | LCC QB86 (ebook) | DDC
522/.2—dc23
LC record available at https://lccn.loc.gov/2021009251
LC ebook record available at https://lccn.loc.gov/2021009252

ISBN: 9781439871898 (hbk)
ISBN: 9781032040035 (pbk)
ISBN: 9781315374659 (ebk)

DOI: 10.1201/9781315374659

Typeset in Times
by codeMantra

Contents

Preface

This book originated from a series of lectures given to undergraduate and MSc students at Queen Mary University of London and Cardiff University on experimental and observational astronomy and on the detection of electromagnetic radiation. Over the years, it became clear that for the target audience there were two themes to these lectures: understanding the physical principles behind the detection and measurement of radiation, and the continuing evolution and improvement of instrumentation to measure the characteristics of astronomical sources. The former topic has a wider relevance as it covers the fundamentals of electromagnetic radiation types and characteristics and is of interest to most physics-based students. The latter is of more specific interest to students of observational or experimental astronomy.

Astronomical technology continues to evolve at a fast pace. Today's research facilities and instruments are sophisticated and complex, usually built by large consortia of institutes and experts, and indeed not comprehensively understood by any single individual. Nevertheless, there remain underlying key principles that motivate and dictate the essential features of an observing system, and that is what we have sought to focus on. Contemporary astronomy also requires access to the whole spectrum, provided by a combination of Earth-based and space-borne observatories. Whether or not they become involved in instrumentation, today's observational astronomers usually deal with multi-wavelength observations and data from across the spectrum and obtained from many different observatories, and it is always an advantage to have a good conceptual understanding of the data and how these were acquired.

This book is designed to provide an overview, encompassing the whole of the electromagnetic spectrum, of the relevant physical principles and how they are applied. It is structured to cover, in the first part, the underlying physics of radiation and detection, and in the second part to provide a run-through of the electromagnetic spectrum outlining the application of those principles to astronomical instruments and measurements. Brief case studies are presented in the later chapters to illustrate the application of the key principles in recent and current instruments. While no individual aspect is treated in great depth, it may serve to provide a broad understanding of the physical principles, techniques, and scientific trade-offs involved in instrument design, and also as a jumping-off point enabling deeper investigation of any particular spectral region or technology.

It is not only suitable for undergraduates but should also be useful to MSc and PhD students, and to observational astronomy students who seek to attain an understanding of how their data are obtained. Undergraduate physics, engineering, or medical physics students may also find the material presented here to be a valuable supplement to any course on experimental techniques, as would MSc or PhD students in experimental or environmental science who need to understand how to interpret data from laboratory or remote-sensing instrumentation.

Inevitably, considering the range of wavelengths and physics involved, and the technological complexity of advanced instrumentation, much has been left out or only covered briefly, and, except for brief descriptions of neutrino and cosmic ray detection in the final chapter, we have concentrated on electromagnetic astronomy.

There is also some material that may be in common with other courses undertaken by physics or astrophysics students, but it is treated here in the context of astronomical detection and presented in one place in a way that we hope will prove useful and convenient.

It is a pleasure to acknowledge the extremely helpful comments on chapter drafts by Tom Brien, Peter Clegg, Dave Clements, Jim Emerson, Roger Emery, Will Grainger, Wayne Holland, Ian Robson, and especially Gary Davis who reviewed the whole manuscript.

Authors

Peter A. R. Ade received his PhD from Queen Mary College, London, in 1973, where he continued to build up a submillimetre wave instrumentation group specialising in producing state-of-the-art instruments for use in both atmospheric and astronomical research. In 2001 he relocated to Cardiff with other colleagues to form a larger instrumentation group. He is a member of the Royal Astronomical Society and is a chartered physicist with the Institute of Physics. He has over 40 years of experience in instrumentation design and manufacture whilst pursuing his observational astrophysics and atmospheric science interests. He has been involved with the development and deployment of many astronomical instruments including ISO-LWS, Cassini-CIRS, Mars-PMIRR, SPT-pol, ACT-pol, EBEX, Pilot, BLAST, SCUBA, SCUBA-2, Spitzer, *Herschel*-SPIRE, and *Planck*-HFI. In 1994, he was awarded a NASA Public Service Medal for his contributions to fundamental advances in far-infrared detector and sensor systems, which enabled critical measurements of atmospheric ozone chemistry. In 2009, he was presented with the Royal Astronomical Society Jackson-Gwilt Medal for contributions to astronomical instrumentation.

Matthew J. Griffin studied electrical engineering at University College Dublin and astrophysics at Queen Mary College London, receiving his PhD in 1985. His research work has included the development of instruments for both ground-based and space-borne observatories, and their use in the study of planetary atmospheres, star formation, and galaxy evolution. He remained at Queen Mary until 2001, and was involved in various ground-based submillimetre instruments and in ESA's Infrared Space Observatory. Since 2001, he has been with the Astronomy Instrumentation Group at Cardiff University. In addition to participating in the SCUBA, SCUBA-2, and *Planck*-HFI instruments, he was the principal investigator for the *Herschel*-SPIRE satellite instrument, for which he was awarded the Royal Astronomical Society Jackson-Gwilt Medal in 2011. He is a fellow of the Institute of Physics and of the Learned Society of Wales.

Carole E. Tucker studied physics and maths at Reading University, then medical radiation physics at Queen Mary and Westfield College, London, completing her PhD in 2001. Having undertaken a great deal of cleanroom device fabrication and spectroscopy work, she took her first postdoctoral position with the Astronomy Instrumentation Group at QMW, working on hardware provision and characterisation for the *Herschel* and *Planck* satellite missions. In 2001, she moved with the instrumentation group to Cardiff University, where she took up her first academic position in 2006. Prof. Tucker manages the quasi-optical filter production facility at Cardiff, which leads to involvement with a great number of international FIR space and ground-based instrument teams. In addition, she works with industry, supplying technology to scientific disciplines outside of astronomy. She is a member of the Institute of Physics, and a fellow of the Royal Astronomical Society and of the Learned Society of Wales.

1 Review of Electromagnetic Radiation

1.1 INTRODUCTION

In this chapter, we review the basic properties of electromagnetic (EM) radiation that are pivotal to our understanding of the radiation detected from astronomical sources and the design of instrumentation to characterise it.

EM radiation originates from the acceleration of charged particles. First, consider a charged particle that is made to oscillate back and forth in simple harmonic motion around some location in space. In forcing this motion, the particle is being continually accelerated. As it moves, the electric field which it produces oscillates and, as a moving charge constitutes a current, the oscillation of the charge also leads to an oscillating magnetic field. Therefore, an oscillating charge produces oscillating electric and magnetic fields, which are the basis of EM radiation.

While EM radiation can be described as a travelling oscillation of the electric and magnetic fields (the wave picture), it can also be described as a stream of discrete packets of energy (the photon picture). As we shall see, the wave picture is usually adopted for long-wavelength/low energy radio waves while the photon picture for higher energy radiation in the visible to γ-ray regions of the spectrum. In between, at millimetre to infrared wavelengths, both conceptualisations can be used, depending on the circumstances. We begin with the wave picture, which can be derived from Maxwell's equations for the electric and magnetic fields.

1.2 MATHEMATICAL DESCRIPTION OF WAVES

A transverse travelling wave involves oscillation in a direction perpendicular to the direction of propagation; EM waves are transverse waves. A snapshot of a travelling electric field wave moving in the z direction is illustrated in Figure 1.1.

E is the magnitude of the electric field. The wave's frequency, v, and wavelength, λ, are related through $v = v\lambda$, where v is the wave speed. Frequency and wavelength are thus linked by the constant speed of the wave. However, the speed depends on the physical nature of the medium in which the wave propagates, so the relationship between wavelength and frequency is particular to a given physical medium. When a wave moves from one medium to another (for instance, light moving from air into glass), it is the frequency that remains constant, not the wavelength.

We can then represent the electric field wave by a sinusoidal function with amplitude E_o and angular frequency $\omega = 2\pi v$:

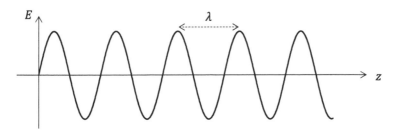

FIGURE 1.1 Sinusoidal electric field travelling wave of wavelength λ.

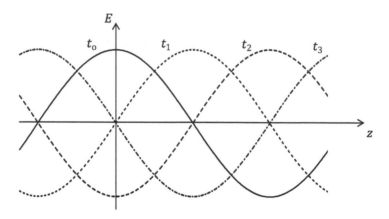

FIGURE 1.2   Snapshots of a travelling wave at different times.

$$E(t) = E_o \cos(2\pi v t) = E_o \cos(\omega t). \tag{1.1}$$

The wave travels in space at velocity, v, so in a time interval, t, a particular wave crest moves through a distance $z = vt$. Now the motion of a wave crest at $z = 0$ is the same as that for the crest at distance z so we can replace t with $(t - z/v)$ to characterise the wave motion for one complete oscillation. With this formalism, we can account for the wave oscillation at any point in time and space by

$$E(z, t) = E_o \cos\left(\omega\left(t - z/v\right) + \phi\right), \tag{1.2}$$

where we have also included an arbitrary initial phase, ϕ.

Defining a quantity known as the wave vector, $\boldsymbol{k}$, which points in the direction of propagation of the wave and has magnitude

$$k = 2\pi/\lambda = 2\pi v/v = \omega/v , \tag{1.3}$$

we arrive at the standard notation for a transverse wave:

$$E(z,t) = E_o \cos(\omega t - kz + \phi). \tag{1.4}$$

Figure 1.2 shows snapshots of this wave for a short distance in space and time over one cycle of oscillation.

1.3   MAXWELL'S EQUATIONS

Maxwell's equations, as derived in many textbooks on electromagnetism, are a set of four equations, which together completely describe the behaviour of electric and magnetic fields in classical physics. They can be expressed either in integral or in differential form, with the latter formulation being a more suitable starting point when considering EM waves:

	Equation	Integral Form	Differential Form
1	Gauss's Law for $\mathbf{E}$	$\oint \mathbf{E} \cdot d\mathbf{s} = \dfrac{1}{\varepsilon} \oint \rho \, dV$	$\text{div } \mathbf{E} = \nabla \cdot \mathbf{E} = \dfrac{\rho}{\varepsilon}$
2	Gauss's Law for $\mathbf{B}$	$\oint \mathbf{B} \cdot d\mathbf{s} = 0$	$\text{div } \mathbf{B} = \nabla \cdot \mathbf{B} = 0$
3	Faraday's Law	$\oint \mathbf{E} \cdot d\mathbf{l} = -\dfrac{d}{dt} \oint \mathbf{B} \cdot d\mathbf{s}$	$\text{curl } \mathbf{E} = \nabla \times \mathbf{E} = -\dfrac{\partial \mathbf{B}}{\partial t}$
4	Maxwell-Ampere Law	$\oint \mathbf{B} \cdot d\mathbf{l} = \mu \oint \mathbf{J} \cdot d\mathbf{s} + \mu\varepsilon \dfrac{d}{dt} \oint \mathbf{E} \cdot d\mathbf{s}$	$\text{curl } \mathbf{B} = \nabla \times \mathbf{B} = \mu\left(\mathbf{J} + \varepsilon \dfrac{\partial \mathbf{E}}{\partial t} \right)$

$$(1.5)$$

Here, $\mathbf{E}$ and $\mathbf{B}$ are the electric and magnetic field vectors, respectively; ρ is the volume charge density; ε and μ are the electric permittivity and magnetic permeability of the medium, respectively; $\mathbf{J}$ is the current density vector; $d\mathbf{s}$ is a surface element normal vector; $d\mathbf{l}$ is a line element vector; and dV is a volume element. All of the integrals are over closed surfaces or paths.

Gauss's Law for $\mathbf{E}$ states that the total electric flux through a closed surface is equal to the enclosed charge divided by ε and is equivalent to the statement that electric fields are due to electric charges and obey the inverse square law: $E \propto 1/r^2$, where r is distance from a point charge. Gauss's Law for $\mathbf{B}$ states that the total magnetic flux through a closed surface is zero and is equivalent to stating that there are no magnetic equivalents of electric charges (no magnetic monopoles), so that when magnetic field lines are used to depict the field they must form closed loops.

Maxwell's third equation is Faraday's Law of induction, representing the fact that a changing magnetic flux gives rise to an electric field (with the causality going from right to left in the equation). Likewise, the Maxwell-Ampere Law represents the generation of magnetic fields either by charge motion (current) or by a changing electric field.

1.4 ELECTROMAGNETIC WAVES

Initially, we consider the simplest case for $\mathbf{E}$ and $\mathbf{B}$ fields in free space or in a medium with no free charges. In this situation, $\rho = 0$, and there is no electrical conduction, so $\mathbf{J} = 0$. Maxwell's equations in differential form then reduce to

$$\nabla \cdot \mathbf{E} = 0$$

$$\nabla \cdot \mathbf{B} = 0$$

$$\nabla \times \mathbf{B} = \mu\varepsilon \frac{\partial \mathbf{E}}{\partial t} \qquad (1.6)$$

$$\nabla \times \mathbf{E} = -\frac{\partial \mathbf{B}}{\partial t}.$$

We can merge these equations by using the vector triple product theorem:

$$\mathbf{A} \times (\mathbf{B} \times \mathbf{C}) = \mathbf{B}(\mathbf{A} \cdot \mathbf{C}) - \mathbf{C}(\mathbf{A} \cdot \mathbf{B}). \qquad (1.7)$$

Applying this to the $\nabla \times \mathbf{E}$ term in Maxwell's fourth equation, we get

$$\nabla \times (\nabla \times \mathbf{E}) = \nabla(\nabla \cdot \mathbf{E}) - \mathbf{E}(\nabla \cdot \nabla) = \nabla(\nabla \cdot \mathbf{E}) - \nabla^2 \mathbf{E}. \qquad (1.8)$$

However, we know that $\nabla \cdot \mathbf{E} = 0$ from Maxwell's first equation, so

$$\nabla^2 \mathbf{E} = -\frac{\partial(\nabla \times \mathbf{B})}{\partial t}. \tag{1.9}$$

We also have $\nabla \times \mathbf{B}$ from the third equation, giving

$$\nabla^2 \mathbf{E} = \mu\varepsilon \frac{\partial^2 \mathbf{E}}{\partial t^2}. \tag{1.10}$$

Similarly, it can be shown that for the magnetic field

$$\nabla^2 \mathbf{B} = \mu\varepsilon \frac{\partial^2 \mathbf{B}}{\partial t^2}. \tag{1.11}$$

Any equation with this differential form in position and time is a wave equation.

1.5 PLANE EM WAVES

The simplest solutions to the EM wave equations are plane-polarised waves, for which $\mathbf{E}$ and $\mathbf{B}$ are uniform in the plane perpendicular to the direction of propagation, and with the two field oscillations perpendicular to each other. More complex situations can be represented by superpositions of many plane-polarised waves.

For a plane-polarised wave propagating in the z direction, the electric field can be taken to oscillate in the x direction, with the field vector denoted by $\mathbf{E}_x(z,t)$, and the magnetic field in the y direction, with the field vector denoted by $\mathbf{B}_y(z,t)$, as illustrated in Figure 1.3, so that

$$\frac{\partial^2 E_x}{\partial z^2} = \mu\varepsilon \frac{\partial^2 E_x}{\partial t^2} \text{ and } \frac{\partial^2 B_y}{\partial z^2} = \mu\varepsilon \frac{\partial^2 B_y}{\partial t^2}. \tag{1.12}$$

Solutions to these equations are

$$E_x(z,t) = E_o \cos(\omega t - kz + \phi) \text{ and } B_y(z,t) = B_o \cos(\omega t - kz + \phi). \tag{1.13}$$

Differentiating the electric field equation with respect to t and z gives

$$\frac{\partial^2 E_x}{\partial t^2} = \omega^2 \cos(\omega t - kz + \phi) \text{ and } \frac{\partial^2 E_x}{\partial z^2} = k^2 \cos(\omega t - kz + \phi). \tag{1.14}$$

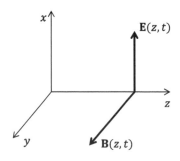

FIGURE 1.3 Electric and magnetic field vectors for a plane-polarised wave propagating in the z direction.

Therefore,

$$\omega^2 = \mu\varepsilon k^2. \tag{1.15}$$

Since the speed of the wave is ω/k, we have, for the speed of propagation of EM waves

$$v = \left(\frac{1}{\mu\varepsilon}\right)^{1/2} = \left(\frac{1}{\mu_r\mu_o\varepsilon_r\varepsilon_o}\right)^{1/2}. \tag{1.16}$$

where μ_o and ε_o are the permeability and permittivity constants (permeability and permittivity of free space) and μ_r and ε_r are the relative permeability and relative permittivity of the medium, respectively. The speed of propagation of EM waves in free space is thus

$$c = \left(\frac{1}{\mu_o\varepsilon_o}\right)^{1/2} = 2.998 \times 10^8 \text{ m s}^{-1}. \tag{1.17}$$

Figure 1.4 shows a visualisation of the electric and magnetic field waves at a certain instant. We can use Faraday's Law to derive a relationship between the amplitudes of the two fields:

$$\nabla \times \mathbf{E} = -\frac{\partial \mathbf{B}}{\partial t}. \tag{1.18}$$

With $\mathbf{E}$ only having a y component, and $\mathbf{B}$ only an x component this reduces to

$$\frac{\partial E_x}{\partial z} = -\frac{\partial B_y}{\partial t}. \tag{1.19}$$

Differentiating equation (1.13) with respect to z and t thus gives

$$kE_o \sin(\omega t - kz + \phi) = B_o\omega \sin(\omega t - kz + \phi). \tag{1.20}$$

Therefore

$$\frac{E_o}{B_o} = \frac{\omega}{k} = c. \tag{1.21}$$

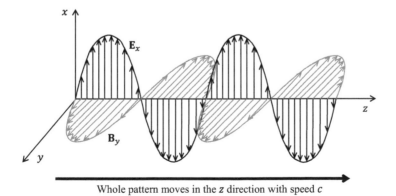

Whole pattern moves in the z direction with speed c

FIGURE 1.4 Illustration of a plane-polarised electromagnetic wave with transverse electric and magnetic fields $\mathbf{E}$ and $\mathbf{B}$, propagating in the z direction.

The amplitudes of the electric and magnetic field oscillations are thus directly related via the wave velocity. Equation (1.21) does not mean that the electric field is more significant than the magnetic field in EM waves. The value of c is only large as a consequence of the definition of the metre in the SI system – the most natural system of units is one in which $\mu_o = \varepsilon_o = c = 1$, but that would not be very suitable for everyday purposes.

For propagation in a (charge-free) physical medium, the speed of propagation of the wave is

$$v = \frac{c}{\left(\mu_r \varepsilon_r\right)^{1/2}}.$$

(1.22)

The refractive index of a medium is defined as the ratio of the wave speed to that in vacuum:

$$n = \frac{c}{v} = \left(\mu_r \varepsilon_r\right)^{1/2}.$$

(1.23)

Most media are not strongly magnetic, with $\mu_r \approx 1$, so that $n \approx \varepsilon_r^{1/2}$. These relationships will become useful later when we consider optical components such as dielectric lenses and wave-plates, and the operation of the acousto-optic spectrometer.

1.6 ENERGY IN EM WAVES

The electric and magnetic fields contain energy, with energy densities (energy per unit volume) given by

$$u_E = \frac{\varepsilon_o E^2}{2} \text{ and } u_B = \frac{B^2}{2\mu_o}.$$

(1.24)

Since $E = cB$ and $c = 1/\left(\mu_o \varepsilon_o\right)^{1/2}$, these two expressions have the same magnitude for electromagnetic waves: $u_E = u_B$. The electric and magnetic fields are therefore equally significant in containing and transporting energy.

The total energy in a region where the two fields have non-zero values is

$$u = u_E + u_B = \varepsilon_o E^2 = \frac{B^2}{\mu_o}.$$

(1.25)

Now consider the energy flowing through a region of area A and thickness dz, as shown in Figure 1.5. In a time dt, the distance travelled by the wave is $dz = cdt$ so the volume swept out is $Acdt$. The electromagnetic energy contained in this volume is thus $Acudt$. The energy flow per unit area per unit time (i.e. power per unit area) is therefore

$$S = \frac{Acudt}{Adt} = cu = c\varepsilon_o E^2 = \frac{cB^2}{\mu_o} = \frac{EB}{\mu_o}.$$

(1.26)

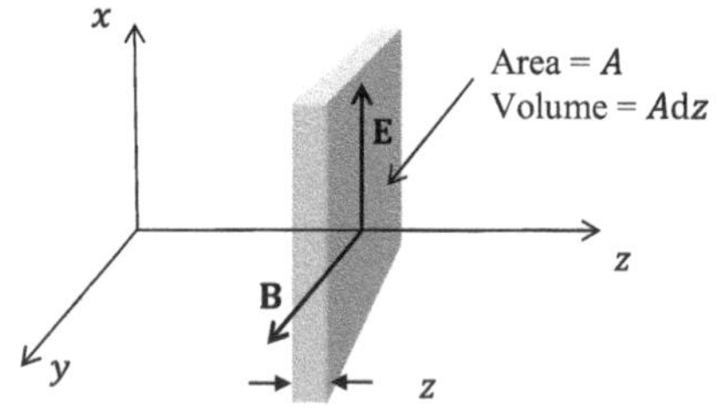

FIGURE 1.5 Electric and magnetic fields within a thin slab.

The electromagnetic power flows in the z direction, which is the direction of $\mathbf{E} \times \mathbf{B}$. The power flow per unit area can thus be represented by a vector, known as the Poynting vector:

$$\mathbf{S} = \frac{\mathbf{E} \times \mathbf{B}}{\mu_0}. \tag{1.27}$$

The magnitude of the Poynting vector, S, with units of power per unit area, is known as the irradiance and is directly proportional to the square of the field amplitude – either E^2 or B^2.

1.7 WAVE PARTICLE DUALITY

Instead of regarding EM radiation as a wave phenomenon, as above, it is more appropriate in many cases to regard the radiation as being quantised – made up of discrete packets of energy called photons.

The energy of a photon of frequency v is

$$E = hv, \tag{1.28}$$

where h is Planck's constant $\left(6.626 \times 10^{-34} \text{ J s}\right)$.

A photon, travelling with speed c, carries an amount of momentum given by

$$p = \frac{hv}{c} = \frac{h}{\lambda}. \tag{1.29}$$

In addition to considering waves to have particle-like properties, we can do the opposite and consider an elementary particle as a matter wave. For a particle of mass m and velocity v, the kinetic energy and momentum, respectively, are

$$K = \frac{1}{2}mv^2 \text{ and } p = mv = \sqrt{2mK}. \tag{1.30}$$

For a particle of momentum p, we can use the same expression as above to define its wavelength (called the de Broglie wavelength):

$$\lambda = \frac{h}{p}. \tag{1.31}$$

The amplitude of a matter wave corresponding to a particle such as an electron can be interpreted as dictating the probability that the particle will be found at that position. The wavelength of the wave corresponds to the particle's momentum.

In crude terms, if a particle has no extent (i.e. it is a point particle), we can know its position exactly. However, if it has wave properties, then, because waves are extended, we cannot say that the particle is "exactly here"; the best we can do is to say that it is "somewhere around here". The wavelength is only significant for very small (sub-atomic) particles. For macroscopic objects, it is extremely small and the position uncertainty is negligible.

This uncertainty is expressed by Heisenberg's uncertainty principle relating the uncertainties in momentum, p, and position, x, of a particle

$$\Delta x \Delta p \geq \frac{h}{2\pi} = \hbar. \tag{1.32}$$

There is thus a fundamental limit to our ability to measure simultaneously a particle's momentum and position. The better we define its position by narrowing the window in which we look for it, the

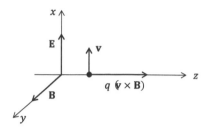

FIGURE 1.6   An EM wave interacting with a point charge.

greater uncertainty we have in its momentum and vice versa. The uncertainty principle can also be formulated in terms of two other complementary quantities – energy and time:

$$\Delta E \Delta t \geq \hbar. \tag{1.33}$$

where E and t are the energy and time uncertainties in a measurement. An important consequence of the uncertainty principle, in this and other forms, is that there are fundamental limitations to the accuracy with which physical quantities can be measured.

1.8   INTERACTION OF EM RADIATION WITH CHARGED PARTICLES

Because the electric and magnetic fields contain energy, and because they exert forces on electric charges, a travelling EM wave transports energy and can interact with a charged particle. Consider an EM wave, propagating in the z direction, interacting with a positive test charge, q, as shown in Figure 1.6.

The transverse electric field exerts a force on the charge causing it to oscillate up and down in the x direction. Due to its motion in the magnetic field of the wave, when it has instantaneous velocity **v**, it now experiences a magnetic force in the direction of $q(\mathbf{v} \times \mathbf{B})$ – i.e., in the z direction. If this force is significant, the charge is pushed forward in the direction of the wave and the wave therefore transfers momentum to the charge.

In the low energy limit ($h\nu \ll$ rest mass energy of the particle), the magnetic force is negligible. The interaction is elastic and is called Thomson scattering. The electric field of the wave causes the particle to oscillate, which, in turn, generates a scattered EM wave with the same frequency, but there is no overall energy exchange between the radiation and the particle. At higher energies, the interaction becomes inelastic, with energy being transferred from the photon to the electron, and is termed Compton scattering. These processes are described in Chapter 3.

1.9   COMPLEX NUMBER REPRESENTATION OF WAVES

For calculations, we can use an exponential notation based on Euler's law for the representation of the quantity, e^{ikx}:

$$e^{ikx} = \cos x + i \sin x, \tag{1.34}$$

so that

$$\cos x = \mathrm{Re}\left(e^{ikx}\right). \tag{1.35}$$

For simplicity of notation, the "Re" is usually omitted, and the electric field wave of equation (1.4) can thus be written as

$$E(z,t) = E_o e^{i(\omega t - kz + \phi)} = E_o e^{-ikz} e^{i\omega t} e^{i\phi}. \tag{1.36}$$

The wave can be considered as having a complex amplitude, $E_o e^{-ikz}$, which takes account of any absorption as it proceeds. Note that the magnitude of the wave vector, k, is also complex with its imaginary part representing any absorption of the wave by the medium and the real part representing the velocity of propagation of the wave in that medium. In the case of no absorption, k is real. The harmonic time factor, $e^{i\omega t}$, describes the oscillatory behaviour. The energy of the electric field is proportional to the square of the amplitude carried by the wave, which is given by the product of $E(z,t)$ with its own complex conjugate: $E(z,t)E(z,t)^*$. The arbitrary phase of the wave at $z = 0$ can be set to zero for convenience.

1.10   SUPERPOSITION AND WAVE INTERFERENCE

To examine the interference of waves, consider a plane wave source which is propagated through the system shown in Figure 1.7. One portion of the wave is transmitted through the two beam splitters and at point P, at coordinate z, leads to a a wave which is described at time t by

$$E_{o1} e^{i(\omega t - kz + \phi_1)}. \tag{1.37}$$

The other portion is reflected by both beam splitters as shown, so that at point P it has travelled through a longer distance, resulting in a difference in phase with respect to the first wave. Allowing for this additional phase shift, and for the possibility of unmatched attenuation affecting the wave amplitude, we can express the second wave at P as

$$E_{o2} e^{i(\omega t - kz + \phi_2)}. \tag{1.38}$$

The resultant wave at P is thus the superposition of the two waves:

$$E(z,t) = E_{o1} e^{i(\omega t - kz + \phi_1)} + E_{o2} e^{i(\omega t - kz + \phi_2)}. \tag{1.39}$$

As noted in Section 1.6, the irradiance of the field is proportional to the square of the field amplitude, given by the product of $E(z,t)$ with its complex conjugate: $E(z,t)E(z,t)^*$:

$$I \propto \left(E_{o1} e^{i(\omega t - kz + \phi_1)} + E_{o2} e^{i(\omega t - kz + \phi_2)} \right) \left(E_{o1} e^{-i(\omega t - kz + \phi_1)} + E_{o2} e^{-i(\omega t - kz + \phi_2)} \right). \tag{1.40}$$

Multiplying this out gives

$$I \propto E_{o1}^2 + E_{o2}^2 + E_{o1} E_{o2} \left(e^{i(\phi_2 - \phi_1)} - e^{-i(\phi_2 - \phi_1)} \right). \tag{1.41}$$

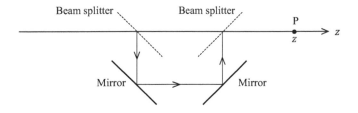

FIGURE 1.7   A beam of radiation following paths of different lengths before interfering.

Therefore,

$$I \propto E_{o1}^2 + E_{o2}^2 + E_{o1}E_{o2}\cos(\phi_2 - \phi_1). \tag{1.42}$$

The resultant irradiance is thus the sum of the irradiances of the two input waves together with an interference term that depends on the relative phase difference between the two waves, $\phi_2 - \phi_1 = \Delta\phi$.

If $\Delta\phi = 0$, $\pm 2\pi$, $\pm 4\pi$, etc. so that $\cos(\Delta\phi) = 1$, the resultant amplitude is a maximum (constructive interference) and if $\Delta\phi = \pi$, $\pm 3\pi$, etc. so that $\cos(\Delta\phi) = -1$, it is a minimum (destructive interference). If the two waves have the same irradiance I_o, then they cancel out completely when they interfere destructively, and they combine to produce an irradiance of $4I_o$ when they interfere constructively.

Letting z_1 and z_2 be the distances travelled by the two beams from the source to point P, the phase difference is

$$\Delta\phi = k(z_2 - z_1) = \frac{2\pi}{\lambda}(z_2 - z_1). \tag{1.43}$$

The wavelength depends on the refractive index, n, of the medium through which the wave propagates. With λ_o being the free-space wavelength,

$$\lambda = \frac{\lambda_o}{n}, \tag{1.44}$$

so

$$\Delta\phi = \frac{2\pi}{\lambda_o}n(z_2 - z_1) = k(\mathrm{OPD}), \tag{1.45}$$

where $n(z_2 - z_1)$, the physical path difference multiplied by the refractive index, is known as the optical path difference, OPD.

When two beams having unequal intensities combine, the maximum (in-phase; $\cos(\Delta\phi) = 1$) and minimum (out of phase; $\cos(\Delta\phi) = -1$) irradiances are given by

$$I_{\max} = I_1 + I_2 + 2(I_1 I_2)^{1/2}, \quad I_{\min} = I_1 + I_2 - 2(I_1 I_2)^{1/2}. \tag{1.46}$$

The visibility, V, of the interference fringes observed as a result of this superposition is defined as

$$V = \frac{I_{\max} - I_{\min}}{I_{\max} + I_{\min}}. \tag{1.47}$$

In spatial interferometry (see Chapter 8), the fringe visibility is important in determining whether or not we can observe the interference.

1.11 RADIATIVE TRANSFER

Before discussing astrophysical radiation mechanisms and their spectral characteristics in Chapter 2, we need to introduce the terminology and method used to quantify the radiation and how it may change as it propagates through a physical medium.

1.11.1 FLUX, FLUX DENSITY, AND INTENSITY

Three basic quantities are used to describe the strength, or brightness, of EM radiation:

Radiant flux, F, is defined as the power, in some specified band of frequencies, crossing a unit area. In the SI system, it has units of W m^{-2}.

Flux density, S_v, is defined as the power per unit area per unit frequency. Its SI units are thus W m^{-2} Hz^{-1}.

Intensity, I_v, is the flux density per unit solid angle in a given direction, with SI units W m^{-2} Hz^{-1} sr^{-1}. (The concept of solid angle is defined below.)

Consider an object emitting EM radiation as shown in Figure 1.8. The emitted power per unit area could vary with: (i) position on the surface of the object; (ii) frequency – the power emitted will depend on the spectral properties of the object; and (iii) direction – for instance, the power emitted in directions 1 and 2 could be different. Alternatively, the object could be absorbing incoming power, with potential variation of the intercepted power with frequency, position, and direction.

Let dP be the total power emitted/absorbed by some small area dA, in a frequency range $v \rightarrow v + dv$. The flux emitted/absorbed in frequency range dv is then

$$F = \frac{dP}{dA}. \tag{1.48}$$

The flux density emitted/absorbed by dA is

$$S_v = \frac{dP}{dA\, dv}. \tag{1.49}$$

The conventional unit of flux density in astronomy is the Jansky (Jy) with 1 Jy $= 10^{-26}$ W m^{-2}Hz^{-1}.

To define the directional dependence, we need to specify the amount of power flowing through a narrow cone about a particular direction, such as direction 1 as shown in Figure 1.8. Such a cone defines a solid angle.

The definition of solid angle is illustrated in Figure 1.9. Let θ be the half-angle defining a cone. Let A_{cap} be the area of the surface of a sphere of radius R intercepted by the cone. The solid angle is defined as

$$\Omega = \frac{A_{cap}}{R^2}. \tag{1.50}$$

Just like a two-dimensional angle, a solid angle is dimensionless. Solid angles are usually expressed in steradians (sr).

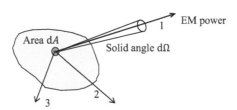

FIGURE 1.8 Emission from a small area, dA, on a source.

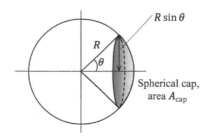

FIGURE 1.9   Geometry for definition of solid angle.

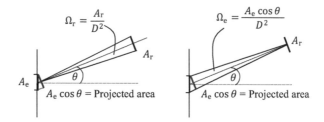

FIGURE 1.10   Geometry for propagation of radiation from an emitting area A_e to a receiving area A_r.

For a hemisphere, $A_{cap} = 2\pi R^2 \Rightarrow \Omega = 2\pi$ sr
For a whole sphere, $A_{cap} = 4\pi R^2 \Rightarrow \Omega = 4\pi$ sr

For a small solid angle, $d\theta$, $\sin(d\theta) \approx d\theta$ and the cap is nearly flat so that we can approximate its area by the area of a circle,

$$A_{cap} \approx \pi \left(R\sin(d\theta) \right)^2 \approx \pi(Rd\theta)^2 \Rightarrow \Omega \approx \pi(d\theta)^2. \tag{1.51}$$

If we specify the power emitted/absorbed per unit solid angle in a particular direction, we can take any variation with direction into account. The specific intensity, I_ν, is defined as the power emitted/absorbed by the source per unit area, per unit bandwidth, and per unit solid angle,

$$I_\nu = \frac{dP}{dA d\nu d\Omega}, \tag{1.52}$$

with SI units W m^{-2} sr^{-1} Hz^{-1}.

A graph of I_ν vs ν for a source, or some region of it, is referred to as its spectral emission.

We can use this formula to determine how much power will pass through some "receiving area", for instance a telescope aperture, of area A_r, at distance D from an emitting area, A_e, radiating with a certain intensity, $I_{\nu-e}$, as shown in Figure 1.10. We assume that there is no absorption or emission occurring between the source and receiving areas. Let θ be the angle between the normal to A_r and the centre of A_e. Then, the emitting area seen by A_r is $A_e \cos\theta$. The solid angle subtended by the receiving area as viewed from the emitter is $\Omega_r = A_r/D^2$, and the solid angle subtended by the emitter as seen from the receiving area is $\Omega_e = A_e \cos\theta/D^2$.

By definition of intensity, the power passing through an area from within a given solid angle is the product of the intensity, the area, the solid angle, and the radiation bandwidth (equation 1.52). Therefore, in frequency band $d\nu$, the power passing through A_r from A_e is equal to the power emitted by A_e into the solid angle Ω_r:

$$P_r = I_{\nu-e}(A_e \cos\theta)(\Omega_r) = I_{\nu-e}(A_e \cos\theta)\left(\frac{A_r}{D^2}\right). \tag{1.53}$$

P_r decreases as $1/D^2$, as one would expect. However, even though the power declines with distance along the direction of propagation, the specific intensity does not. The specific intensity at A_r is

$$I_{v-r} = \frac{\text{Power passing through } A_r}{(A_r dv)(\text{solid angle from which it comes})} = \frac{I_{v-e}(A_e \cos\theta)\left(\dfrac{A_r}{D^2}\right)dv}{(A_r dv)\left(\dfrac{A_e \cos\theta}{D^2}\right)} = I_{v-e}. \quad (1.54)$$

Therefore, the specific intensity is constant along the path between the two areas, as long as there is no emission or absorption in between. For this reason, it is the most useful quantity to use to quantify the effects of emission or absorption along the direction of propagation.

1.11.2 THE EQUATION OF RADIATIVE TRANSFER

In a physical medium, emission and/or absorption can occur so that I_v may not be constant. Consider radiation as it propagates through a region from positions s_0 to s_1 in a medium as shown in Figure 1.11. As the radiation passes through a small region of thickness ds at distance coordinate s, the intensity may change due to emission and/or absorption in the medium. The medium's emission and absorption properties can be characterised by an emission coefficient, ε_v, and an absorption coefficient, k_v, such that we can write the net increase in intensity due to emission as

$$dI_{v+} = \varepsilon_v ds, \quad (1.55)$$

and the net decrease due to absorption as

$$dI_{v-} = -k_v I_v ds. \quad (1.56)$$

The SI units of ε_v are thus W m^{-2} sr^{-1} Hz^{-1} m^{-1}, and those of k_v are m^{-1} (although it is often quoted in cm^{-1}).

We assume here that (i) the amount of emission is independent of the intensity – which is valid as long as the incident radiation does not cause the medium to emit (e.g. this would not hold in a laser) and (ii) the amount of absorption, in absolute terms, is proportional to the intensity.

For the thin slab of material, thickness ds, we have

$$I_v(s + ds) = I_v(s) + \varepsilon_v ds - k_v I_v ds, \quad (1.57)$$

so that

$$\frac{dI_v}{ds} = \varepsilon_v - k_v I_v. \quad (1.58)$$

This is the equation of radiative transfer and a very important equation in astrophysics. There are a number of particular situations where simplifying approximations are useful.

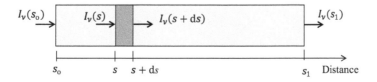

FIGURE 1.11   Change in intensity of radiation in propagating through an absorbing and/or emitting material.

1.11.3 EMISSION ONLY

When there is no absorption, $k_v = 0$ and the equation of radiative transfer becomes

$$dI_v = \varepsilon_v(s)ds, \tag{1.59}$$

where we allow for the possibility that ε_v might vary with position. The emergent intensity is then

$$I_v(s_1) = I_v(s_0) + \int_{s_0}^{s_1} \varepsilon_v(s)ds. \tag{1.60}$$

If ε_v is constant throughout the region, then

$$I_v(s_1) = I_v(s_0) + \varepsilon_v(s_1 - s_0), \tag{1.61}$$

so the incident intensity has been boosted by the thickness times the emission coefficient of the medium, or, if there is no incident emission the emergent intensity is just ε_v times the thickness. An example of this case would be a region of hot tenuous plasma that is transparent to its own radiation.

1.11.4 ABSORPTION ONLY

With no emission, $\varepsilon_v = 0$ so that

$$dI_v = -k_v I_v ds. \tag{1.62}$$

Allowing the absorption coefficient to vary with position, we get

$$\frac{dI_v}{I_v} = -k_v(s)ds. \tag{1.63}$$

Integrating this from s_0 to s_1 gives

$$\left[\ln(I_v)\right]_{s_0}^{s_1} = -\int_{s_0}^{s_1} k_v(s)ds. \tag{1.64}$$

The right-hand side represents the integrated absorption of the medium. It is convenient to represent this as a new parameter, the optical depth, τ_v:

$$\tau_v = \int_{s_0}^{s_1} k_v(s)ds. \tag{1.65}$$

The optical depth thus corresponds to the physical distance weighted by the absorption coefficient. Equation (1.64) then gives

$$I_v(s_1) = I_v(s_0)e^{-\tau_v}. \tag{1.66}$$

Therefore, the intensity decreases exponentially with distance into the medium, at a rate that is determined by τ. If k_v is uniform throughout the medium, then,

$$\tau_v = k_v(s_1 - s_0) \quad \text{and} \quad I_v(s_1) = I_v(s_0)e^{-k_v(s_1-s_0)}. \tag{1.67}$$

This is a very useful relationship when considering, for instance, attenuation due to a cloud of cold gas and dust in the interstellar medium absorbing radiation from a star or other source behind it, or the absorption efficiency of a detector of EM radiation – clearly, the optical thickness of a detector must be large enough to ensure that most of the incident radiation gets absorbed by the detector rather than passes through it.

1.11.5 THERMAL EQUILIBRIUM

For a region of space in thermal equilibrium, the intensity is the same everywhere – if it were not, then there would be a net flow of energy from one place to another. Thus, from the equation of radiative transfer,

$$\varepsilon_v = k_v I_v. \tag{1.68}$$

In the case of thermal equilibrium, the radiation intensity as a function of frequency is described by the Planck Function, $B_v(T)$ (see Chapter 2 for details), giving rise to a relationship between the emission and absorption coefficients (Kirchhoff's Law):

$$\varepsilon_v = k_v B_v(T). \tag{1.69}$$

1.11.6 THERMAL EMISSION FROM A SEMI-TRANSPARENT REGION

Consider emission from a partially absorbing/partially emitting region of material at uniform temperature T, and with ε_v and k_v constant throughout the region. This is an important case for astrophysics, Earth observing, plasma physics, and laboratory spectroscopy. In Figure 1.12, we let the incoming intensity be zero, so we only see the emitting region, and we redefine the coordinates so that the front edge of the region corresponds to $s = 0$. The intensity emitted by a thin slab of thickness ds is then

$$dI_v = \varepsilon_v ds = k_v B_v(T) ds. \tag{1.70}$$

This emission then has to be transmitted through the lossy region between ds and the observer so that the corresponding outgoing intensity is

$$dI_v(\text{out}) = \left[k_v B_v(T) ds \right] e^{-k_v s}. \tag{1.71}$$

The total intensity seen by the observer is obtained by integrating this over the whole length of the emitting region ($s=0$ to s_0):

$$I_v(\text{out}) = \int_0^{S_0} k_v B_v(T) e^{-k_v s} ds. \tag{1.72}$$

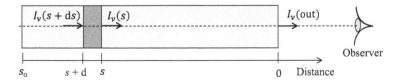

FIGURE 1.12 Observation of emission from a semi-transparent region.

$B_\nu(T)$ is a constant because T is constant, so that

$$I_\nu(\text{out}) = B_\nu(T)k_\nu \int_0^{s_0} e^{-k_\nu s} ds = B_\nu(T)k_\nu \left[\frac{-1}{k_\nu} e^{-k_\nu s} \right]_0^{s_0} = B_\nu(T)\left(1 - e^{-k_\nu s_0}\right). \qquad (1.73)$$

Since $k_\nu s_0 = \tau_\nu$, the optical depth of the region, the intensity seen by the observer is

$$I_\nu(\text{out}) = B_\nu(T)\left(1 - e^{-\tau_\nu}\right). \qquad (1.74)$$

There are two limiting cases.

If $\tau_\nu \ll 1$, then $I_\nu(\text{out}) \approx B_\nu(T)\left(1 - (1 - \tau_\nu)\right) \approx \tau_\nu B_\nu(T)$ and the region is said to be optically thin. Its spectrum is that of a black body, but diluted and modified in shape by τ_ν.

If $\tau_\nu \to 1$, then $I_\nu(\text{out}) \approx B_\nu(T)$ and the region is said to be optically thick – the spectrum of a perfect absorber is thus the black body spectrum. Note that we have made no assumptions about the original mechanism by which the radiation is emitted. Regardless of the physical process involved, the spectrum of an opaque object is described by the Planck function.

1.11.7 TRANSMISSION OF THE EARTH'S ATMOSPHERE

Ground-based telescopes view the cosmos through a partially transmitting atmosphere and usually track sources as they move across the sky. The slant path traversed by the radiation from the source thus varies as the Earth rotates. Figure 1.13a shows the geometry of a typical observation of a source at angle θ_z from the zenith. We can use the equation of radiative transfer (absorption-only case) to determine the overall attenuation of the incoming astronomical radiation experiences before it reaches the telescope. Equation (1.66) states

$$I_\nu(s_1) = I_\nu(s_0)e^{-\tau_\nu}, \qquad (1.75)$$

where s_1 refers to the ground and s_0 to the top of the atmosphere, and we assume that the optical depth, τ_ν, accounts for both absorption and scattering losses in the atmosphere.

It is common to assume that the atmosphere has a plane parallel structure with layers near the ground being more dense (higher loss) than layers near the top.

We can define a zenith opacity, $\tau_\nu(0)$ and an opacity along the line of sight to the source of $\tau_\nu(0)\sec\theta_z$. The transmission, $t(\theta_z)$, along the line of sight is then

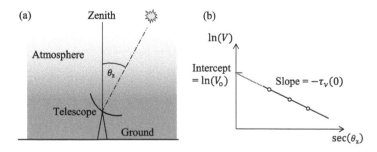

FIGURE 1.13 (a) A telescope observing a source at zenith angle θ_z; (b) example of a plot of measured signal voltage vs source zenith angle, allowing the atmospheric optical depth and the unattenuated source voltage to be calculated from the slope and intercept of the plot.

$$t(\theta_z) = \frac{I_v(s_1)}{I_v(s_o)} = e^{-\tau_v(0)\sec\theta_z}, \tag{1.76}$$

or,

$$\ln(I_v(s_1)) = \ln(I_v(s_o)) - \tau_v(0)\sec\theta_z. \tag{1.77}$$

Therefore, from a series of measurements of, for instance a detector voltage, V, proportional to the intensity, at different zenith angles, we can create a secant plot, as in Figure 1.13b – a log-linear plot of measured voltage vs $\sec(\theta_z)$, to calculate both the signal that would be observed at the top of the atmosphere, $V_o \propto I_v(s_o)$, and the vertical opacity of the atmosphere, $\tau_v(0)$. This is a standard procedure used by astronomers to correct for atmospheric losses.

1.12 POLARISATION

1.12.1 LINEAR, CIRCULAR, AND ELLIPTICAL POLARISATION

We have observed that the oscillating electric and magnetic fields are orthogonal to each other and to the direction of propagation of an electromagnetic wave. The polarisation state depends on the physical conditions in the radiation source and can yield important astronomical information. It is therefore useful to construct instruments – polarimeters – which are capable of measuring the state of polarisation of received radiation.

For a plane-polarised wave, the direction of oscillation of the electric field vector is termed the direction of polarisation of the wave. In general, any arbitrary wave can be decomposed into two orthogonal components, $\mathbf{E}_x$ (horizontal) and $\mathbf{E}_y$ (vertical). Figure 1.14 shows two such linearly polarised waves, with a phase delay, δ, between the two components.

These waves can be described as

$$\mathbf{E}_x(z,t) = E_{ox}e^{i(\omega t - kz)}\,\hat{\mathbf{i}}, \tag{1.78}$$

$$\mathbf{E}_y(z,t) = E_{oy}e^{i(\omega t - kz + \delta)}\,\hat{\mathbf{j}}. \tag{1.79}$$

where $\hat{\mathbf{i}}$ and $\hat{\mathbf{j}}$ are unit vectors in the x and y directions, respectively.

The resultant wave, as observed at some position on the z axis, is thus given by

$$\mathbf{E}(z,t) = \mathbf{E}_x(z,t) + \mathbf{E}_y(z,t) = \left(E_{ox}\hat{\mathbf{i}} + E_{oy}e^{i\delta}\hat{\mathbf{j}}\right)e^{i(\omega t - kz)}. \tag{1.80}$$

FIGURE 1.14 Orthogonal electric field waves travelling in the $+z$ direction with arbitrary phase delay δ.

If the amplitude of the x component, E_{ox}, is zero, there is only one component and we have a linearly polarised wave $\mathbf{E}_y$ with so-called vertical polarisation. Similarly if E_{oy} is zero, then only $\mathbf{E}_x$ exists, and the wave is said to be horizontally polarised. If the two components have equal amplitudes and there is no phase difference $\delta = 0$, then the two waves combine to create a single polarised wave at 45° to both x and y. These cases are shown in Figure 1.15.

If the two components have equal amplitudes and the phase difference is $\delta = \pi/2$, then

$$\mathbf{E}(z,t) = \left(E_o \hat{\mathbf{i}} + E_o e^{i\pi/2} \hat{\mathbf{j}} \right) e^{i(\omega t - kz)}. \tag{1.81}$$

The y-component thus lags the x-component by a constant value of $\pi/2$. The amplitude of the resultant is therefore constant. The resultant wave vector of constant amplitude rotates in an anticlockwise direction as viewed by an observer looking along the z-axis towards the origin. If the phase difference is $\delta = -\pi/2$, then the electric field vector sweeps out a circle in the clockwise direction. These cases are known as left-handed and right-handed circular polarisation, respectively, and are depicted in Figure 1.16.

In the more general case of unequal amplitudes, the result is an elliptically polarised wave, in which the resultant amplitude is described by the equation of an ellipse as its direction in the $x - y$ plane rotates. The ellipse has a constant form in the $x - y$ plane, as shown in Figure 1.17, since the wave amplitudes, E_{ox} and E_{oy}, and the phase difference, δ, are constant.

In general, an arbitrary wave can be considered as the combination of a linear wave and an elliptically polarised wave. In astronomy, the observed radiation is sometimes partially polarised in a way that depends on the emission mechanism or on effects that have occurred between emission and detection.

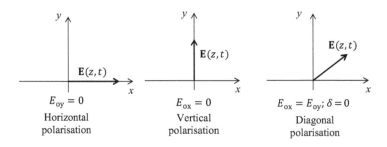

FIGURE 1.15   Horizontal, vertical, and diagonal polarisation states.

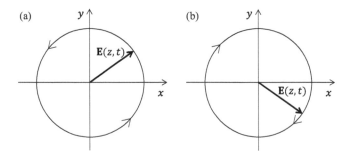

FIGURE 1.16   (a) Anticlockwise rotation of the electric field vector of a left circularly polarised wave as observed at a position on the z axis; (b) clockwise rotation for a right circularly polarised wave.

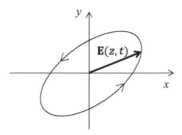

FIGURE 1.17 Elliptical polarisation.

1.12.2 STOKES PARAMETERS

The polarisation ellipse parameters provide an instantaneous representation of the polarised wave but are not actually directly measurable. Stokes identified four parameters that can be derived from simple measurements of intensity, and which characterise the polarisation ellipse as follows:

$$I = E_{ox}^2 + E_{oy}^2 = I_0 + I_{90}$$
$$Q = E_{ox}^2 - E_{oy}^2 = I_0 - I_{90}$$
$$U = 2E_{ox}E_{oy}\cos\delta = I_{45} - I_{-45}$$
$$V = 2E_{ox}E_{oy}\sin\delta = I_{C-R} - I_{C-L}.$$

(1.82)

where I_0 and I_{90} are the intensities measured in any pair of orthogonal polarisation directions, I_{45} and I_{-45} are intensities measured in the corresponding diagonal directions, and I_{C-R} and I_{C-L} are the intensities measured in right and left circular polarisations. I characterises the total intensity, Q the degree of polarisation, U the direction of polarisation, and V the degree of ellipticity. They are often represented as a column vector, known as the Stokes vector:

$$\mathbf{S} = \begin{bmatrix} I \\ Q \\ V \\ V \end{bmatrix}.$$

(1.83)

Only three of the parameters are independent since they are trigonometrically related by

$$V^2 = I^2 - Q^2 - U^2,$$

(1.84)

so they can be evaluated from linear polarisation measurements alone. The Stokes vector allows unpolarised, partially polarised, and fully polarised light to be described. Stokes vectors for some common polarisation states are as follows:

$$\begin{bmatrix} 1 \\ 1 \\ 0 \\ 0 \end{bmatrix} \Rightarrow \text{Linearly polarised (horizontal)} \qquad \begin{bmatrix} 1 \\ -1 \\ 0 \\ 0 \end{bmatrix} \Rightarrow \text{Linearly polarised (vertical)}$$

$$\begin{bmatrix} 1 \\ 0 \\ 1 \\ 0 \end{bmatrix} \Rightarrow \text{Linearly polarised} (+45°) \qquad \begin{bmatrix} 1 \\ 0 \\ -1 \\ 0 \end{bmatrix} \Rightarrow \text{Linearly polarised} (-45°)$$

$$\begin{bmatrix} 1 \\ 0 \\ 0 \\ 1 \end{bmatrix} \Rightarrow \text{Right hand circularly polarised} \qquad \begin{bmatrix} 1 \\ 0 \\ 0 \\ 1 \end{bmatrix} \Rightarrow \text{Left hand circularly polarised}$$

$$\begin{bmatrix} 1 \\ 0 \\ 0 \\ 0 \end{bmatrix} \Rightarrow \text{Unpolarised}$$

2 Astrophysical Radiation

2.1 ASTROPHYSICAL RADIATION MECHANISMS

A fundamental consequence of Maxwell's equations is that if a charged particle undergoes acceleration – changes its state of motion – it emits or absorbs EM radiation. Objects at non-zero temperature emit radiation due to the random thermal motions of the electric charges within them. Other radiation mechanisms include quantum state transitions in atoms and molecules, which generally result in narrow spectral features, and radiation emitted due to charged particles undergoing deflections as they move in magnetic fields (cyclotron or synchrotron radiation). These radiation phenomena are described in this chapter along with a description of the polarisation state. The role of an observational astronomer is to measure the spatial, spectral, or polarimetric intensity from a source and then infer information on the physical nature of the source.

2.2 BLACK BODY RADIATION AND THE PLANCK FUNCTION

A black body is defined as an object that absorbs all of the radiation incident on it (absorption coefficient, $k_\nu = 1$). The physical origin of black body radiation is the thermal motion of the charges in a material at non-zero temperature. Its brightness and spectral distribution are dictated only by the temperature of the object – its chemical composition or physical state make no difference.

It was shown by Max Planck that the intensity, at frequency ν, of the EM radiation emitted by a black body at temperature T is given by

$$I_\nu = B_\nu(T) = \frac{2h\nu^3}{c^2 \left(e^{\frac{h\nu}{k_B T}} - 1 \right)}, \tag{2.1}$$

where h is Planck's constant, c is the speed of light, and k_B is Boltzmann's constant. As a particular example of specific intensity, the SI units of I_ν are W m^{-2} sr^{-1} Hz^{-1}. The intensity of black body radiation can also be expressed as a function of wavelength, λ:

$$B_\lambda(T) = \frac{2hc^2}{\lambda^5 \left(e^{\frac{hc}{k_B \lambda T}} - 1 \right)}, \tag{2.2}$$

with SI units W m^{-2} sr^{-1} m^{-1}.

Either one of these formulae may be used when working with the Planck function – it is purely a matter of choice. Two limiting cases are frequently invoked: the Rayleigh–Jeans and the Wien regions of the spectrum.

2.2.1 RAYLEIGH–JEANS AND WIEN REGIONS

If $h\nu \ll k_B T$ (i.e. the photon energy is low compared to characteristic thermal energy), then the spectrum is well-described by the Rayleigh–Jeans approximation:

$$e^{\frac{h\nu}{k_B T}} - 1 \approx \left(1 + \frac{h\nu}{k_B T} - 1\right) \approx \frac{h\nu}{k_B T}, \tag{2.3}$$

so that the Planck function becomes

$$B_\nu(T) \approx \frac{2\nu^2 k_B T}{c^2}. \tag{2.4}$$

In this low-frequency limit, the intensity is proportional to the square of the frequency and linearly proportional to the temperature. Radio astronomy systems usually operate in this regime, and the detected signals are thus directly proportional to the temperature of the source if it emits thermally.

If $h\nu \gg k_B T$ (i.e. the photon energy is high compared to characteristic thermal energy), then the spectrum is well described by the Wien approximation:

$$e^{\frac{h\nu}{k_B T}} - 1 \approx e^{\frac{h\nu}{k_B T}}, \tag{2.5}$$

giving

$$B_\nu(T) = \frac{2h\nu^3}{c^2 e^{\frac{h\nu}{k_B T}}}. \tag{2.6}$$

The exponential term in the denominator is much stronger than the ν^3 term in the numerator, so the intensity falls off very rapidly with increasing frequency. The characteristic form of the black body spectrum is shown in Figure 2.1, which shows $B_\nu(T)$ vs ν for temperatures of 100 and 200 K. The left-hand plot is with linear scales, and on the right is a log–log plot, which allows the curves to be plotted over much wider ranges of frequency and intensity. The power law (ν^2) dependence on frequency in the Rayleigh–Jeans region, appearing as a straight line on the log–log plot, and the rapid decrease with frequency in the higher-frequency Wien regime are also apparent. In between the Rayleigh–Jeans and Wien regimes is the peak of the black body emission curve. The $T = 200$ K curve has a higher peak value, which also occurs at a higher frequency, but otherwise the two curves have the same shape. This too is characteristic of all black body spectra – Planck functions for different temperatures do not intersect, have similar shapes, and the peak in emission shifts to higher intensity and higher frequency as the temperature increases.

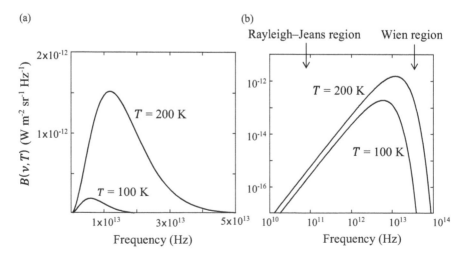

FIGURE 2.1   Blackbody spectra for temperatures of 100 and 200 K with linear scales (a) and logarithmic scales (b).

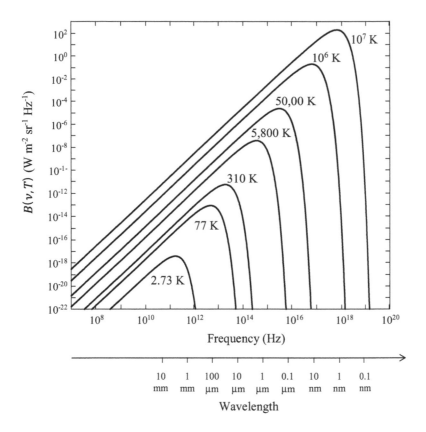

FIGURE 2.2   Blackbody spectra for temperatures between 2.73 and 10^7 K.

Figure 2.2 shows a nest of Planck functions for a wide range of temperatures: 2.73 K (the temperature of the cosmic microwave background (CMB) radiation that fills the Universe and is a relic of the Big Bang); 77 K (the temperature of liquid nitrogen, which is used to cool some kinds of detectors, and also characteristic of dust and gas in star-forming regions); 310 K (the approximate temperature of a person); 5800 K (the temperature of the Sun's photosphere); 50,000 K (the typical photospheric temperature of a very hot massive star); 10^6 K (characteristic of the accretion disc around a black hole); and 10^7 K (characteristic of the temperature in the core of a star like the Sun).

For any black body, most of the radiation is emitted at frequencies close to the peak. Therefore, for example, the cosmic microwave background radiation is strongest at millimetre wavelengths; planets, with temperatures of 100 to a few hundred K, radiate predominantly in the infrared; star-forming clouds have temperatures of ~ 10–100 K and so radiate primarily in the far infrared, and stars, which have surface temperatures of a few × 1000 to a few × 10,000 K, exhibit peak emission at visible or UV wavelengths.

2.2.2   Wien's Displacement Law

The relationship between the temperature and the frequency or wavelength of peak emission is known as Wien's displacement law. The frequency of the peak emission, ν_{peak}, can be found by differentiating $B_\nu(T)$ and setting the result equal to zero, giving

$$\nu_{peak} = \left(5.88 \times 10^{10}\right)T \ \ \text{Hz}. \tag{2.7}$$

The same can be done for $B_\lambda(T)$ to get

$$\lambda_{peak} = \frac{2.90 \times 10^{-3}}{T} \text{ m.} \qquad (2.8)$$

These do not coincide: that is, $v_{peak} \neq c/\lambda_{peak}$. This is because the units of the two versions of the Planck functions are not the same: one is in terms of a unit frequency interval (Hz) and the other in terms of a unit wavelength interval (m).

2.2.3 STEFAN'S LAW

By integrating the Planck function over all frequencies and emission angles, we can derive an expression for the total power per unit area emitted by a black body of temperature T:

$$B(T) = \sigma T^4, \qquad (2.9)$$

where, σ, the Stefan–Boltzmann constant, is given by

$$\sigma = \frac{2\pi^5 k_B{}^4}{15 c^2 h^3} = 5.67 \times 10^{-8} \text{ W m}^{-2} \text{ K}^{-4}. \qquad (2.10)$$

2.2.4 EMISSIVITY

Some objects emit or absorb imperfectly, but they have a spectral distribution the same as or similar to a black body. The departure of the emitted intensity from the black body function is often characterised by the emissivity, ε_v:

$$I_v = \varepsilon_v B_v(T) \quad \text{where} \quad \varepsilon_v \leq 1. \qquad (2.11)$$

If ε_v is constant across the spectral region of interest, then the emission is sometimes referred to as grey body emission. The optically thin approximation to equation (1.74) for the thermal emission from a semi-transparent region, $I_v \approx \tau_v B_v(T)$, is similar to equation (2.11), with τ_v replacing ε_v. An example in astrophysics is the radiation from dust grains in interstellar clouds. In this case, the emissivity decreases at low frequency (longer wavelength), and the term "modified black body" is often used to describe the source emission.

Another important example is emission from the Earth's atmosphere. Neglecting the possible effects of scattering, at any wavelength for which the transmission, t_{Atm}, is less than unity, the emissivity will be given by $\varepsilon_{Atm} = 1 - t_{Atm}$. For ground-based observations, as well as the background from atmospheric emission, there is often a background from the telescope and the instrument itself – for instance, a mirror that reflects 99% of the incident radiation and absorbs 1% will radiate as a black body with an emissivity of 1%. As we shall see in later chapters, this phenomenon poses problems for astronomical telescopes in that the source may be viewed through an unwelcome background of sky and/or telescope and instrument emission.

2.3 FREE–FREE RADIATION

Free–free radiation (also known as thermal Bremsstrahlung – the literal meaning is "braking radiation") is emitted by ionised gas. This is due to the accelerations experienced by electrons as they encounter positively charged ions, as shown schematically in Figure 2.3.

The intensity and its spectral distribution depend on the number of electrons and ions and on the distribution of their speeds (i.e. on the temperature). Over most of the EM spectrum, the emission

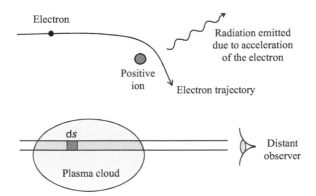

FIGURE 2.3   (a) Free–free emission from an electron encountering a positive ion; (b) Observation of optically thin free–free emission from a plasma cloud. Emission from any path elements ds is observed with negligible attenuation.

is optically thin and the plasma is close to being transparent – the probability of an emitted photon being re-absorbed before escaping from the source is small, so an observer sees emission from a column through the source. The emission coefficient of the plasma can then be shown to be given by

$$\varepsilon_v = (\text{Constant})\, g(T,v)\, n_e n_i Z^2 T^{-0.5}\, e^{-\frac{hv}{k_B T}}, \tag{2.12}$$

where n_e and n_i are the number densities of electrons and ions, respectively, Z is the atomic number of the material, and $g(T,v)$ is called the Gaunt factor and has a weak dependence on frequency and temperature. The exponential term arises from the Maxwell–Boltzmann energy distribution of the particles.

In thermal equilibrium, the absorption coefficient is $k_v = \varepsilon_v / B_v(T)$, from equation (1.69), and for optically thin radiation,

$$I_v = B_v(T)\tau_v = B_v(T) \int_{\text{Path}} k_v ds = \int_{\text{Path}} \varepsilon_v ds. \tag{2.13}$$

Assuming that conditions are uniform throughout the source, we can distinguish three different spectral regimes, which are illustrated in Figure 2.4.

i. Rayleigh–Jeans region ($hv \ll k_B T$): using the Rayleigh–Jeans approximation, gives

$$I_v \propto \int_{\text{Path}} g(T,v)\, n_e n_i Z^2 T^{-0.5} ds. \tag{2.14}$$

Taking the Gaunt factor dependencies on frequency and temperature $\left(g(v) \propto T^{0.15} v^{-0.1} \right)$ into account gives

$$I_v \propto \int_{\text{Path}} n_e n_i Z^2 T^{-0.35} v^{-0.1} ds. \tag{2.15}$$

The intensity is nearly independent of frequency (i.e. the spectrum is close to flat), and it decreases with increasing temperature. This arises because as the temperature increases, the typical electron velocity increases, leading to shorter electron–ion encounter durations and smaller deflections.

ii. Wien region ($hv \gg k_B T$):High-frequency emission is associated with encounters of short duration, so with very energetic electrons. Because of the exponentially declining term in

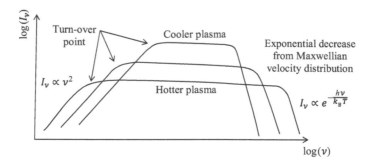

FIGURE 2.4 Spectrum of free–free emission for different plasma temperatures.

the Maxwell–Boltzmann distribution, there is a sharp decrease in the intensity due to the lack of high-energy electrons.

iii. Very low frequencies: For very low frequencies, the free electrons can respond to the free–free radiation produced by other electrons and reabsorb it. The emitting region thus becomes opaque to its own radiation – i.e., the optical thickness becomes large. As described in Section 2.2, this means that the spectrum is described by the Rayleigh–Jeans black body function at lower frequencies. For simple singly ionised plasma $(n_e = n_i = n)$, the plasma frequency (see Chapter 3) defines a break point from an optically thick to an optically thin region and is defined as

$$\nu_p = \frac{1}{2\pi}\left(\frac{ne^2}{\varepsilon_0 m_e}\right)^{\frac{1}{2}}. \tag{2.16}$$

The turn-over frequency from a flat spectrum to that of a black body gives a direct measure of the electron density in the plasma.

Astrophysical free–free radiation is observed from regions containing tenuous ionised gas, such as gaseous nebulae (at radio frequencies) and the hot ($> 10^7$ K) intergalactic gas in galaxy clusters (at X-ray frequencies).

2.4 NON-THERMAL EMISSION MECHANISMS: CYCLOTRON AND SYNCHROTRON RADIATION

A charged particle moving in a magnetic field experiences magnetic force, and so undergoes an acceleration, leading to the emission of EM radiation. This is an important emission mechanism in astrophysics.

Consider an electron of mass m_e, moving at constant velocity, $\mathbf{v}$, in a uniform magnetic field, $\mathbf{B}$. It experiences a magnetic force given by

$$\mathbf{F}_m = -e(\mathbf{v} \times \mathbf{B}). \tag{2.17}$$

with magnitude

$$F_m = ev_\perp B, \tag{2.18}$$

where $\mathbf{v}_\perp$ is the component of the electron velocity that is perpendicular to the magnetic field. The direction of the magnetic force is perpendicular to $\mathbf{v}_\perp$, as shown in Figure 2.5, and there is no force due to the component of motion parallel to the field, so the speed of the particle is not affected, only

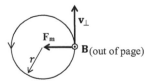

FIGURE 2.5   Cyclotron motion of an electron in a uniform magnetic field.

its direction of motion. The magnitude of the force, F_m, is therefore also constant. This constant force perpendicular to the direction of motion leads to motion in a circle in a plane perpendicular to the magnetic field (or a spiral around the magnetic field direction if the component of velocity parallel to **B** is non-zero).

The magnetic force constitutes a centripetal force given by

$$F_m = ev_\perp B = \frac{m_e v_\perp^2}{r},$$ (2.19)

where r is the radius of the circular orbit.

So

$$r = \frac{m_e v_\perp}{eB},$$ (2.20)

and the period and frequency of the orbit are

$$T = \frac{2\pi r}{v_\perp} = \frac{2\pi m_e}{eB} \text{ and } \nu = \frac{1}{T} = \frac{eB}{2\pi m_e}.$$ (2.21)

ν is known as the cyclotron frequency. In the non-relativistic case, the electron radiates EM waves continuously at this frequency, which is directly related to the strength of the magnetic field so that the field strength can be inferred directly from the cyclotron frequency.

If the electron moves with a relativistic speed, v, then, as it spirals along the magnetic field direction, it emits radiation into a narrow forward cone, as shown in Figure 2.6. The half-angle of the cone is $\theta \approx 1/\gamma$, where γ is the Lorentz factor of the electron:

$$\gamma = \left[1 - \frac{v^2}{c^2}\right]^{-1/2}.$$ (2.22)

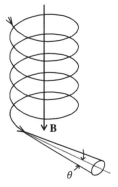

FIGURE 2.6   Synchrotron emission by a relativistic electron.

Therefore, the more relativistic the electrons, the narrower the emission cone.

Such emission is referred to as synchrotron radiation. An observer who happens to be viewing in the right direction will see a burst of radiation each time the emission cone sweeps past. In the electron rest frame, the time for one orbit is the cyclotron period. However, in the observer's frame, the orbital period is increased due to time dilation (moving clocks run slow) by a factor of γ:

$$T_{\text{obs}} = \frac{2\pi\gamma m_e}{eB}. \tag{2.23}$$

The observer only sees emission from a fraction, $2\theta/(2\pi) = 1/(\pi\gamma)$, of the orbital period, so it is visible for a time

$$\Delta t = \frac{2m_e}{eB}. \tag{2.24}$$

However, this interval is made shorter still because the radiation emitted at the end of the interval has a shorter distance to travel than the radiation emitted at the beginning of the interval (Doppler effect), thus compressing further the pulse of radiation observed.

Referring to Figure 2.7, assume that the radiation pulse comes into view at position A, at $t = 0$, when the distance to the observer is x. The observer thus first sees this pulse at time $t_1 = x/c$. The electron goes out of view at point B, having travelled distance L with speed v, at time $t \approx L/v$. It therefore goes out of view at time

$$t_2 = \frac{L}{v} + \frac{x - L}{c}. \tag{2.25}$$

The length of time for which the electron's radiation is observed is therefore

$$\Delta t_{\text{obs}} = t_2 - t_1 = \frac{L}{v} - \frac{L}{c}. \tag{2.26}$$

Since $L \approx v\Delta t$, we have

$$\Delta t_{\text{obs}} = \left[1 - \frac{v}{c}\right]\frac{2m_e}{eB}. \tag{2.27}$$

If the speed is highly relativistic $(v \approx c)$, then

$$1 - \frac{v^2}{c^2} \approx \left(1 + \frac{v}{c}\right)\left(1 - \frac{v}{c}\right) \approx 2\left(1 - \frac{v}{c}\right), \tag{2.28}$$

so

$$\left(1 - \frac{v}{c}\right) \approx \frac{1}{2\gamma^2}. \tag{2.29}$$

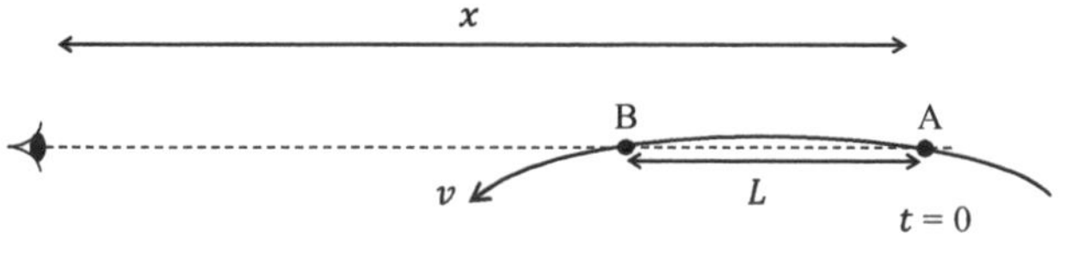

FIGURE 2.7   Electron path during synchrotron emission.

The length of time for which emission is observed is thus

$$\Delta t_{\text{obs}} = \frac{m_e}{\gamma^2 eB}. \tag{2.30}$$

Since γ can be very large, Δt_{obs} can be many orders of magnitude less than Δt. As the electron orbits around the magnetic field lines, the observer sees a sharp pulse every time its "lighthouse beam" sweeps past, as shown in Figure 2.8. The corresponding spectrum is found by taking the Fourier transform of this pulse train and has the shape illustrated in Figure 2.9. The frequency of peak emission is proportional to $1/\Delta t_{\text{obs}}$ and can be shown to be given by

$$\nu_{\text{peak}} = \frac{1}{4\pi} \frac{\gamma^2 eB}{m_e}. \tag{2.31}$$

Note that the frequency of the peak depends on both the magnetic field strength and the electron energy, so it is not possible to evaluate B without knowing the energy or vice versa. Astrophysical synchrotron radiation is generally composed of the superposition of many single-electron spectra. The observed spectrum will thus depend on the energy distribution of the emitting electrons, which is often characterised as a power law: $N(E)dE \propto E^{-m}dE$. In this case, the synchrotron spectrum also turns out to be a power law, with a spectral index, α, that depends on the electron power law index, m:

$$I_\nu \propto B^{\alpha+1} \, \nu^{-\alpha} \quad \text{with } \alpha = \frac{m-1}{2}. \tag{2.32}$$

The spectrum is characterised by a straight line on a log–log plot as shown in Figure 2.10.

FIGURE 2.8   Regular synchrotron emission pulses from a relativistic electron.

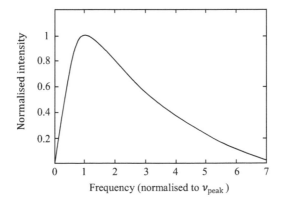

FIGURE 2.9   Single-electron synchrotron spectrum.

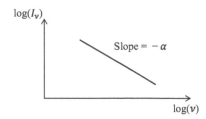

FIGURE 2.10   Typical astrophysical synchrotron emission spectrum.

Synchrotron radiation is generally strongly polarised (see Section 1.12), due to regularity of the magnetic field direction within the emitting region, and this can be a clear signature distinguishing it from other potential mechanisms such as thermal or free–free radiation.

Astrophysical examples of synchrotron emission are (i) active galactic nuclei (AGN) which contain very strong magnetic fields and highly energetic electrons, (ii) radiation from the hot ionised interstellar medium which pervades our own galaxy resulting in a background of synchrotron radiation in the microwave-millimetre region produced by the charged particles spiralling along the Galactic magnetic field lines, and (iii) aurorae produced by solar wind particles interacting with the magnetic fields of the Earth and other planets.

2.5   QUANTUM TRANSITIONS

Quantum mechanical systems have discrete energy states, and transitions between these involve the emission or absorption of precise amounts of energy. The energy change ΔE can be either emitted or absorbed in the form of a photon with frequency given by $h\nu = \Delta E$, resulting in distinct spectral features – emission or absorption lines. The use of the word "line" arose originally from the appearance of atomic absorption or emission features appearing as dark or bright lines within the continuum spectrum of the Sun. The frequency of a spectral line depends on the energy change involved, and different physical systems tend to have features in different regions of the EM spectrum.

An enormous amount of information can be determined from the study of spectral line measurements: atomic or molecular composition, isotopic ratios, temperature, density, velocity, etc. Spectroscopy is a standard tool in almost all areas of science including physics, chemistry, materials science, medicine, astronomy, and remote sensing.

A detailed discussion of the various line emission/absorption mechanisms requires a quantum treatment of the energy states (e.g., Griffiths 2005; Hanel et al. 2003; Hollas 2004; Unger 1970). Here, we merely review the basic mechanisms and the information which may be derived from spectral measurements at different wavelengths.

2.5.1   ATOMIC TRANSITIONS

The simplest atomic system, a hydrogen atom, illustrates some of the key features of atomic spectra. Using the concept of a trapped electron having only certain allowed energy states, in the early 20th century, Bohr devised a model of the hydrogen atom, solving what were two problems at the time: (i) why does the negatively charged electron not get pulled into the nucleus by the positively charged proton and (ii) why does hydrogen only emit or absorb radiation at specific wavelengths?

In Bohr's model, the electron, of mass m_e, is considered to be in a circular orbit around the nucleus (a single proton) with radius r and speed v, with the electric force providing the necessary centripetal acceleration:

$$\frac{e^2}{4\pi\varepsilon_0 r^2} = \frac{m_e v^2}{r} \quad \text{giving} \quad v = \left[\frac{e^2}{4\pi\varepsilon_0 m_e r}\right]^{1/2}. \tag{2.33}$$

The total energy (positive kinetic plus negative potential) of the electron is

$$E = \frac{1}{2}\frac{m_e e^2}{4\pi\varepsilon_0 m_e r} - \frac{e^2}{4\pi\varepsilon_0 r} = -\frac{e^2}{8\pi\varepsilon_0 r}. \tag{2.34}$$

Bohr's quantum mechanical constraint on the allowed energies was based on assuming that the angular momentum, $m_e vr$, of the electron can only have certain discrete values:

$$m_e vr = n\frac{h}{2\pi}, \quad n = 1, 2, 3 \ldots \tag{2.35}$$

This requirement corresponds to the circumference of the orbit being equal to an integer number times the electron wavelength: $2\pi r = nh/(m_e v)$ so that the electron wave function can only have one unique value at any position.

Imposing this condition results in allowed radii given by

$$lr_n = \frac{nh}{2\pi m_e}\left[\frac{4\pi\varepsilon_0 m_e r}{e^2}\right]^{1/2}, \tag{2.36}$$

and corresponding total energies of

$$E_n = -\frac{e^4 m_e}{8\varepsilon_0^2 h^2}\left(\frac{1}{n^2}\right) = -2.18\times10^{-18}\left(\frac{1}{n^2}\right) \text{ J or } -\frac{13.62}{n^2} \text{ eV}. \tag{2.37}$$

Here we have introduced the electron volt, eV, a unit of energy commonly used in atomic spectroscopy and usually used instead of frequency or wavelength to characterise photons in the X-ray and γ-ray regions. One eV is defined as the work done on an electronic charge $(1.6\times10^{-19}$ C$)$ in moving it through a potential difference of 1 V, so 1 eV = 1.6×10^{-19} J. A photon of wavelength λ has an energy of $hc/(\lambda e)$ eV – for example 1 nm $\equiv 1.24$ keV.

The energies of the first few levels in the hydrogen atom are: $E_1 = -13.62$ eV, $E_2 = -3.41$ eV, $E_3 = -1.51$ eV, $E_4 = -0.85$ eV, and so on. The electron can be

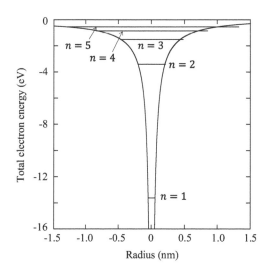

FIGURE 2.11 Total electron energy vs Bohr radius for the hydrogen atom, with the first five allowed energy levels.

regarded as trapped in the potential well of the proton, with allowed energies restricted to particular values. Figure 2.11 shows the total electron energy vs radius, with the first five allowed energy levels indicated.

The first seven energy levels of the hydrogen atom are shown in Figure 2.12. The vertical arrowed lines indicate the possible transitions from upper levels down to the ground state $(n = 1)$ and the first and second excited states $(n = 2, 3)$, with the emission of a photon of energy equal to the corresponding energy difference. Transitions to or from the ground state correspond to the Lyman spectral series, with wavelengths given in Table 2.1. All of these wavelengths are in the UV part of the spectrum. Similarly, transitions involving $n = 2$ and higher levels are associated with the Balmer series of lines in the visible region, and transitions up from or down to $n = 3$ give rise to the Paschen series of lines, in the infrared region.

Hydrogen gas, which is very common in the Universe, can emit or absorb at these wavelengths giving rise to corresponding emission or absorption lines.

The spectral series continues into the FIR and radio regions. Highly excited H ions form when free electrons recombine with protons to form H atoms, and the electrons cascade down to the ground state with photon emission at each downward transition. High-n transitions (e.g. $n = 110 - 109$) are observable at radio wavelengths and are referred to as radio recombination lines. Astrophysical

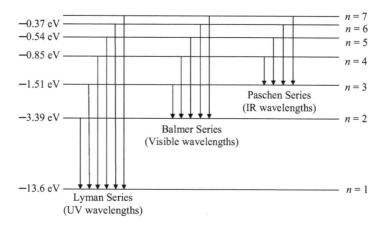

FIGURE 2.12 Energy-level diagram for hydrogen showing Lyman transitions to the ground state, $(n = 1)$, Balmer transitions to the first excited state $(n = 2)$, and Paschen transitions to the second excited state $(n = 3)$. The corresponding energy levels, referenced to the energy of a free electron, defined to be zero, are indicated on the left-hand scale.

TABLE 2.1

Lyman Spectral Series of H Atom Transitions from Higher-Energy Levels to the Ground State

Transition	Wavelength (nm)
2 – 1	121.6
3 – 1	102.6
4 – 1	97.3
5 – 1	95.0
6 – 1	93.8
-	-
∞ -1	91.2

observations of these lines are valuable since the interstellar medium is relatively transparent at radio wavelengths, being free of dust absorption which masks observations at optical/IR wavelengths. For example, measurements of recombination lines from H and other atoms can be used to determine the temperature of deeply embedded stellar objects in obscured star-forming regions.

The ground state of the hydrogen atom is split into two not-quite-equal energy levels due to the two possible configurations of the electron and nuclear spins. The lowest energy state occurs when the spins are antiparallel, with total spin=0, with a slightly higher-energy state when the spins are parallel, with total spin=1. This is referred to as hyperfine splitting, as the energy difference between the two states is very small, $\Delta E = 5.87 \times 10^{-6}$ eV. This type of splitting is present in all energy levels but is so small that, in general, it is not observable even in high-resolution spectra. However, the spin-flip transition from the higher to the lower ground state is observed at a frequency of $v = \Delta E/h = 1.43$ GHz, corresponding to a wavelength of 21 cm. The higher-energy state can be populated through the interaction of hydrogen atoms with the cosmic microwave background 2.7-K radiation field in which the photons have typical energies of $\sim 2 \times 10^{-4}$ eV. Thus, we can observe neutral hydrogen clouds in emission throughout our galaxy, and in other galaxies, unperturbed by dust extinction which is negligible at this wavelength. The line intensity provides an estimate of the cloud mass and the observed Doppler shift traces the motion along the line of sight. In dense interstellar clouds, where stars generally form, most of the hydrogen is in molecular form, H_2, which involves the spins of the two orbital electrons being locked in an antiparallel configuration, so there is no 21-cm emission. Molecular hydrogen can only be observed via infrared vibrational transitions, as described in the next section.

Electron energy levels within more complex atoms are also quantised, their states being described by four quantum numbers: the principal quantum number, n; the angular momentum quantum number, l; the magnetic quantum number, m; and the electron spin quantum number, s.

n	specifies the main energy level (orbit or "shell")	$n = 1, 2, 3 \ldots$
l	specifies the orbital angular momentum states	$l = 0, 1, 2 \ldots (n-1)$
m	specifies the direction of the orbital angular momentum vector	$m = -l, \ldots 0 \ldots + l$
s	specifies the electron spin	$s = \pm \ 1/2$

Transitions between the numerous allowed energy levels result in photon emission or absorption, with wavelengths in the optical-UV-X-ray region, that are characteristic of the particular atom or ion, and provide information on the composition of the gas and on its ionisation state, and thus on its temperature. Atomic spectra exhibit complex structure due to the principal energy levels being split by the interaction of the electron orbital and spin vectors, giving rise to intrinsic fine structure in the spectrum. The state energy is different for the spin-up and spin-down cases (i.e. electron spin reinforcing the orbital angular momentum or opposing it). Thus, transitions associated with these states appear as pairs of very closely spaced spectral lines, referred to as the "fine structure".

If there is an external magnetic field, this allows only a quantised set of particular relative orientations of an atom's angular momentum vector with respect to the field, resulting in spectral line splitting known as the Zeeman effect. The degree of splitting depends on the magnetic field strength, something that can be exploited to measure astrophysical magnetic fields.

2.5.2 MOLECULAR VIBRATIONAL TRANSITIONS

Molecules are composed of atoms bound together by Coulomb forces arising from the sharing of electron orbitals. This can lead to charge asymmetries within the molecule and hence electric dipoles. The thermal vibration of the molecule is quantised such that only particular vibrational states are allowed, and transitions between the vibrational energy levels occur with photon emission or absorption. The energy levels are lower than for atomic transitions (typically ~ 0.1 eV), resulting in radiation at longer wavelengths, in the infrared part of the spectrum.

A simple diatomic molecule can be modelled as an oscillator with quantised energy levels within the potential well defined by the energy vs distance curve for the molecule, as depicted in Figure 2.13. The equilibrium separation between the atoms, r_e, corresponds to the bond length (typically 100–200 pm). The difference between the energy at r_e and the zero level for large separation is the bond strength (typically a few eV). The energy levels are described by the vibrational quantum number, v.

More complex molecules can be asymmetric and can have many possible vibrational modes, each of which has a characteristic frequency. The number of modes depends on the number of atoms. In a molecule with N atoms, each atom can move in three dimensions, and so there are $3N$ degrees of freedom. Six of these correspond to the motion of the molecule as a whole, with no associated change in shape (three for the position in space and three for the angles defining its orientation). Thus, the total number of degrees of freedom (vibrational modes) is $3N - 6$. For linear triatomic molecules, this becomes $3N - 5$ as one degree of freedom is removed. If two or more of the frequencies are identical, the modes are said to be degenerate. The astrophysically common water molecule, H_2O, provides an example of a non-linear triatomic molecule. It has $(3 \times 3) - 6 = 3$ normal modes of vibration. These vibrational modes are conventionally labelled v_1, v_2, and v_3, as illustrated in Figure 2.14.

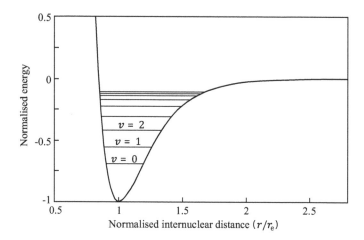

FIGURE 2.13 Potential energy (normalised to the bond strength) vs internuclear distance (normalised to the bond length, r_e) and vibrational energy levels for a diatomic molecule.

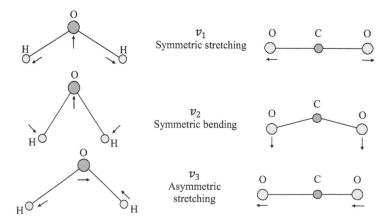

FIGURE 2.14 Fundamental vibrational modes, v_1, v_2, v_3 for the water and carbon dioxide molecules.

The vibrational modes of the linear triatomic molecule CO_2 are also shown in Figure 2.14. There are four vibrational modes, but the two bending modes are degenerate due to molecular symmetry. In addition, the symmetric stretch vibration is spectrally inactive as there is no dipole moment associated with it, so only two CO_2 infrared vibrational spectral features are seen.

2.5.3 MOLECULAR ROTATIONAL TRANSITIONS

Molecules with a finite electric dipole moment can undergo transitions between different rotational energy levels. The energy differences are typically ~ 1 meV, corresponding to the radio or millimetre/submillimetre spectral regions. The energy of the rotating molecule is quantised in terms of its total angular momentum quantum number, J:

$$E = BJ(J+1), \tag{2.38}$$

where the parameter B is known as the rotational constant of the molecule and is related to the molecule's moment of inertia, I, by

$$B = \frac{h^2}{8\pi^2 I}. \tag{2.39}$$

For a transition between two adjacent levels defined by upper level $J+1$ and lower level J, the energy difference is

$$\Delta E = 2B(J+1). \tag{2.40}$$

For the CO molecule, $B = 3.837 \times 10^{-23}$ J, corresponding to 57.636 GHz. The $J = 1 - 0$ transition thus occurs at a frequency of 115.27 GHz, the $J = 2 - 1$ transition is at 230.6 GHz, and so on. Rotational lines of CO provide a very important probe of the density and temperature of molecular hydrogen gas in the cold dense clouds from which stars form. CO molecules are well mixed with molecular hydrogen, and their rotational states are collisionally excited by hydrogen molecules. For optically thin lines, the measured CO line intensity depends on the total mass of molecular hydrogen within the telescope beam. This provides an indirect tracer of the cold molecular hydrogen, which is not directly detectable because the temperature is too low to excite H_2 vibrational emission. In addition, the relative intensities of the rotational lines in the sequence depend on the population of the different rotational energy levels, which in turn depends on the temperature of the gas – at low temperatures, low-J lines are most prominent, while at higher temperatures higher-J lines dominate – and by plotting the line intensity against the rotational level, the gas temperature can be inferred.

2.5.4 VIBRATIONAL-ROTATIONAL TRANSITIONS

Molecular vibration and rotation can occur together, so transitions can involve changes in both vibrational and rotational quantum numbers, v and J. This gives rise to spectra in which the vibrational transitions (coarse structure) have fine rotational structure around each line.

For a certain vibrational transition, the spectrum is centred at the corresponding vibrational frequency and has two series of rotational lines on either side of the fundamental vibrational frequency. On the higher-energy/frequency side is the R branch ($\Delta J = 1$), and the lower energy/frequency set is called the P branch ($\Delta J = -1$). Figure 2.15 shows the spectral signature exhibited by a CO molecule undergoing vibration/rotation transitions. Note that $\Delta J = 0$ is not allowed for this molecule, so the fundamental vibrational transition (labelled the Q branch) does not appear.

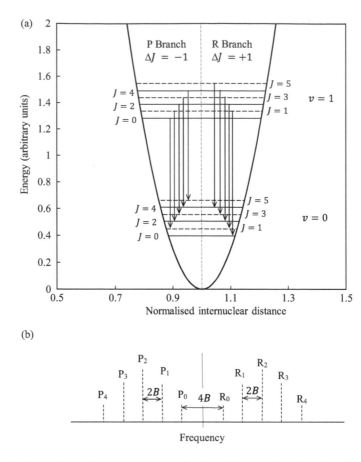

FIGURE 2.15   (a) Energy-level diagram showing the P and R branches for CO; (b) the corresponding observed distribution of spectral lines.

2.5.5 LINE BROADENING

The natural line width (spread of frequencies) of a spectral line depends on the lifetime of the state from which the transition occurs. The exact time at which a transition occurs is probabilistic, and in a population of emitting atoms or molecules, there will thus be a distribution of state durations. For a state with typical lifetime, τ, the uncertainties in the energy, E, or corresponding frequency, ν, of the emitted photon, are given by the uncertainty principle (equation 1.33):

$$\Delta E \Delta t = \Delta E \tau = h \Delta \nu \tau \sim \hbar \Rightarrow \Delta \nu \sim \frac{1}{2\pi\tau}. \tag{2.41}$$

This spread in frequencies is known as natural line broadening. With random transitions, the distribution of state durations is exponential (analogous to the exponential decline in the quantity of a radioactive substance). Under these conditions, the normalised intensity of the line profile can be shown to be given by the Lorentzian function:

$$L(\nu) = \frac{I(\nu)}{I(\nu_o)} = \frac{1}{\pi} \frac{\nu_L/2}{(\nu_L/2)^2 + (\nu - \nu_o)^2}. \tag{2.42}$$

where ν_o is the central frequency of the line and $\Delta\nu_L$ is the full width at half maximum (FWHM) line width.

Natural line widths are usually very narrow compared to the emitting frequency. For example, for atomic species emitting at optical wavelengths $(\sim 500 \text{ nm} \equiv 6 \times 10^{14} \text{ Hz})$, with lifetimes $\sim 10^{-9}$ s, the natural line widths are $\sim 10^9$ Hz, so $\nu/\Delta\nu \sim 6 \times 10^5$. However, the observed line widths in astronomical spectra are usually much broader than the natural line widths. This is because other factors influence the distribution of frequencies, including Doppler shifts due to random thermal or turbulent velocity distributions within the observed region, or collisions between atoms/molecules reducing the lifetimes of energy states. These processes modify both the line shape and the line width in ways that depend on the physical parameters of the region.

2.5.5.1 Collisional (Pressure) Broadening

If gas is sufficiently dense, it becomes significantly probable that an atom or molecule in the process of making a transition collides with another molecule. The collision disrupts the phase continuity of the transition by effectively reducing its lifetime and hence increasing its line width, $\Delta\nu$. If collisions are very frequent, they may also reduce the lifetime of the states, also leading to broadening of the line. Such conditions are common in astrophysics, for instance in the atmospheres of stars and planets, and collisional broadening can often lead to line widths several orders of magnitude larger than the natural line width.

The collisionally broadened line profile also has a Lorentzian shape, except that the width is now determined by the collision rate, and therefore the pressure, of the gas. The collisional linewidth, $\Delta\nu_c$, depends on the mean time between collisions and, for a given temperature, is proportional to the pressure.

2.5.5.2 Thermal (Doppler) Broadening

Natural and collisional broadening are examples of homogeneous broadening, in which each atom or molecule behaves similarly and can contribute to any frequency across the line profile. In contrast, Doppler broadening is an example of what is termed inhomogeneous broadening. In this case, the overall line shape consists of contributions from many radiating particles, or many segments of trajectories between collisions of one particle, some moving towards and some away from the observer.

If a particular quantum transition results in emission at frequency, ν_{em}, the frequency observed, ν_{obs}, when the particle is moving along the line of sight with non-relativistic speed, v, is Doppler shifted by an amount $\Delta\nu$ given by

$$L\Delta\nu = \nu_{\text{obs}} - \nu_{\text{em}} = \left(\frac{v}{c}\right)\nu_{\text{em}}, \tag{2.43}$$

where c is the speed of light. The observed frequency is blueshifted $(\nu_{\text{obs}} > \nu_{\text{em}})$ if the emitter is moving towards the observer (positive, v) or redshifted $(\nu_{\text{obs}} < \nu_{\text{em}})$ if the emitter is moving away from the observer (negative v).

For a gas of particles of mass m at temperature, T, the thermal velocities along any line of sight have a Maxwellian velocity distribution, $f(v)$, given by

$$f(v) = \left(\frac{m}{2\pi k_B T}\right)^{1/2} e^{-\frac{mv^2}{2k_B T}}. \tag{2.44}$$

The velocity distribution is thus Gaussian with zero mean and a typical thermal velocity along the line of sight of $v_{\text{th}} = \sqrt{2k_B T/m}$, corresponding to a Doppler frequency shift of

$$\Delta\nu_D = \frac{\sqrt{2k_B T/m}}{c} \nu_{\text{em}}. \tag{2.45}$$

In addition to the thermal motions of the microscopic gas particles, there may also be macroscopic turbulent motions within the emitting region, resulting in additional broadening with

$$\Delta v_{\mathrm{D}} = \frac{v_{\mathrm{em}}}{c}\left(\frac{2k_{\mathrm{B}}T}{m} + \Delta v_{\mathrm{t}}^2\right)^{1/2},$$

(2.46)

where Δv_{t} characterises the typical line of sight turbulent velocity.

This distribution of speeds results in a Gaussian distribution of observed intensity given by

$$\frac{I(v)}{I(v_0)} = G(v) = \frac{1}{\sqrt{\pi}\Delta v_{\mathrm{D}}} e^{-\left(\frac{v - v_{\mathrm{em}}}{\Delta v_{\mathrm{D}}}\right)^2}.$$

(2.47)

Note that the collisional broadening linewidth is independent of the frequency of the line, but the Doppler line broadening is proportional to frequency. Thus, for a gas of given temperature, density, and composition, at low frequencies the Doppler width will be much less than the collisional width, while at higher frequencies the opposite may be the case. For atomic transitions in the optical/UV region, Doppler broadening is the dominant effect.

When both Doppler and collisional broadening are present, the function that describes the composite line profile is a convolution of the Gaussian, $G(v)$, and Lorentzian, $L(v)$, functions

$$I(v) = \int_{v'} G(v)L(v - v')\mathrm{d}v'.$$

(2.48)

This generalised line shape, introduced by Voigt, has no analytical expression, but it can easily be computed numerically. The core of the line (within the FWHM) can be closely approximated by a Gaussian, while Lorentzian or Voigt profiles are needed to match the wings of the line. Figure 2.16 shows the shapes of the Gaussian, Lorentzian, and Voigt profiles. All three curves have been normalised to their peak values. The Gaussian and Lorentz profiles each have the same FWHM of unity, and the Voigt profile, being the convolution of those two, is broader.

In addition to the information gleaned from observing the frequencies and intensities of spectral lines (atomic and molecular composition, excitation, and number densities), measurements of line profiles can be used to determine the physical conditions of a region (gas velocities, density, and pressure). Extracting line shape information requires using a spectrometer with sufficient resolution

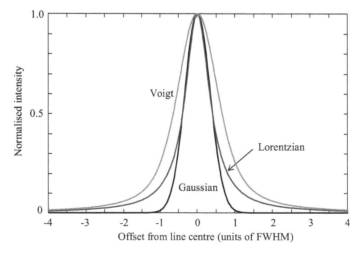

FIGURE 2.16 Lorentzian, Voigt, and Gaussian line shapes. The Voigt line shape is the convolution of the Gaussian and Lorentzian profiles.

to resolve the line features of interest, and this is often one of the main drivers of astronomical instrumental design.

2.5.6 SPECTRAL LINE DATABASES

Modelling of the spectral emission from astronomical sources, and the identification of lines and interpretation of observational data, requires accurate knowledge of the spectral characteristics of many atomic and molecular species. Spectral line databases provide compilations of the key parameters of spectral lines (energy levels, wavelengths, transition probabilities, etc.). Commonly used databases include the CHIANTI (Dere et al. 1997) and NIST (Kramida et al. 2019) databases for atomic and ionic spectroscopy, and the GEISA (Jacquinet-Husson et al. 2016) and HITRAN (Hill et al. 2016) databases for molecular spectroscopy.

2.5.7 γ-RAY LINE EMISSION

Spectral lines in the γ-ray regions can be produced by excited atomic nuclei or by the annihilation of electron-positron pairs.

The allowed energies of a proton within a nucleus are quantised, and a nucleus in an excited state can emit a photon in transitioning to a lower energy state. The forces involved are much greater than for electronic transitions, resulting in energy levels of a few MeV rather than the few eV typical of electrons in atoms. The photons emitted in nuclear transitions are thus in the γ-ray part of the spectrum. For example, unstable nuclei created in large quantities by supernova explosions emit γ-ray spectral lines.

Electron-positron annihilation results in a pair of γ-ray photons each with energy equal to the electron rest mass (0.511 MeV). Positrons can be created in nuclear reactions in supernovae and on the surfaces of neutron stars, and through photon–photon interactions in the intense gravitational fields near black holes.

2.6 POLARISED RADIATION IN ASTRONOMY

As well as the polarised synchrotron radiation from a variety of astrophysical sources mentioned in Section 2.4, several other mechanisms can produce polarised radiation.

The Zeeman effect discussed in Section 2.5.1 also involves polarised emission or absorption. The alignment of the angular momentum vector with respect to the source magnetic field leads to the emitted or absorbed light possessing an inherent linear or circular polarisation state that depends on the field magnitude, the specific transitions and the viewing geometry. Zeeman polarisation measurements are used to characterise the magnetic field structure in the Sun's atmosphere and in astrophysical maser sources.

Interstellar dust grains are mostly non-spherical, and can become preferentially aligned with respect to the local magnetic field. This leads to polarisation of scattered starlight in the visible region, and also to polarised thermal emission by the dust particles, appearing in the far infrared and submillimetre region. These phenomena can be used to trace the structure and strength of the magnetic field in the interstellar medium and in the dusty molecular clouds in which stars form.

Rotation of the direction of polarisation of a polarised beam occurs when it travels through a physical medium in which there is a magnetic field, a phenomenon known as Faraday rotation. In astronomy, this occurs when a polarised wave travels through ionised gas in the interstellar medium which is influenced by the Galactic magnetic field. A plane polarised wave can be represented as the sum of two circularly polarised components, one left-handed and one right-handed. These two waves have slightly different speeds of propagation due to the interaction between the waves and free electrons in the presence of the local magnetic field. This means that the refractive index of the interstellar medium is slightly different for the two components. The amount of rotation depends on the electron number density, the path length, and the magnetic field strength along the line of sight. It also increases as the square of the wavelength, so the effect is most important at radio wavelengths. The wavelength dependence can be used to determine that Faraday rotation has taken place and to quantify the effect. Radio polarimetry thus provides a means of measuring the interstellar plasma properties and/or source properties and if a number of sources are studied across the sky, the 3-D structure of the Galactic magnetic fields can be investigated.

There are anisotropies in the 2.7-K CMB radiation, with a typical magnitude of 1 part in 10^5, which are also inherently polarised at the level of about 10% (so that the polarised component is only about 1 part in 10^6). Linear polarisation is induced by density inhomogeneities and metric perturbations (gravitational waves) during the epoch of recombination, through the elastic scattering of CMB photons by free electrons (Thomson scattering; see Chapter 3). Polarimetry of the CMB anisotropies is very challenging experimentally but is a very powerful probe of the physical conditions during the big bang.

3 Interaction of Electromagnetic Radiation with Matter

In this chapter, we review some of the most important processes by which electromagnetic radiation interacts with matter, with particular emphasis on the detection of the radiation and on the effects of the Earth's atmosphere on the incoming radiation from astronomical sources. The implications for astronomical observations in different parts of the spectrum will be considered in later chapters. The interaction between the radiation and the medium in which it propagates can include absorption, scattering, refraction, and reflection. In the case of a detector, we are mainly interested in absorption, with the additional requirement that the absorbed energy be readily measurable in some way. In general, EM radiation interacts with matter by transferring energy to charged particles (almost always electrons, except at γ-ray energies), in a way that depends strongly on the photon energy (frequency) of the radiation and on the nature and conditions of the medium.

3.1 RADIO WAVELENGTHS $\left(\lambda >\sim 1\,\text{mm}\right)$

If an electron in the material is bound to an atom, it can only gain energy in amounts equal to or greater than the difference between its energy and that of the next highest quantum state. At radio wavelengths, the photon energy is too small to generate new charge carriers in any material. The radiation can therefore only interact if there are free carriers already in the absorbing material, so it must be a conductor, such as a metal, an electrolyte, or a plasma. Radio waves can propagate through significant thicknesses of non-conducting media such as air and concrete but are strongly affected by metal structures.

In the radio region, it is more useful to regard the incident radiation not as a stream of photons, but as a wave – a travelling oscillation of the electric and magnetic fields $\mathbf{E}$ and $\mathbf{B}$. These oscillating electric and magnetic fields can influence the motion of free charge carriers. In a metal, the electrons are not bound to the lattice atoms and form a dense "free electron gas". A radio wave incident on a metal will therefore induce oscillating electric currents, as shown in Figure 3.1. An electric field oscillation exerts an oscillating force on the electrons in any metal structure, leading to an oscillating current. An oscillation of the magnetic field leads to an oscillating magnetic flux through the loop, resulting in an oscillating current. These induced currents will, in turn, generate EM waves, resulting in the reflection of the incident power. However, some of the power in the induced currents can be extracted by connecting up the metal structure to a suitable detector.

3.1.1 THE EARTH'S IONOSPHERE

In the case of radio waves propagating in a plasma, the behaviour depends on the frequency of the wave and on the electron density in the plasma. An important case for astronomical observations is the influence on radio waves of the Earth's ionosphere, an ionised layer in the atmosphere at an

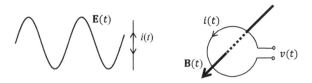

FIGURE 3.1   Generation of currents by the interaction of electric and magnetic fields with metal wires.

altitude between 80 and several hundred kilometres, created and maintained by photoionisation due to UV light from the Sun. As a layer of low-density plasma, it is different from the neutral atmosphere in that electrons and ions are free to move under the influence of electromagnetic fields. The electrons, being much lighter, are much more responsive to electromagnetic waves than the ions and exhibit oscillatory motion in response to the electric field vector, producing a current at the same frequency as the incident electric field.

Consider a typical electron in the plasma and a corresponding ion. We can assume that because the ion is so much heavier, it remains stationary while the electron oscillates in response to the electric field. The electron-ion pair thus constitutes an electric dipole with a sinusoidally varying separation – i.e., an oscillating dipole moment.

Let the incident electric field, with amplitude E_0 in the x-direction and angular frequency ω, be represented by

$$E = E_0 e^{i\omega t}. \tag{3.1}$$

The force exerted on an electron, of mass m_e, is

$$F = m_e \frac{d^2 x}{dt^2} = -eE = -eE_0 e^{i\omega t}. \tag{3.2}$$

Integrating this expression gives

$$x = -\left[\frac{eE_0}{m_e \omega^2}\right] e^{i\omega t} = -\left[\frac{e}{m_e \omega^2}\right] E. \tag{3.3}$$

The electron thus exhibits simple harmonic motion out of phase with the electric field. Taking the static ion into account, the magnitude of the corresponding dipole moment (charge multiplied by distance) is

$$p = -\left[\frac{e^2}{m_e \omega^2}\right] E. \tag{3.4}$$

If the number density of electrons is n_e, and if the density is not so high that the dipoles interact significantly with each other, then the magnitude of the total dipole moment (electric polarisation) is

$$P = -\left[\frac{n_e e^2}{m_e \omega^2}\right] E. \tag{3.5}$$

For a given field strength, the magnitude of the polarisation decreases as the frequency increases. This is because the force on the electrons is proportional to the fixed field strength but the time for which it can act before the polarity is reversed is smaller for higher frequencies, resulting in a smaller amplitude of oscillation.

In any medium, the electric displacement vector, $\mathbf{D}$, is related to the polarisation vector, $\mathbf{P}$, by

$$\mathbf{D} = \varepsilon_0 \varepsilon_r \mathbf{E} = \varepsilon_0 \mathbf{E} + \mathbf{P}, \tag{3.6}$$

where ε_r is the relative permittivity of the medium.

Therefore

$$\varepsilon_0 \varepsilon_r E = \varepsilon_0 E \left[1 - \frac{n_e e^2}{\varepsilon_0 m_e \omega^2}\right], \tag{3.7}$$

Giving

$$\varepsilon_r = 1 - \frac{n_e e^2}{\varepsilon_0 m_e \omega^2} = 1 - \frac{\omega_p^2}{\omega^2}. \tag{3.8}$$

The plasma frequency is defined as

$$\nu_p = \frac{\omega_p}{2\pi} = \frac{1}{2\pi} \left(\frac{n_e e^2}{\varepsilon_0 m_e} \right)^{1/2} \approx 9 n_e^{1/2}, \tag{3.9}$$

with n_e in electrons/m³.

For frequencies much higher than the plasma frequency, the electrons undergo simple harmonic motion with low amplitude because they cannot respond quickly enough to the incident wave. The relative permittivity is close to but less than unity, and so the refractive index of the plasma $\left(n = \varepsilon_r^{1/2} \right)$ is also just less than unity. The electrons are not interacting strongly with each other or the ions, and there is no significant mechanism for dissipation of the wave's energy, so it is transmitted with a small amount of refraction. As the frequency decreases and approaches ω_p, the amplitude of the electron oscillations increases as they become more strongly coupled to the field oscillations, the refractive index decreases, and the angle of refraction increases. When the refraction angle becomes $\sim 90°$, the wave is no longer transmitted. A signal coming up from below is reflected back down towards the ground (allowing low-frequency radio transmission over long distances), and a signal coming from outside the atmosphere is reflected back into space.

The electron density in the ionosphere varies considerably from day to night, and with altitude. A typical value of 2×10^{12} m⁻³ for the F-layer (altitude 150–300 km) corresponds to a plasma frequency of ~13 MHz (wavelength ~24 m), and most of the time ground-based radio astronomy observations are not possible at wavelengths longer than this.

3.1.2 ANTENNAS

An antenna is a particular metal structure designed to transmit or receive EM waves of a particular direction and frequency. A detailed description of the many kinds of antennas that are used for communications and remote sensing is beyond the scope of this book. Here, we will only define some of the key parameters that characterise all antennas. It can be shown that these properties of an antenna are the same whether it is being used to transmit or to receive. This equivalence of properties for transmission and reception is known as the principle of reciprocity.

The gain, G, of an antenna is defined as

$$G = \frac{\text{Maximum power radiated per unit solid angle}}{\text{Power radiated per unit solid angle if the radiation were isotropic}}. \tag{3.10}$$

An isotropic radiator has no preferred direction and power is radiated uniformly. The power per unit solid angle is equal to $P/(4\pi)$ where P is the total power output. A directional antenna radiates preferentially in a certain direction with a maximum value of power per unit solid angle of $GP/(4\pi)$.

The full width half maximum (FWHM) beam-width, $\Delta\theta_{FWHM}$, is defined as the angle between the directions in which the gain has dropped to half of the maximum value, as shown in Figure 3.2.

An antenna beam will have some angular profiles, $G(\theta, \phi)$, where θ is the angle with respect to the direction of peak emission and ϕ is the angle around that direction (with $G(0,0)$ denoted by G). Typically, a beam profile will have a main lobe, into which most of the power is radiated, and smaller sidelobes with sizes and shapes depending on the details of the design.

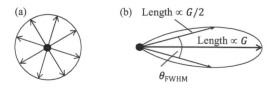

FIGURE 3.2 Depiction of beam patterns of an isotropic radiator (a) and an antenna with gain G (b).

The beam solid angle, Ω_{Beam}, is defined as

$$\Omega_{\text{Beam}} = \oint_{4\pi} B(\theta, \phi)\,d\Omega, \tag{3.11}$$

where $B(\theta, \phi)$ is the normalised gain,

$$B(\theta, \phi) = \frac{G(\theta, \phi)}{G}. \tag{3.12}$$

For an isotropic radiator, $\Omega_{\text{Beam}} = 4\pi$. The smaller the value of Ω_{Beam}, the more strongly directional is the antenna.

It is often a reasonable approximation to regard the beam as being cylindrically symmetrical (i.e., G is independent of ϕ), and as having a Gaussian profile in which the contribution of the main lobe dominates the solid angle so that

$$B(\theta) = e^{-\left(\frac{\theta}{\theta_0}\right)^2}. \tag{3.13}$$

The parameter θ_0 determines the width of the main lobe, with $\Delta\theta_{\text{FWHM}}$ being given by the difference between the positive and negative angles at which $B(\theta) = \frac{1}{2}$:

$$\Delta\theta_{\text{FWHM}} = \left[4\ln(2)\right]^{1/2}\theta_0 = 1.665\theta_0. \tag{3.14}$$

In the case of a Gaussian beam, the beam solid angle is given by

$$\Omega_{\text{Beam}} = \int_0^\infty e^{-\left(\frac{\theta}{\theta_0}\right)^2} 2\pi\theta\,d\theta. \tag{3.15}$$

Using the standard integral

$$\int_0^\infty e^{ax^2} x\,dx = \frac{1}{2a}e^{ax^2}, \tag{3.16}$$

this gives

$$\Omega_{\text{Beam}} = \pi\theta^2 = \frac{\pi(\theta_{\text{FWHM}})^2}{4\ln(2)} \approx 1.13(\theta_{\text{FWHM}})^2. \tag{3.17}$$

An antenna has an effective area, A_e, corresponding to the area over which EM radiation is emitted or received. It depends on the size and geometry of the antenna. Consider an antenna viewing a source of intensity I_ν that fills its beam. If the system is sensitive over a frequency range $\Delta \nu$, then, from the definition of intensity, the power intercepted is given by

$$P = A_e \Omega_{\text{Beam}} \Delta \nu \, I_\nu, \tag{3.18}$$

where in SI units we have $W \equiv (m^2)\,(sr)\,(Hz)\,(W\,m^{-2}\,sr^{-1}\,Hz^{-1})$. Note that P is proportional to the product, $A_e \Omega_{\text{Beam}}$, of the effective area and the solid angle. This product is known as the throughput of the antenna.

For a dish-type antenna of diameter D, operating at wavelength λ, diffraction at the primary aperture (see Section 4.6) limits the beam-width to

$$\theta_{\text{FWHM}} \approx \frac{\lambda}{D}. \tag{3.19}$$

If we take the effective area to be equal to the geometrical area, then

$$A_e \approx \frac{\pi D^2}{4}, \tag{3.20}$$

so that

$$A_e \Omega_{\text{Beam}} \approx \frac{\pi D^2}{4} \frac{\pi \left(\dfrac{\lambda}{D}\right)^2}{4 \ln(2)} \approx \frac{\pi^2}{16 \ln(2)} \lambda^2 \approx 0.9 \lambda^2. \tag{3.21}$$

A more rigorous calculation yields the result that $A_e \Omega_{\text{Beam}} = \lambda^2$ exactly, and furthermore, it can be shown that this result holds for *any* kind of antenna (Kraus 1986).

The gain and beam solid angle are related: if we assume for simplicity that the total emitted power is radiated into solid angle Ω_{Beam}, then the antenna gain is

$$G = \frac{P/\Omega_{\text{Beam}}}{P/(4\pi)} = \frac{4\pi}{\Omega_{\text{Beam}}}. \tag{3.22}$$

Consequently, the gain can be expressed as

$$G = \frac{4\pi A_e}{\lambda^2}. \tag{3.23}$$

So for high gain (i.e., radiation in a well-defined direction), we need a large area or a short wavelength or both. The characteristics of antennas are further discussed in Chapter 8.

3.1.3 DETECTION OF THE INTERCEPTED RADIATION

The currents induced in an antenna are proportional to the magnitude of the electric or magnetic fields intercepted by the antenna. The corresponding electromagnetic power is proportional to the square of the field amplitude. To produce an electrical signal proportional to the incident power, the signal current from the antenna, $i(t)$, can be applied to a device with a non-linear voltage–current characteristic (e.g. a diode) to generate a voltage $v(t) \propto E^2(t)$, as shown in Figure 3.3.

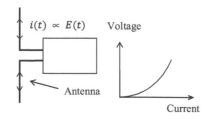

FIGURE 3.3 An antenna produces a current proportional to the incident electric field, and a square-law detector produces a voltage proportional to electromagnetic power.

This approach to detection is valid for long wavelengths (λ > a few millimetres; v > ~ 100 GHz). At shorter wavelengths, problems arise because the size of practical components gets large in relation to the wavelength, and because transmission line or waveguide losses between the antenna and the detector, and losses in the detector itself, get worse with increasing frequency. For this reason, most high-frequency radio receivers use a technique known as mixing, which involves combining the input signal with a local oscillator to convert the information to a more convenient lower frequency. This will be described in detail in Chapter 8.

3.2   INFRARED-OPTICAL-UV WAVELENGTHS ($\lambda = 1$ mm–30 nm)

As the wavelength decreases and enters the millimetre regime, it becomes smaller than the size of practical detectors and circuit components, and the photon energy, $h\nu$, becomes large enough to interact directly with bound electrons. Hence, it is more convenient to regard the radiation as a stream of discrete photons rather than a wave. As the first step in the detection process, the photons must be absorbed and converted to either a measurable rise in temperature (a bolometric detector) or to charge carriers that can move under the influence of an electric field (a photodetector).

3.2.1   PHOTON ABSORPTION MECHANISMS

The manner in which a photon interacts with a material depends strongly on the conducting properties of the material. High-conductivity materials, such as metals, do not absorb – they are good reflectors and so are unsuitable for use as detectors (but are useful as mirrors). In the case of a resistive material, the currents flowing in the material as a result of interaction with the incident wave result in $i^2 R$ dissipation and consequent heating. The degree of heating is proportional to the absorbed power. This is the principle of the bolometric detector (described in Chapter 9). Dielectric materials (insulators) may be transparent or absorbing, depending on the material properties and thickness, and the wavelength. If the photon energy is great enough, then ionisation can occur through the photoelectric effect, and the liberated electron may remain inside the material or may escape. The photoelectric effect is also important as a mechanism for X-ray absorption (see Chapter 11).

Many photon detectors use semiconductors, materials for which the conductivity can be increased by the absorption of photons through the promotion of electrons to higher energy levels.

3.2.2   ELECTRON PROMOTION IN SEMICONDUCTORS

There are two basic types of semiconductors: intrinsic (pure material), and extrinsic (doped material), for which the energy band structures are depicted in Figure 3.4. Intrinsic semiconductors have no available energy states between the valence band (VB) and the conduction band (CB). If the energy of an incoming photon is greater than the energy gap between the valence and conduction bands, ΔE, it can be absorbed by a valence band electron which thus

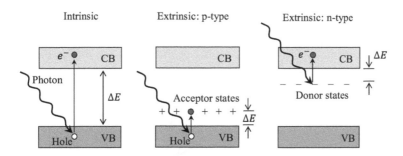

FIGURE 3.4   Electron energy band structure and photon absorption in intrinsic and doped semiconductors.

TABLE 3.1

Energy Gaps at 300 K and Corresponding Cut-Off Wavelengths for Some Semiconductor Materials

Material	$\Delta E\,(eV)$		$\lambda_c\,(\mu m)$	
	300 K	**77 K**	**300 K**	**77 K**
Si	1.12	1.16	1.11	1.07
Ge	0.66	0.73	1.88	1.69
InSb	0.17	0.23	7.19	5.31
GaAs	1.42	1.51	0.87	0.82
InP	1.35	1.41	0.92	0.88

Source: Data from Siklitsky & Tolmatchev (2019).

---

gains enough energy to move up to the conduction band. The condition for absorption is therefore $h\nu \geq \Delta E$ or $hc/\lambda \geq \Delta E$. Therefore,

$$\lambda \leq \lambda_c = hc/\Delta E. \qquad (3.24)$$

λ_c is called the cut-off wavelength: charge promotion cannot occur for wavelengths greater than λ_c. Both the electron in the conduction band and the positive hole left in the valence band can move under the influence of an applied electric field.

The energy gaps and cut-off wavelengths, at room temperature and 77 K, of a few commonly used intrinsic semiconductors are listed in Table 3.1. The band gap tends to increase slightly (leading to shorter cut-off wavelength) when the temperature is reduced. This is a consequence of the effective interatomic spacing getting smaller as the vibrational energy of the lattice atoms decreases.

Figure 3.5 shows the absorption coefficient as a function of wavelength in the optical and near infrared for some commonly used semiconductor materials. In the vicinity of the cut-off wavelength, the strength of the absorption changes by several orders of magnitude as the material changes from being essentially transparent at wavelengths longer than the cut-off to being strongly absorbing for shorter wavelengths. Except for Si and Ge, the transition is quite sharp. Si and Ge are indirect gap semiconductors, for which promotion of an electron from the valence to the conduction band involves a change in momentum as well as in energy. This complicates the interaction process since a phonon (quantised lattice vibration) must also be involved, and the interaction probability depends on wavelength. The other materials are direct gap semiconductors, which involve no change in electron momentum on promotion to the conduction band. Interaction is then highly favoured as long as

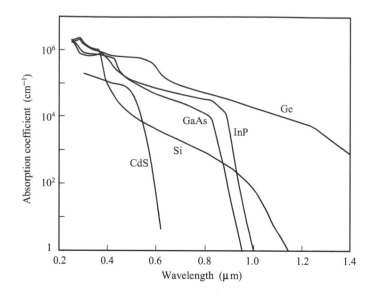

FIGURE 3.5 Absorption coefficient in the optical-infrared region for several semiconductor materials. (Credit: C. B. Honsberg and S. G. Bowden, Absorption coefficient page at www.pveducation.org, 2019.)

the photon energy is high enough, leading to a much sharper transition. Note that germanium also has a direct gap feature at 1.6 μm as well as the more gradual cut-off at around 0.7 μm.

Extrinsic p-type material is doped with impurities that create acceptor states just above the valence band. A photon can promote a valence band electron up to one of these levels, leaving a hole, which is then available for conduction. Extrinsic n-type material is doped with impurities that create donor states just below the conduction band A photon can promote an electron from one of these states up to the conduction band.

Extrinsic semiconductors have much smaller energy gaps than intrinsic silicon or germanium, and so can be used as detectors at much longer wavelengths. For example, arsenic-doped silicon (Si:As) has a cut-off wavelength of around 25 μm and usable response out to ~28 μm. Gallium-doped germanium (Ge:Ga) has a cut-off around 115 μm. Semiconductor-based detectors for infrared and optical wavelengths will be described in Chapters 9 and 10.

3.2.3 THE PHOTOELECTRIC EFFECT

More energetic photons can liberate bound electrons in a semiconductor or an insulator through the photoelectric effect:

$$h\nu + \text{atom} \rightarrow \text{positive ion} + e^- \tag{3.25}$$

as shown in Figure 3.6. The electron, with binding energy W (the work function), completely absorbs the energy of the incident photon, and photon energy over and above the work function goes to provide kinetic energy to the electron. Depending on the nature of the material and the depth at which the photon is absorbed, the free electron can remain in the material (perhaps fixed in position or able to drift slowly under the influence of an applied electric field), or it may escape from the surface (photoemission). The photoelectric effect is the basic interaction mechanism involved in many optical, X-ray and γ-ray detectors. Silicon has a work function of ~4.5 eV, corresponding to λ ~270 nm, and can be used as a UV or soft X-ray detector. Other optical/UV detectors, such as photomultiplier tubes, and microchannel plates use photo-emitting materials to generate currents in vacuo. Such detectors are described in Chapter 10.

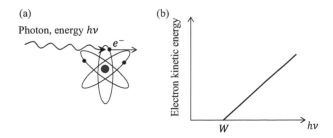

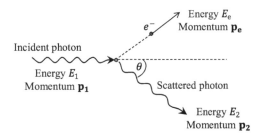

FIGURE 3.6 The photoelectric effect (a) and the liberated electron kinetic energy vs photon energy (b).

FIGURE 3.7 The Compton effect.

3.3 X-RAY AND γ-RAY WAVELENGTHS

There are three main photon absorption mechanisms for X-ray and γ-ray photons: the photoelectric effect, as described above, the Compton effect, and electron-positron pair production. Which of these dominates depends on the photon energy and the nature of the material.

3.3.1 PHOTOELECTRIC INTERACTION

For the photoelectric effect, the interaction probability per unit mass is roughly proportional to $Z^5/E^{7/2}$ where E is the photon energy (which must be at least as great as the work function) and Z is the atomic number of the material. High-Z materials are thus much more efficient absorbers, and the absorption probability declines rapidly with increasing photon energy.

3.3.2 THE COMPTON EFFECT

As we move up in photon energy, photoelectric absorption becomes increasingly ineffective, but another process, the Compton effect, starts to dominate. This is the elastic scattering of a photon by an electron:

$$h\nu_1 + e^- \rightarrow h\nu_2 + e^- (\nu_2 < \nu_1). \tag{3.26}$$

as shown in Figure 3.7. Here the electron absorbs some, but not all, of the photon energy. We are now in an energy regime in which the binding energy of the electron to its parent atom is small compared to the photon energy so that we can ignore it – it makes no difference whether the electron is initially bound or not. The direction of the scattered photon is at an angle θ to the incident photon direction, and the incident and scattered photons have energies E_1 and E_2 and momenta vectors $\mathbf{p}_1$ and $\mathbf{p}_2$, while the electron energy and momentum are E_e and $\mathbf{p}_e$.

We can apply the laws of conservation of energy and momentum to such an encounter. The energies and momenta of the incident and scattered photons are given by

$$E_1 = p_1 c = \frac{hc}{\lambda_1} \quad \text{and} \quad E_2 = p_2 c = \frac{hc}{\lambda_2}. \tag{3.27}$$

Using the relativistic expression for the energy of the scattered electron,

$$E_e = \left[(p_e c)^2 + (m_e c^2)^2 \right]^{1/2}, \tag{3.28}$$

and applying the law of conservation of energy gives

$$\frac{hc}{\lambda_1} + m_e c^2 = \frac{hc}{\lambda_1} + \left[(p_e c)^2 + (m_e c^2)^2 \right]^{1/2}. \tag{3.29}$$

This can be written as

$$(p_e c)^2 = \left(\frac{hc}{\lambda_1} \right)^2 + \left(\frac{hc}{\lambda_2} \right)^2 - 2 \frac{h^2 c^2}{\lambda_1 \lambda_2} - 2 m_e c^2 \left(\frac{hc}{\lambda_2} - \frac{hc}{\lambda_1} \right). \tag{3.30}$$

Conservation of momentum requires that

$$\mathbf{p_1} = \mathbf{p_e} + \mathbf{p_2}, \tag{3.31}$$

so that

$$p_e^2 = (\mathbf{p_1} - \mathbf{p_2}) \cdot (\mathbf{p_1} - \mathbf{p_2}) = p_1^2 + p_2^2 - 2\mathbf{p_1} \cdot \mathbf{p_2} = p_1^2 + p_2^2 - 2 p_1 p_2 \cos\theta. \tag{3.32}$$

Multiplying by c^2, we get another expression for $p_e^2 c^2$:

$$(p_e c)^2 = \left(\frac{hc}{\lambda_1} \right)^2 + \left(\frac{hc}{\lambda_2} \right)^2 - 2 \frac{h^2 c^2}{\lambda_1 \lambda_2} \cos\theta. \tag{3.33}$$

Equating (3.30) and (3.33), we find

$$2 \frac{h^2 c^2}{\lambda_1 \lambda_2} + 2 m_e c^2 \left(\frac{hc}{\lambda_2} - \frac{hc}{\lambda_1} \right) = 2 \frac{h^2 c^2}{\lambda_1 \lambda_2} \cos\theta, \tag{3.34}$$

which reduces to

$$\Delta\lambda = \lambda_2 - \lambda_1 = \frac{h}{m_e c}(1 - \cos\theta). \tag{3.35}$$

$h/(m_e c)$ is called the Compton wavelength, λ_C, and is a constant for a given elementary particle. For the electron, $\lambda_C = 2.4 \times 10^{-12}$ m. The maximum change in wavelength that a photon can undergo in a Compton encounter with an electron is therefore $2\lambda_C = 4.8 \times 10^{-12}$ m, which is a very small value except at X-ray and γ-ray wavelengths.

The maximum energy that can be transferred to the electron is, from equation (3.35) with $\theta = \pi$,

$$E_{e-max} = E_1 \left(\frac{2a}{1 + 2a} \right), \tag{3.36}$$

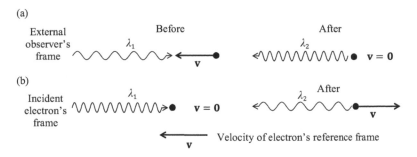

(a)

External
observer's λ_1 After
frame λ_2
                                                                          v = 0

Before

(b)

Incident λ_1 After
electron's   v = 0 λ_2
frame   v

                    ←——   Velocity of electron's reference frame
                        v

FIGURE 3.8   Equivalence of the inverse Compton effect in an external observer's frame (a) to the Compton effect in the rest frame of the electron (b).

where

$$a = \frac{E_1}{m_e c^2} = \frac{\lambda_C}{\lambda_1}. \tag{3.37}$$

If a is large (photon energy large compared to the electron rest mass), most of the photon energy is transferred to the electron, and if a is small, hardly any is transferred. For long wavelengths, therefore, the photon is scattered elastically (negligible change in energy), a process termed "Thomson scattering" (as noted in Section 1.8).

For a significantly greater than unity, the interaction probability per unit mass for the Compton effect is largely independent of the atomic number of the material and declines roughly linearly with photon energy. When the incident photon energy is high, the scattered photon may have sufficient energy to lead to secondary Compton encounters, producing further interactions that result in a cascade of photons and electrons.

In the Compton effect, a high-energy photon loses energy in collision with low-energy electron. The inverse Compton effect is the reverse of this process: a low-energy photon gains energy in collision with high-energy electron. However, by changing our frame of reference, we can see that inverse Compton scattering as seen by an external observer is the same process as Compton scattering as viewed in the rest frame of the electron. To see this, consider the "head-on" case $(\theta = 0)$. Inverse Compton scattering in an external observer's frame is shown in the upper panel of Figure 3.8. The same event as viewed in the rest frame of the electron, which moves with velocity **v** from right to left throughout the encounter, is shown in the lower panel. In the electron's rest frame, it has zero initial velocity, but the incident photon is strongly blueshifted. After the encounter, the electron now has a high velocity and the photon is strongly redshifted. Therefore, in this electron's frame, a high-energy photon has given up energy to a low-energy electron – the Compton effect.

3.3.3   Electron-Positron Pair Production

If the photon energy is greater than twice the rest-mass energy of the electron $\left(2m_e = 1.02\ \text{MeV}\right)$, then, in an encounter with the intense electric field of an atomic nucleus, this energy can be converted to an electron/positron pair, as illustrated in Figure 3.9. Any additional energy that the photon has will be converted to the kinetic energy of the electron and the positron, which usually move off

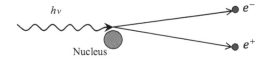

$h\nu$ e^-

Nucleus e^+

FIGURE 3.9   Electron-positron pair production.

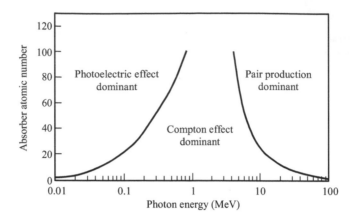

FIGURE 3.10   Regimes of photon energy and absorber atomic number in which the photoelectric effect, Compton effect, and pair production dominate. (From Evans 1955; © McGraw-Hill Education, reproduced with permission.)

at relativistic speeds. The interaction probability increases rapidly above the threshold energy and then flattens off with only a weak energy dependence above ~ 10 MeV. It is also roughly proportional to Z^2, so high-Z materials are more efficient absorbers.

The relative importance of the three interactions – photoelectric, Compton, and pair production – for different energy regimes and absorber atomic weights is illustrated in Figure 3.10.

3.3.4   Mass Attenuation Coefficient

As the photon energy increases, radiation becomes more and more penetrating. The attenuation of a beam of radiation passing through a uniform absorbing medium is, as shown in Chapter 2,

$$I(s) = I(0)e^{-ks}, \tag{3.38}$$

where k is the absorption coefficient and s is the distance travelled through the medium. At X- and γ-ray energies, the attenuation is often given in terms of the mass attenuation coefficient, $\mu_m = k/\rho$ where ρ is the density of the material. The conventional units for μ_m are cm^2g^{-1}. For an individual photon passing through a piece of material of thickness s, the probability of its being absorbed is thus

$$p = 1 - e^{-\mu_m \rho s}. \tag{3.39}$$

The dependence of the mass attenuation coefficient on the incident photon energy is illustrated in Figure 3.11, for silicon $(Z = 14)$ and lead $(Z = 82)$. The energy dependencies of the three processes are indicated and have a similar form for both materials. However, the attenuation is greater and the cross-over energy between the photoelectric and Compton effects is much higher for lead (~0.5 MeV) than for the lower-Z silicon (~60 keV). The discontinuities correspond to wavelengths at which the photon energy exceeds the binding energy of a particular electron shell in the atom, and so ionisation becomes more probable.

3.4   THE EFFECTS OF THE EARTH'S ATMOSPHERE

3.4.1   Atmospheric Absorption

Based on the processes described above, we can consider how the Earth's atmosphere affects the propagation of incoming astronomical radiation across the whole electromagnetic spectrum. The

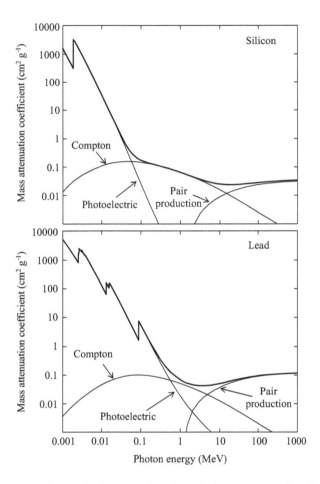

FIGURE 3.11 Mass attenuation coefficient as a function of photon energy for silicon ($Z = 14$) and lead ($Z = 82$). The thick line represents the sum of the contributions from the photoelectric effect, the Compton effect, and electron-positron pair production. (Data from NIST Standard Reference Database 8 (XGAM), Berger et al. 2010.)

atmosphere is composed mainly of molecular nitrogen, N_2, and oxygen, O_2. Other species which strongly influence atmospheric transmission are ozone (O_3), carbon dioxide (CO_2), and water (H_2O).

As well as absorption, other atmospheric processes such as emission, scattering, and turbulence, can have important effects on the radiation reaching an Earth-based telescope from a source and can produce an undesired background of atmospheric radiation against which the source must be detected. The effects of the atmosphere across the spectrum, and the measures adopted to cope with them, will be discussed in more detail in the subsequent chapters dealing with detection in different spectral regions.

In the radio region, wavelengths above ~20 m (frequencies below ~ 15 MHz) are reflected by the ionosphere, so longer-wavelength Earth-based radio astronomy is impossible. As noted in Section 3.1, for the bulk of the Earth's atmosphere, which is neutral and non-conducting, there are no significant mechanisms for electromagnetic radiation at radio wavelengths to interact with the air, although the presence of significant amounts of water in vapour or liquid form can introduce some attenuation since water is a partially conducting. The atmosphere is effectively transparent at radio wavelengths, except in very wet conditions, down to millimetre wavelengths. Consequently, radio astronomy observations can be carried out from sea level in temperate climates and even in cloud.

Millimetre and submillimetre-wavelength photons can excite rotational modes (Section 2.5.3) of molecules in the atmosphere, mainly H_2O, O_2, and O_3, producing absorption. The corresponding spectral line widths are strongly pressure-broadened (Section 2.5.5.1). Intense pressure broadening in the troposphere results in many overlapping absorption features completely blocking large portions of the spectrum. Water vapour is the most strongly absorbing species, and unlike oxygen, its concentration varies enormously with geographical location, altitude, and local weather conditions. The key parameter that dictates the amount of attenuation due to water vapour is the precipitable water vapour, or PWV, which corresponds to the depth of the layer of water that would result if the entire water content of the atmosphere were to precipitate out as rain. The best places for astronomical observatories working in this part of the spectrum are thus dry, high mountain-top sites such as Maunakea in Hawai'i (altitude 4.2 km), the Atacama plateau in Chile (altitude typically 5 km), and the Antarctic plateau (altitude 2.8 km).

Figure 3.12 shows a model of the atmospheric transmission between 1 μm and 10 mm at Maunakea for 1 mm PWV. There are strong pressure-broadened absorption features at 5 and 2.5 mm due to O_2, and several absorption lines due to H_2O at shorter wavelengths, with features at 540 and 400 μm being very intense and defining a gap between the transparency windows around 450 and 350 μm. Shortward of around 300 μm (frequencies above 1000 GHz), the atmosphere is largely opaque, due to intense water vapour absorption, until one gets to around 25 μm wavelength.

Transmission in the submillimetre region depends strongly on the quality of the observing site and is best at the highest and driest observatories. Figure 3.13 shows a higher-resolution model of the atmospheric transmission at the Chajnantor, altitude 5060 m on the Atacama plateau, between 300 μm and 1 mm (1000–300 GHz). Transmission plots are shown for different amounts of PWV (0.2, 0.5, 1.0, and 2.0 mm), with the lowest PWV corresponding to the highest transmission. As well as the broad absorption features, there are many narrower lines due to O_3 and O_2. The transmission below 1 mm wavelength declines in a manner that depends very strongly on the PWV due to the wing of the pressure-broadened 540-μm H_2O feature, and the transmission in the 350- and 450-μm window regions also varies considerably with PWV.

In the infrared, photons can excite vibrational modes in atmospheric molecules (see Section 2.5.4), mainly CO_2, and H_2O. Figure 3.14 shows a model of the transmission of the atmosphere between 1 and 25 μm at Maunakea, for 1 mm PWV. The transmission is high for large parts of the infrared region, but some parts of the spectrum are completely blocked (e.g. between ~5.5 and 7 μm, due to water vapour, and above 14 μm, due to CO_2) and others contain numerous deep

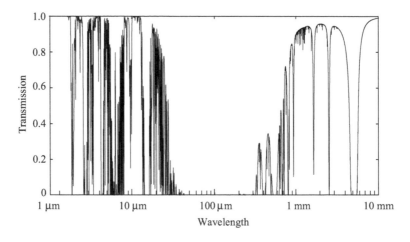

FIGURE 3.12   A model of the zenith atmospheric transmission at the summit of Maunakea between 1 μm and 10 mm wavelength, for 1 mm PWV. Data from the BTRAM (Version 3), atmospheric model (Chapman et al. 2009). (Courtesy of Blue Sky Spectroscopy Inc.; available at https://blueskyspectroscopy.com/?page_id=21.)

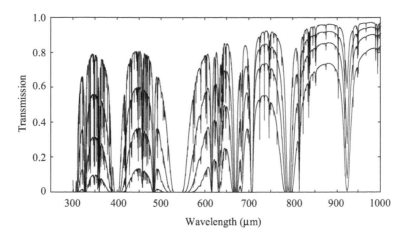

FIGURE 3.13 A model of the zenith atmospheric transmission at Chajnantor, on the Atacama plateau, in the 300-μm–1 mm range, for four different values of PWV: 0.2, 0.5, 1.0, and 2.0 mm, with transmission decreasing with increasing PWV. Data from the BTRAM (Version 3), atmospheric model (Chapman et al. 2009). (Courtesy of Blue Sky Spectroscopy Inc.; available at https://blueskyspectroscopy.com/?page_id=21.)

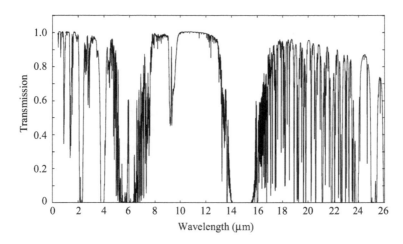

FIGURE 3.14 Zenith infrared transmission of the atmosphere between 1 and 26 μm wavelength at Maunakea, for 1 mm PWV. (Data generated using NASA's ATRAN website https://atran.arc.nasa.gov/cgi-bin/atran/atran. cgi, based on the ATRAN model of Lord 1992.)

pressure-broadened absorption features (e.g. 4–5.5 μm and 16–25 μm). For ground-based infrared observations, standard filters are used to define photometric bands within the highly transmissive parts of the spectrum. These are denoted J, H, K, L, M, N, and Q and are centred at 1.22, 1.63, 2.19, 3.45, 4.75, 10.4, and 21 μm, respectively (and are further discussed in Chapter 5).

Although high dry sites such as Maunakea, the Atacama plateau, or the Canary island of La Palma, do not have such a dramatic advantage in infrared atmospheric transmission over lower altitude sites as they do at submillimetre wavelengths, performance is also superior in terms of atmospheric temperature (affecting the background) and image quality.

In the visible region, there are only weak features due to electronic transitions (Section 2.5.1). However, scattering by molecules and aerosols becomes significant and strongly affects the atmospheric transmission. Scattering due to particles, such as molecules, which are much smaller than the wavelength, is termed Rayleigh scattering, and occurs even in the clear atmosphere. The Rayleigh scattering cross section is proportional to λ^{-4}, and so the effect increases greatly as the

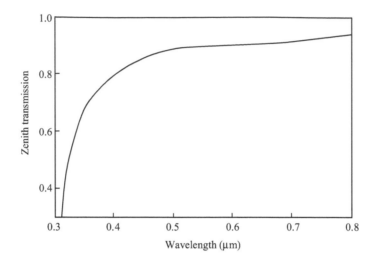

FIGURE 3.15 Median zenith transmission in the visible region at Maunakea. (Data from Bèland, Boulade & Davidge (1988) and the Canada France Hawai'i Telescope (CFHT) Observatory Manual 2003.)

wavelength decreases towards the blue end of the spectrum, resulting in significant attenuation of astronomical sources in the short-wavelength part of the visible band. (Rayleigh scattering is also what gives the sky its blue colour.) The median zenith transmission at Maunkea is shown in Figure 3.15 with the increasing attenuation at shorter wavelengths being largely due to Rayleigh scattering. In addition to diminishing the signal from an astronomical object, scattering of moonlight or sunlight increases the general sky brightness (even with the Sun below the horizon), which reduces the contrast between a source and the surrounding sky, further compromising sensitivity.

UV photons have enough energy to dissociate oxygen and ozone molecules. Absorption by ozone in the ozone layer (altitude ~ 80 km) is particularly strong below 300 nm wavelength, resulting in complete blockage of shorter UV wavelengths. This is due to X-ray photons ionising atmospheric gas atoms by the photoelectric effect, knocking out inner electrons. The whole of this range is blocked by the atmosphere, and observations are carried out from space.

Very high-energy γ-ray photons (and also cosmic rays, mainly high-energy protons) can interact with nuclei in the atmosphere resulting in the production of relativistic electron-positron pairs which, in turn, lead to further interactions and cascades of secondary elementary particles and photons. With the relativistic particles travelling at greater than the speed of light in air, Čerenkov radiation is produced in the visible region, which can be detected and used to infer the energy and direction of the incoming photon.

3.4.2 Atmospheric Emission

In addition to absorbing or scattering incoming astronomical radiation, the atmosphere emits electromagnetic radiation, thus constituting a source of additional radiation (known as background radiation, although in this case it is literally a foreground) against which sources must be detected. This background is particularly problematic in the infrared and submillimetre regions, with the troposphere, by virtue of its 200–300 K temperature range, emitting as a grey body in those parts of the spectrum in which the transmission is partial. In the near infrared and visible region, there is also always some residual sky emission, called airglow, even on moonless nights and in the absence of terrestrial light pollution. Photoionisation and chemical reactions due to solar UV photons during the day leave some atoms and molecules at high altitudes in excited states during the nighttime, and these states can decay with the emission of visible photons. For some sensitive optical and near-infrared observations from the ground, it is this residual sky emission that acts as the ultimate limit to sensitivity.

3.4.3 Atmospheric Turbulence and Seeing

Another important effect that the atmosphere has, particularly at visible and infrared wavelengths, but also at longer wavelengths even into the radio region, is the rapid and random modulation of the refractive index due to atmospheric turbulence. This causes fluctuating distortions of the (ideally plane) wavefront at the telescope aperture and results in an image that is smeared and unstable.

This is the phenomenon that causes stars to twinkle, and it is characterised by the astronomical seeing. It limits the angular resolution of ground-based telescopes to around $1''$, regardless of telescope size, and sensitivity is also severely degraded by the smearing of the image. However, adaptive optics systems, which are now routinely used on large ground-based optical and infrared telescopes, can sense and correct the wavefront distortions introduced by the atmosphere, resulting in major gains in spatial resolution and sensitivity. Such systems will be described in Chapter 10.

In the case of interferometric observations in the radio region, unwanted phase variations can also be introduced by atmospheric turbulence (see Chapter 8).

3.4.4 Earth-Based and Space-Borne Observatories

For those parts of the spectrum in which the atmospheric transmission is high, observations can be made from good (preferably high and dry) sites on the ground. Telescopes can also be operated on aircraft, at altitudes of around 20 km, and on stratospheric balloons flying at around 40 km altitude, getting above much of the atmosphere, but still with residual emission and absorption compromising performance. Most γ-ray, X-ray, and UV observations are made from space, with recent and current facilities including the ESA INTEGRAL and NASA Swift γ-ray satellites, NASA's Chandra and ESA's XMM-Newton observatories in the X-ray region, and NASA's Hubble Space Telescope (HST) in the UV. For some spectral regions in the infrared-millimetre range, space observation is essential, and for others, it provides improved sensitivity and stability compared to what can be achieved with an Earth-based platform, although with smaller telescopes than can be built on the ground. Recent examples of space-borne telescopes in this region are NASA's *Spitzer* and ESA's *Herschel* observatories.

For observations of the cosmic background radiation in the millimetre regime, although transmission to the ground is high, observation from or near the Earth can result in systematic effects making it very challenging to achieve the high precision needed, and space-borne platforms such as NASA's COBE and WMAP and ESA's *Planck* missions have achieved much-improved performance.

At visible and near-infrared wavelengths that are accessible from the ground, operation in space above the distorting effects of the atmosphere can also provide superior performance over ground-based observatories, even with a smaller telescope size, as with the HST. Observations that require great stability or the ability to map large areas of sky with very uniform characteristics are also often best carried out from space – for example, ESA's GAIA satellite, launched in 2013, which is measuring the astrometric positions of stars in the Milky Way with great precision, and the ESA Euclid mission, with launch planned in 2022, to study the nature of dark energy by measuring the small effects on observed shapes of galaxies of weak gravitational lensing.

Although space observatories avoid the unwelcome effects of the atmosphere, compared to ground-based observatories they are much more expensive, limited in angular resolution, usually have shorter operating lifetimes, and are impossible to upgrade with improved instruments (except for the HST). Most observational research themes require access to a broad range of wavelengths and benefit from the range of ground-based and space-borne facilities that is available today.

4 Telescopes and Optical Systems

4.1 INTRODUCTION

The function of a telescope is to collect astronomical radiation and direct it into an instrument for further manipulation and detection. The instrument can take one of many different forms depending on the wavelength and the scientific purpose, but the principles underlying the telescope optics are the same. It is not our intention here to describe telescope systems in the detail presented in other texts, but to review the principles and phenomena that dictate their performance, and to discuss which configurations are commonly used for particular applications.

In this chapter, we outline the physical optics considerations that dictate the performance of telescope systems. We start by describing the most commonly used telescope configurations and the optical principles underlying telescope optics, before considering imperfections from aberrations and diffraction. As we progress, we show that other approaches using Fourier optics can enhance our understanding of instrument design and provide valuable tools for designing and analysing telescopes. At shorter wavelengths, in the X-ray and γ-ray regimes, a different approach is necessary because photons can penetrate deeply into most materials making conventional focusing difficult. X- and γ-ray telescopes will be described in Chapter 11.

4.2 TELESCOPE CONFIGURATIONS

A simple telescope has two lenses or mirrors, a large diameter (primary) that collects and focuses the source flux and a secondary (termed eyepiece if the eye views the image) which is positioned to couple the radiation collected by the primary onto the astronomical detector. A telescope with a lens as the primary is known as a refracting telescope or refractor, and one with a mirror as the primary is a reflecting telescope or reflector. Most telescopes used for astronomical research today are reflectors because lenses can suffer from one or more disadvantages: they are difficult to manufacture and to support with the large diameters often required for modern telescopes, and they are generally more massive than mirrors of equivalent size. They also tend to be more lossy than mirrors and can be opaque in some parts of the spectrum, and they can suffer from chromatic aberration (Section 4.5.6) arising from variation of refractive index with wavelength. The largest refracting telescope still in operation is the 102-cm diameter telescope at the Yerkes Observatory in Wisconsin, USA, originally installed in 1887.

Some typical telescope configurations are shown in Figure 4.1. The Newtonian, which is rarely adopted today, has a parabolic primary and flat secondary which moves the viewing port out of the telescope beam. The Cassegrain system has a parabolic primary and a convex hyperbolic secondary in front of the prime focus to focus the beam to an image behind the primary through a hole at its centre. The Gregorian is similar, but with a concave secondary behind the prime focus. All of these involve some obscuration of the primary mirror by the secondary, resulting in some loss of light collecting power and potentially some problems due to diffraction caused by the secondary obscuration affecting the telescope beam profile. A variant that avoids the effects of obscuration is the offset Cassegrain or Gregorian configuration (offset Cassegrain shown here).

The Cassegrain has become the most common astronomical telescope type for optical to millimetre wavelengths. Large focal plane instruments can be located near the centre of gravity of the

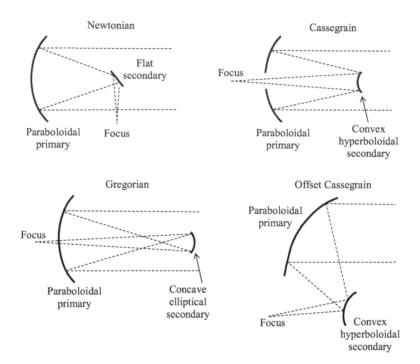

FIGURE 4.1   Common telescope configurations.

primary without mechanical distortion of the telescope structure. Due to the symmetry about the optical axis, the telescope is easily balanced, and the instruments are conveniently located near floor level when the telescope is parked pointing at the zenith. The Gregorian shares these advantages but is less compact along the optical axis, requiring a larger dome and/or a more substantial support structure for the secondary.

4.3   TELESCOPE MOUNTING METHODS

An Earth-based astronomical telescope must be mounted so that it can be pointed at the source to be observed, and so that its pointing can be adjusted as the source moves on the sky due to the Earth's rotation. The absolute position of an object on the sky is conventionally described by its right ascension (RA) and declination (Dec), using the coordinate system shown in Figure 4.2. The celestial equator is the projection of the Earth's equator onto the celestial sphere. The declination of an object is equivalent to the latitude of the intersection of a line joining the object to the centre of the Earth and can have values between −90° and +90°. RA is equivalent to the longitude of the intersection of a line joining the object to the centre of the Earth and is measured with respect to a standard reference point, chosen to be the direction of the Sun at the March equinox. It can have values between 0° and 360° and is often quoted in hours, minutes, and seconds, with a full circle (360°), corresponding to 24 hours.

The two most common mounting arrangements for ground-based telescopes, known as the equatorial and the altitude-azimuth (alt-az) mounts, are illustrated in Figure 4.3. In both configurations, the telescope can rotate about two perpendicular axes. In the case of the equatorial mount, one axis, the polar axis, is parallel to the Earth's axis of rotation – i.e., parallel to a line joining the north and south rotational poles. Rotation about this axis corresponds to varying the RA of the telescope boresight. Rotation about the other (declination) axis adjusts the declination of the boresight. Pointing at and tracking an object on the sky are straightforward. The RA and Dec axes can be fixed at the appropriate value at the start of an observation. The declination angle can thereafter be left fixed, and it is only necessary to rotate the telescope about the polar axis at a fixed angular rate equivalent

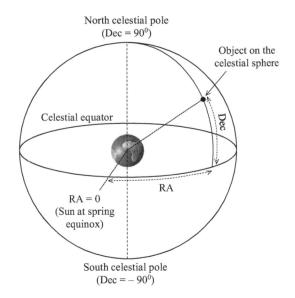

FIGURE 4.2 Right ascension and declination coordinate system.

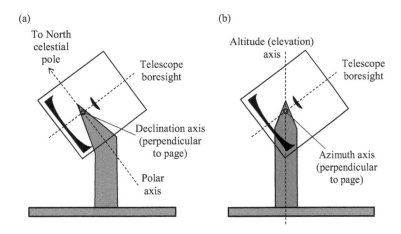

FIGURE 4.3 Equatorial (a) and alt-az (b) telescope mounting configurations.

to the rotation rate of the Earth (known as the sidereal rate). The equatorial mount was the most common and convenient for many years when major telescopes were much smaller than they are today, but it has now been largely superseded by the alt-az mount, which is more suitable for very large telescopes.

In the alt-az configuration, one axis is horizontal, with rotation about this axis allowing the elevation above the horizon to be selected, and the other is vertical (towards the zenith), with rotation about that one allowing the source azimuth angle to be selected. With an alt-az mount, the centre of gravity of the telescope is arranged to be on the vertical axis, allowing the weight of the telescope to be borne along that axis. With large modern telescopes, this simplifies the design of the support structure and allows for a more compact design that can fit into a smaller (and so cheaper) dome. Acquisition and tracking are more complex than for the equatorial mount, with the telescope having to be driven independently and at variable speeds in both axes; however, this is not a problem for modern computerised drive systems. Another complication is that the image of an extended source in the focal plane rotates about the telescope optical axis as the source is tracked across the sky. This is usually inconvenient for instrument design, and an image rotation system is therefore introduced

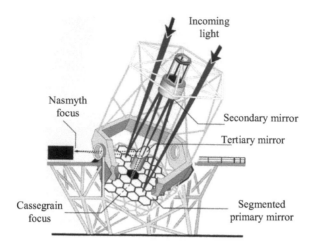

FIGURE 4.4   Keck telescope Nasmyth focus configuration. (Credit: W. M Keck Observatory and California Association for Research in Astronomy.)

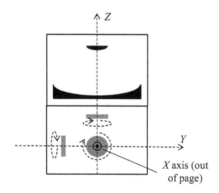

FIGURE 4.5   Principle of three-axis spacecraft stabilisation.

in the optical train to compensate exactly for the effect so that the focal plane image is fixed in the instrument coordinate system.

With alt-az telescopes, a variant of the Cassegrain configuration, which is often adopted for instruments that are physically large and massive, involves locating the instrument at the Nasmyth focus. The beam is diverted by a flat tertiary mirror through a hole in the telescope elevation bearing to a fixed platform on which heavy equipment can be mounted. The image remains in a fixed location on the Nasmyth platform. Figure 4.4 shows a schematic of the Nasmyth focus of the Keck 10-m telescope, located at Maunakea Observatory, Hawai'i. Two Nasmyth platforms are often available, one on either side, and can be selected simply by rotating the tertiary mirror.

In the case of space-borne telescopes, stable pointing is achieved by means of variable-speed rotating wheels, known as reaction wheels. A typical system involves three such reaction wheels rotating about perpendicular axes, as shown schematically in Figure 4.5. Changing the speed of rotation of a wheel changes its angular momentum along its rotation axis, leading to a torque on the spacecraft causing it to rotate in the opposite sense. Angular momentum can thus be exchanged in a controlled manner between the wheels and the spacecraft.

The pointing direction of the telescope must be known and controlled with an accuracy that depends on the angular resolution of the system. With a telescope of diameter D operating at wavelength λ, the radiation from a point source is spread by diffraction over a region of the focal plane corresponding to an angle of $\sim\lambda/D$ radians on the sky (Section 4.6) so that a detector in the focal plane is sensitive to a patch on the sky of roughly that extent, known as the beam size. For accurate calibration and astrometry, the telescope control system needs to be capable of pointing to within a small fraction of the beam.

4.4 TELESCOPE OPTICS

4.4.1 LENS OPTICS

Most modern astronomical telescopes use a reflective mirror as the primary optical element. Instruments can have mirrors or lenses or both. Here we consider the basic properties of both, starting by considering the refractive properties of a thin plano-convex lens with a spherical surface, of radius R, for rays near the optical axis (the paraxial approximation), as illustrated in Figure 4.6.

The object is at distance s_o from the lens and has height h_o. Ray 1 from the top of the object is parallel to the optical axis and is refracted by lens, passing through the lens focus F, at an angle α to the optical axis. The focus is at a distance f, the focal length, from the lens. The normal to the surface of the lens at the point of refraction passes through the lens centre of curvature, C, at an angle γ to the optical axis. The angle of refraction is thus $(\alpha + \gamma)$.

Ray 2 passes through the lens at the vertex, making an angle β, with the optical axis. For small β and a thin lens, the two refracting surfaces are parallel resulting in no deflection. The image, of height h_i, is formed where these rays intersect, at distance s_i from the lens.

We have, for small angles

$$\tan \beta \approx \beta = \frac{h_o}{s_o} = \frac{h_i}{s_i}, \tag{4.1}$$

and

$$\tan \alpha \approx \alpha = \frac{h_o}{f} = \frac{h_i}{s_i - f}. \tag{4.2}$$

The magnification, M, is the ratio of the image and object sizes:

$$M = \frac{h_i}{h_o} = \frac{s_i}{s_o}. \tag{4.3}$$

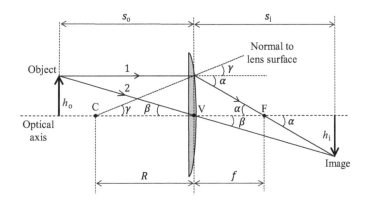

FIGURE 4.6 Ray tracing through a refracting thin lens.

The relation between the focal length and the object and image distances is given by substituting for h_i/h_o in the equation for α:

$$\frac{s_i}{s_o} = \frac{s_i - f}{f},$$
(4.4)

which reduces to

$$\frac{1}{s_o} + \frac{1}{s_i} = \frac{1}{f}.$$
(4.5)

Equation (4.5) is known as the Gaussian thin lens formula after Gauss, who first described the formation of images. The refractive power of the lens is dictated by its index of refraction, n. We can apply Snell's law to the refraction, giving

$$n\sin\gamma = \sin(\gamma + \alpha),$$
(4.6)

where n is the refractive index of the lens, and we take the refractive index of the air to be unity.

Since the angles are small,

$$n\gamma \approx \gamma + \alpha.$$
(4.7)

Noting that $\alpha = h_o/f$ and $\tan\gamma \approx \gamma = h_o/R$, we have

$$n\frac{h_o}{R} \approx \frac{h_o}{R} + \frac{h_o}{f},$$
(4.8)

which reduces to

$$f \approx \frac{R}{n-1}.$$
(4.9)

A more general relation for a biconvex lens is obtained by noting that for two plano-convex lenses of radii R_1 and R_2, and focal lengths f_1 and f_2, which are joined together, the focal length of the combination is given by

$$\frac{1}{f_1} + \frac{1}{f_2} = \frac{1}{f}.$$
(4.10)

Combining equations (4.9) and (4.10) gives the lens maker's formula:

$$\frac{1}{f} = (n-1)\left(\frac{1}{R_1} + \frac{1}{R_2}\right).$$
(4.11)

4.4.2 MIRROR OPTICS

The formation of an image by a concave spherical mirror of radius R is shown in Figure 4.7. Ray 1 from the top of the object, height h_o, at distance s_o from the mirror vertex, V, passes through the centre of curvature, C, of the mirror, making an angle α with the optical axis, and is retro-reflected by the mirror surface. Ray 2 from the top of the object strikes the vertex of the mirror and is reflected with the same angle of incidence, β.

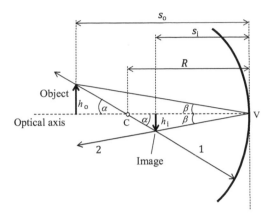

FIGURE 4.7 Concave mirror ray optics.

The image is formed where these rays intersect. For small angles

$$\tan \beta \approx \beta = \frac{h_o}{s_o} = \frac{h_i}{s_i}, \tag{4.12}$$

and

$$\tan \alpha \approx \alpha = \frac{h_o}{s_o - R} = \frac{h_i}{R - s_i}. \tag{4.13}$$

As for the lens, the magnification is given by

$$M = \frac{h_i}{h_o} = \frac{s_i}{s_o}. \tag{4.14}$$

From the equation for α,

$$\frac{s_i}{s_o} = \frac{R - s_i}{s_o - R}, \tag{4.15}$$

which reduces to the mirror equation,

$$\frac{1}{s_o} + \frac{1}{s_i} = \frac{2}{R}. \tag{4.16}$$

For a very distant object, always a valid approximation for astronomical observations, $1/s_o$ is negligible and the image is formed at the mirror focal distance $s_i = f = R/2$ from the vertex.

4.4.3 THE PLATE SCALE

The relationship between angle in the image and distance in the focal plane is given by the plate scale equation. We can represent the primary aperture of the telescope system by a single optical element (either a lens or a mirror) with diameter D and focal length, f, as illustrated in Figure 4.8 which shows the aperture represented as a lens. A source on the sky subtending an angle θ produces

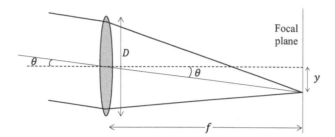

FIGURE 4.8 The relationship between angle on the sky and distance in the focal plane (plate scale).

an image in the focal plane with linear extent y. For small θ, $\tan\theta \approx \theta = y/f$ so that the plate scale, which has units of angle per linear distance (e.g. arcsec/mm), is

$$\frac{\text{Angular size subtended by the source}}{\text{Size of the geometric image in the focal plane}} = \frac{\theta}{y} = \frac{1}{f}. \tag{4.17}$$

4.4.4 Telescope Throughput

There are two important planes in any optics system where apertures can be used to limit the intensity and/or the extent of the image formed by the system. These are known as aperture stops and field stops and are illustrated in Figure 4.9. An aperture stop limits the amount of radiation from the source that is collected, defining the used telescope area, A_{tel}. In many cases, the aperture stop is simply defined by the diameter of the telescope primary itself. An aperture stop can also be placed at any plane in the system where there is an image of the telescope aperture. This reduces the effective collecting area of the telescope, but can be advantageous, for instance to avoid stray light from just beyond the primary mirror being diffracted or scattered into the instrument. A field stop is placed at a real image plane and so limits the extent of the observed image (the field of view), but it does not change the image brightness. For example, in many photographic cameras, there is an adjustable iris behind the main lens that is used to control the aperture diameter and the size of the multi-pixel detector chip (in a CCD camera) serves as a field stop.

It is useful to define another parameter for the optical system – the focal ratio, or F-number, given by $F = f/D$. For an optical system of fixed focal length focusing light from a distant object onto the focal plane, the image brightness is proportional to D^2 and thus proportional to $1/F^2$. This is why the stops on a standard camera have the sequence $F/2$, $F/2.8$, $F/4$, $F/5.6$, $F/8$, $F/11$, and $F/16$ – they increase in steps of approximately $\sqrt{2}$ so that the image intensity is approximately halved with each stop. The F-number thus characterises the condensing properties of the telescope – a small F-number results in an intense compact image with low magnification, and a large F-number produces a large image with lower brightness. The intensity in the image plane depends on both the field aperture size (hence on the solid angle, Ω_{Det}, with which the detector views the telescope) and the field stop size (collecting area of the detector, A_{Det}, which could, for instance, be that of a multi-pixel CCD array or a single collecting aperture for a one-pixel system). The throughput, $A\Omega$, is constant in an optical system without losses – the power propagating through the system (given by intensity times throughput times bandwidth, $I_\nu A\Omega\Delta\nu$) is constant – so as the intensity is constant (Section 1.11.1), the throughput is conserved. The conservation of throughput is illustrated for a simple system in Figure 4.10. The throughput at the detector, $A_{\text{Det}}\Omega_{\text{Det}}$ (small area; large solid angle) matches the throughput on the sky, $A_{\text{Tel}}\Omega_{\text{Tel}}$, (large area, small solid angle).

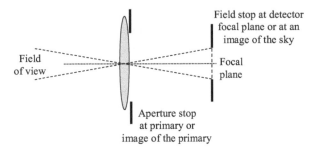

FIGURE 4.9   Aperture and field stops.

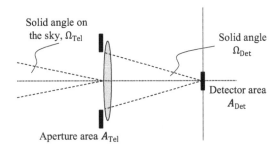

FIGURE 4.10   Conservation of throughput.

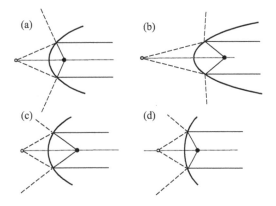

FIGURE 4.11   Spherical and aspheric mirror types; (a) paraboloidal; (b) ellipsoidal; (c) spherical; and (d) hyperboloidal. For parallel light rays coming from the right-hand side, the open circles indicate virtual foci and closed circles real foci.

4.5   OPTICAL ABERRATIONS

As noted in Section 4.2, there are several types of curved surfaces that are used in standard telescope configurations. Figure 4.11 illustrates the four basic types: spherical, paraboloidal, hyperboloidal, and ellipsoidal. Their foci are either real or virtual depending on whether the mirrors are concave or convex.

Although most telescopes use reflecting mirrors, focal plane instrument optics can use either mirrors or lenses to arrive at the most compact and efficient design. Non-ideal effects arise because neither the lenses nor the mirrors produce perfect images as they suffer from optical aberrations and diffraction. For instruments operating in the visible region, the components are many wavelengths in size, and aberrations tend to dominate, while at longer (far infrared to radio) wavelengths, the component size is often not hugely greater than the wavelength, and diffraction is often the fundamental limit to image quality. Here, we will describe the main types of aberration that arise in optical systems before discussing the phenomenon of diffraction in Section 4.6.

Departures from the idealised Gaussian optics described in Section 4.4 are known as aberrations. There are two main types: chromatic aberrations, which arise in refractive systems because the refractive index of a lens material is a function of wavelength, and monochromatic aberrations, which arise even for a single wavelength of illumination.

In deriving the lens maker's formula (equation 4.11), it was assumed that the incidence angle of the incident rays, θ, was small, allowing us to include only the first-order term in the expansion for $\sin\theta$:

$$\sin\theta = \theta + \frac{\theta^3}{3!} + \frac{\theta^5}{5!} + \cdots \tag{4.18}$$

Thus, the thin lens Gaussian theory is referred to as first-order theory, while a more accurate analysis to model more extreme off-axis rays is referred to as third-order theory (first two terms only). We will not describe that here but will identify the nature of the optical imperfections we expect in the image quality. The five aberrations that result are referred to as Seidel aberrations, after Ludwig von Seidel (1821–1896). It should be noted that spherical and aspheric mirrors suffer from the same aberrations as lenses, with the notable exception of chromatic aberration – reflecting surfaces do not disperse light.

4.5.1 SPHERICAL ABERRATION

Spherical aberration, illustrated in Figure 4.12, is a dependence of focal length on aperture. This is the only on-axis aberration. Parallel rays that strike the aperture further from the centre (marginal rays) are refracted more than those that are incident nearer the axis. The longitudinal spherical aberration is the distance between the paraxial focus and the axial intersection of the extreme rays. Similarly, the height above the axis where this extreme ray strikes a screen at the paraxial focus is the transverse spherical aberration. The image of a distant point source is seen as a bright centre spot with a halo. There is a position on the optical axis at which the spot size is minimum – the circle of least confusion. Stopping down the aperture reduces the spherical aberration. For a diverging lens, the aberration is negative (extreme ray bent less), so a combination of the concave and convex elements reduces the overall spherical aberration.

Spherical aberration can be eliminated by altering the profile of the lens or mirror so that the deviation of the rays is progressively reduced with transverse distance from the optical axis. A parabolic shape results in a common focal point for all rays parallel with the axis.

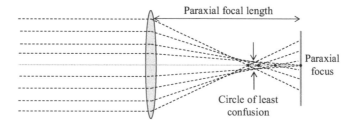

FIGURE 4.12   Spherical aberration.

4.5.2 COMA

Coma is an aberration associated with object points away from the optical axis. The off-axis rays have a different effective focal length, and therefore a different magnification, to the near on-axis rays. Negative coma results when the extreme rays have less magnification, as shown in Figure 4.13, while Figure 4.14 shows how positive coma occurs when the extreme rays have greater magnification.

Coma can be reduced by using non-spherical surfaces to reduce or null out the effect. Alternatively, an aperture stop that eliminates the extreme rays will reduce coma.

4.5.3 ASTIGMATISM

A circular lens or mirror will present an elliptical shape to a bundle of off-axis rays, causing the rays imaged along the longer axis to have a shorter focal length than those imaged by the shorter axis. Figure 4.15 shows the effect of astigmatism in the absence of spherical aberration and coma.

A point source results in orthogonal lines in the short (meridional) and long (sagittal) focal length images. Again, there is a circle of least confusion located between the two foci. In practice, most two-lens or two-mirror systems suffer from astigmatism, reducing the usable focal plane area. The use of additional lens or mirror components or of special surface profiles can reduce the astigmatism to give acceptable image quality over larger areas.

Two final aberrations refer to deformation of images that are free from astigmatism (anastigmatic).

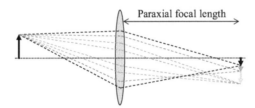

FIGURE 4.13 Negative coma due to extreme rays having less magnification.

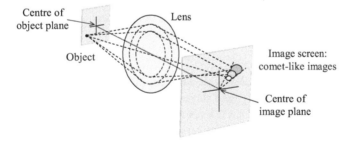

FIGURE 4.14 Positive coma due to extreme rays having greater magnification.

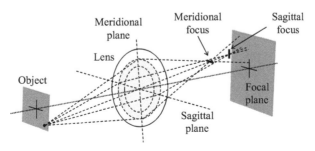

FIGURE 4.15 Astigmatism.

4.5.4 Field Curvature

Field curvature (also referred to as Petzval field curvature) occurs for objects outside of the paraxial condition, for which the image plane becomes curved. The projection of this extended image onto a flat screen or detector array will therefore be blurred at the edges because it is out of focus, as shown in Figure 4.16.

For the single element shown in Figure 4.16, with the focal surface being a sphere of radius f, simple trigonometry gives the longitudinal displacement of the focus, Δx, from the ideal focal plane, at radial distance y from the optical axis:

$$\Delta x = \frac{y^2}{2f}. \tag{4.19}$$

More generally, for a system with N optical elements characterised by focal lengths f_i and indices n_i, equal to the refractive index for a lens or to ± 1 for a mirror (depending on whether it is convex or concave), the overall displacement is

$$\Delta x = \frac{y^2}{2} \sum_{i=1}^{N} \frac{1}{n_i f_i}. \tag{4.20}$$

It is possible, therefore, to cancel out the field curvature effects of positive and negative-index elements to achieve a flat field (the Petzval condition).

4.5.5 Distortion

Distortion arises when the transverse magnification increases or decreases with off-axis position in the image. Depending on whether magnification decreases or increases with radial distance in the focal plane, the image gets radially stretched like a pincushion or compressed like a barrel as illustrated in Figure 4.17, which shows the effects for 20% distortion (far more extreme than would normally apply in practice). Lenses or mirrors with carefully profiled surfaces can reduce this effect, as can relatively simple calibration algorithms which can post-correct digitised images.

4.5.6 Chromatic Aberration

Dispersion (variation of refractive index with wavelength) in a lens results in longer wavelength (red) light being refracted less strongly than shorter wavelengths (blue) so that different colours are focused at different focal distances. For a convex lens, blue light is bent more than red, as shown in Figure 4.18, whereas for a concave lens, the reverse is true.

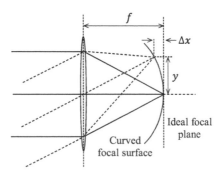

FIGURE 4.16   Petzval field curvature.

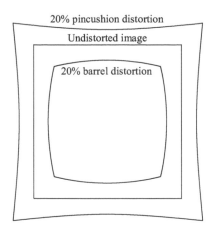

FIGURE 4.17   Image distortion.

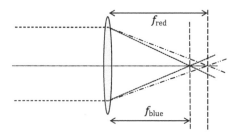

FIGURE 4.18   Chromatic aberration.

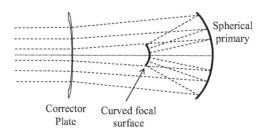

FIGURE 4.19   The Schmidt camera.

Ultra-low-dispersion glass (which is very expensive) can help to reduce this effect. A cheaper solution is to use an achromat, consisting of a cemented concave and convex lens pair, which can give exactly coincident foci at two particular wavelengths to give a better approximation to an undistorted white light image. For even better correction, apochromats can be used which correct the focus at three or more wavelengths. Mirrors do not suffer from chromatic aberration because there is no dependence of the angle of reflection on wavelength, but all the other Seidel aberrations (spherical aberration, coma, astigmatism, field curvature, and distortion) are present.

An interesting variant of the standard reflecting telescope configuration is the Schmidt camera system, which is designed for wide-field imaging. It utilises the advantage of a spherical primary mirror that there is no preferred axis of symmetry so that off-axis rays can be collected with no worse aberration than paraxial rays, eliminating coma and astigmatism. However, the primary exhibits spherical aberration, which must be corrected if an unaberrated image is to be produced over a large field. In the Schmidt system, the spherical aberration is removed by placing a refractive corrector plate before the primary as shown in Figure 4.19. The corrector plate introduces wavefront distortions, which are then exactly cancelled out by the spherical aberration of the primary, resulting in a high-fidelity image over a large (but curved) focal surface. This configuration is referred to as a catadioptric system.

The Schmidt camera has the advantage of having a wide field of view and large aperture and is commonly used for wide-field astronomical surveys.

4.6   DIFFRACTION

A fundamental phenomenon affecting the imaging capabilities of a telescope is Fraunhofer diffraction at the primary aperture, which results in the radiation from a point source being spread out in the focal plane, and so influencing the beam pattern of a detector on the sky. Diffraction is an interference effect that occurs when light passes around or through an obstruction. The various segments of the wavefront that propagate beyond the obstacle interfere creating a particular energy distribution – a diffraction pattern. Here, we outline some of the fundamental diffraction properties that dictate the image quality of telescopes and other optical systems.

We first consider diffraction of a plane wave (for example, from a distant point source) at a single slit type aperture. Figure 4.20 illustrates diffraction of a plane wave incident normally on a slit of width b, resulting in the generation of an interference pattern $I(\theta)$ on a screen at perpendicular distance z. By Huygens' principle, each point on the unobstructed wavefront emits spherical wavelets. The contribution to the electric field at point P from a position within the slit at distance r from P and y from the centre of the screen is given by

$$dE(r,t) = \frac{dE_o}{r}e^{i(\omega t - kr)} = \frac{E_o}{r}\sin(\omega t - kr), \tag{4.21}$$

where dE_o is the elemental field amplitude at the position within the slit and $k = 2\pi/\lambda$ is the magnitude of the wave vector. The radiated field amplitude decreases inversely with distance r.

We can divide up the field at the slit into strips of width dx and set dE_o to be $E_L dx$, where E_L is the source field strength per unit length so that

$$dE(r,t) = \frac{E_L dx}{r}\sin(\omega t - kr). \tag{4.22}$$

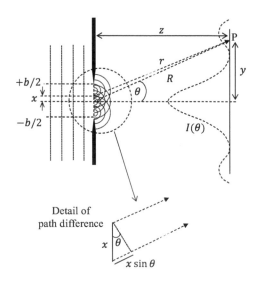

FIGURE 4.20   Diffraction at a single slit.

Fraunhofer noted that we can simplify the calculation of the overall amplitude at points on the screen by assuming that $z \gg b$ so that the distance of P from all points across the aperture is about the same ($r \approx z$), and the amplitude of the field at any point on the screen from all the sub-apertures is the same ($\approx E_L dx/z$). However, the phase is sensitive to small variations in r. The optical path difference between rays emanating from the slit centre and distance x is given by $x \sin \theta = R - r$, where R is the distance from the slit centre to point P. Therefore,

$$dE(r,t) = \frac{E_L dx}{R} \sin\left[\omega t - k(R - x \sin \theta)\right], \tag{4.23}$$

where in the amplitude term, we use the approximation that $R \approx r$.

To find the total electric field at point P from all sub-apertures, we must sum up the field contributions from all elements across the aperture – i.e. we integrate this expression across the aperture of width b:

$$E(r,t) = \frac{E_L}{R} \int_{-b/2}^{b/2} \sin\left[\omega t - k(R - x \sin \theta)\right] dx = \frac{E_L b}{R} \frac{\sin \beta}{\beta} \sin(\omega t - kR), \tag{4.24}$$

where

$$\beta = \frac{kb}{2} \sin \theta. \tag{4.25}$$

For small θ, we have

$$\tan \theta \approx \sin \theta \approx \theta \implies \frac{y}{R} = \frac{2\beta}{kb} = \frac{\lambda \beta}{\pi b}. \tag{4.26}$$

The intensity at the screen is proportional to the time-average of the square of the field amplitude, $\langle E^2 \rangle$. The time-average of the $\sin^2$ term is 1/2, so from equation (4.24), we can write the Fraunhofer diffraction pattern of a 1-D slit as

$$I(\beta) = \frac{1}{2} \left(\frac{E_L b}{R}\right)^2 \left(\frac{\sin \beta}{\beta}\right)^2 = I(0)\mathrm{sinc}^2(\beta), \tag{4.27}$$

This is shown in Figure 4.21 and represents the spatial intensity distribution on the screen produced by a distant point-like source viewed through this aperture.

The sinc^2 function has zeroes at $\beta = \pm\pi, \pm 2\pi, \pm 3\pi \ldots$, when the path difference for a pair of rays from the slit centre and edge differ by integer multiples of $\pi/2$. This is true for all pairs of sub-apertures with separation $b/2$, so we get an intensity minimum from the whole slit.

The first zeroes are at $\beta = \pm\pi$, i.e.,

$$\theta = \pm\frac{\lambda}{b} \quad \text{or} \quad y = \pm\frac{\lambda z}{b}. \tag{4.28}$$

Extending the diffraction calculation to a 2-D aperture simply adds an additional term into the integral for the other coordinate, which, for a rectangular aperture, results in an identical profile in the orthogonal direction. The diffraction pattern for a rectangular aperture is shown in Figure 4.22.

The calculation for a circular aperture follows the same methodology, but adopting spherical polar coordinates for convenience. The result, as we would expect from the above analysis and

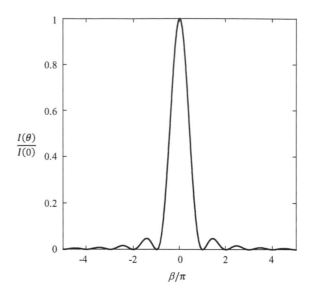

FIGURE 4.21 Single-slit far-field diffraction pattern.

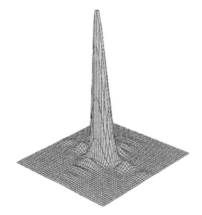

FIGURE 4.22 Rectangular aperture diffraction pattern.

circular symmetry, is circular fringes around a central core. The intensity profile, known as the Airy pattern after Astronomer Royal George Airy who first calculated it, is given by

$$I(\theta) = I(0) \left[\frac{2J_1(u)}{u} \right]^2, \tag{4.29}$$

where J_1 is the first-order Bessel function and $u = (kD\sin\theta)/2$, and D is the aperture diameter.
  For small θ,

$$u = \frac{\pi D \theta}{\lambda}. \tag{4.30}$$

The first minimum of $J_1(u)$ is at $u = 1.22\pi$, i.e.,

$$\frac{\pi D \theta}{\lambda} = 1.22\pi \Rightarrow \theta = \frac{1.22\lambda}{D}. \tag{4.31}$$

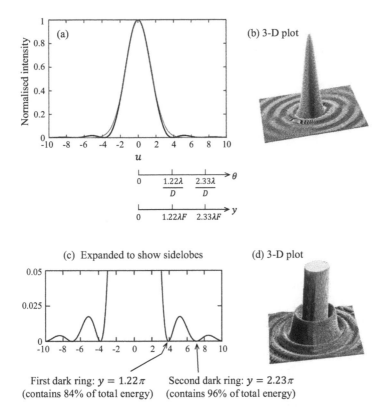

FIGURE 4.23 (a) Circular aperture diffraction pattern (black curve), and a Gaussian approximation with the same FWHM as the central peak (grey line); (b) 3-D plot of the diffraction pattern; (c) expanded version showing sidelobes and zeroes; (d) expanded 3-D plot.

The angular size of the image of a point source is thus proportional to the wavelength divided by the telescope diameter. Using the plate scale equation (4.17), the physical radius of the first minimum of the diffraction pattern can be written as

$$y = f\theta = 1.22\left(\frac{f}{D}\right)\lambda = 1.22F\lambda, \tag{4.32}$$

where $F = f/D$ is the focal ratio. The key features of the Airy diffraction function are shown in Figure 4.23. The central peak can be conveniently approximated by a Gaussian function with the same FWHM. The full width at half maximum of the Airy function is $\theta_{\text{FWHM}} = 1.03\lambda/D$.

4.6.1 THE RAYLEIGH CRITERION

Image spreading due to diffraction limits our ability to discern fine scale details in an image or to distinguish two sources close together. By convention, two sources in an image are said to be resolved (i.e. unambiguously determined to be separate) if the central peak of one source's diffraction pattern lies on or outside the first minimum of the other's – the Rayleigh criterion. This is reasonable, although it is somewhat arbitrary in that if the signal-to-noise ratio (SNR) is low, it might not be possible to resolve the objects, and if it is high, it might be possible to resolve sources that are more closely spaced. The Rayleigh criterion corresponds to a minimum angular separation of $1.22\lambda/D$, as shown in Figure 4.24. In practice, angular resolution can depend on the SNR (to be

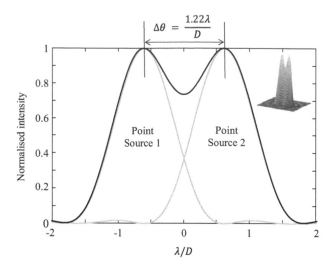

FIGURE 4.24 Graphical representation of the Rayleigh criterion. The inset shows a 3-D representation of the overlapping diffraction patterns.

described in Chapter 5) achieved for the measurement. If the SNR is high and the shape of the beam is known, then reliable information can be derived about the spatial distribution of the source on a size scale smaller than the Rayleigh criterion would predict; if the SNR is poor, then it may not be possible to resolve sources even if their angular separation is larger.

In the case of ground-based optical and infrared observatories, the theoretical angular resolution of the telescope may not be achieved because of atmospheric turbulence. The technique of adaptive optics, in which fluctuating wavefront distortions caused by the atmosphere are sensed and corrected in real time, is now commonly adopted in large ground-based observatories to enable near-diffraction-limited imaging to be carried out (see Chapter 10).

4.6.2 Beam Profile or Point Spread Function

A detector in the focal plane of a telescope has a particular beam profile or point spread function (PSF), which is defined as its response to a point source on the sky as a function of the pointing offset from the position of the source. The PSF depends both on the properties of the optical system and on the detector itself. The coupling of the detector to the sky depends strongly on whether it responds to the intensity distribution in the focal plane or whether it is coupled to the telescope via an antenna, and so responding to the field distribution. In Chapter 8, on radio techniques, we will consider the case of antenna-coupled detectors, but here we assume the former case, in which a detector can be considered as a square pixel that collects an amount of power from the focal plane that depends on its area. Two-dimensional detector arrays with square pixels are commonly used in astronomical instruments at X-ray, UV, optical and infrared wavelengths.

If one imagines a pixel of vanishingly small size and the telescope being scanned across the source in two dimensions, then the response of the pixel will map out the Airy pattern projected onto the sky, which is the diffraction-limited beam profile of the detector. The beam profile in this case is equivalent to the convolution of the intensity distribution and a delta function representing the pixel. The secondary maxima of the Airy function lead to the beam having sidelobes at the corresponding off-axis angles. As noted above, the Airy pattern FWHM is $1.03\lambda/D$, so λ/D is often adopted as an approximation to the diffraction-limited beamwidth.

A real pixel must have finite size to detect any power from the source. The beam profile then corresponds to the convolution of the Airy intensity distribution and the area covered by the pixel, and so is broadened with respect to that of a vanishingly small detector. The average beam profile is shown in Figure 4.25 for different square pixel sides: 0.5, 1.0, and 1.5 times $F\lambda$, corresponding to 0.5, 1.0, and 1.5 λ/D in terms of angle on the sky, in comparison with the narrowest beam profile achievable (for zero pixel size) indicated by the grey line. A pixel size of $0.5\lambda/D$ only broadens the beam by a few percent compared to the Airy pattern. A $1.0\lambda/D$ pixel size results in a modest ~20% loss of angular resolution, but anything larger than $1.0\lambda/D$ starts to cause significant broadening of the beam.

The pixel size also dictates the fraction of the total power from a point source that it collects. This is given by integrating the Airy pattern over the pixel extent and dividing by the total power in the Airy pattern (corresponding to infinite pixel size). Figure 4.26 shows this for the case of an on-axis square pixel, with the pixel size again expressed in units of λ/D. To collect about 70% of the power with a square pixel requires a pixel size of ~$1.3\lambda/D$. There is a thus compromise between the sensitivity of an individual detector and the achieved angular resolution: making the detector small gives the best angular resolution and making it large gives the best sensitivity. In practice, with multi-pixel arrays, a pixel size of ~ $0.5\lambda/D$ (sometimes up to $1.0\lambda/D$) is often used to avoid too much degradation of angular resolution. As long as the inter-pixel gaps are small, this does not

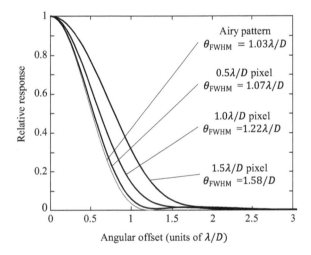

FIGURE 4.25 Beam profiles for square pixels of side 0.5, 1.0, and 1.5 λ/D (black lines) compared with the Airy pattern (grey line), which corresponds to the beam profile for a vanishingly small pixel.

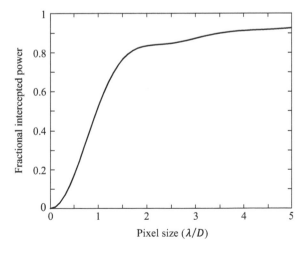

FIGURE 4.26 Fraction of total power from a point source intercepted by an on-axis square pixel as a function of pixel size in units of λ/D.

TABLE 4.1

Diffraction-Limited FWHM Beamwidth of a 10-m Diameter Telescope

Wavelength	Beamwidth
550 nm (optical)	0.012″
2.2 μm (infrared)	0.043″
20 μm (far infrared)	0.43″
300-μm (submillimetre)	6.5″
1 mm (millimetre)	22″
3 cm (microwave)	10.3′
1 m (radio)	6.0°

involve much loss of efficiency, as power missed by one pixel will be collected by neighbouring pixels. It does, however, require the co-addition of pixels to measure the signal, which is less optimal in terms of final signal-to-noise than collecting most of the power with a single pixel.

4.6.3 STREHL RATIO

As for diffracted rays, the Seidel aberrations appear as additional wavefront phase shifts arising from the non-perfect optics, causing off-axis rays to be advanced or delayed with respect to rays arriving at the focus which followed a path closer to the optical axis, so both effects can contribute to the spread in energy from a point-like source. The overall PSF will include both diffraction and Seidel effects. For astronomical telescopes operating at visible wavelengths (size of optical components $\gg$ wavelength), diffraction is usually small compared to the Seidel aberrations, but at far infrared and longer wavelengths, diffraction dominates and the aberrations can often be neglected.

To quantify the image blur caused by non-diffracting effects, the Strehl ratio, S, is defined as the peak image intensity from a point source compared to the peak intensity of an ideal optical system limited only by diffraction. Therefore, a near-perfect optical system which is diffraction-limited would have a Strehl ratio close to unity.

Table 4.1 lists the diffraction-limited beam size (taken as $1.05\lambda/D$) for a 10-m aperture for different wavelengths. As we move from the optical region where the Seidel aberrations or atmospheric turbulence may limit the image clarity, we rapidly become dominated by diffraction.

4.7 FOURIER OPTICS

So far, we have followed the classical geometric physical optics methodology. It is also instructive to analyse optical systems in the Fourier domain. Importantly, this allows a more intuitive approach to be applied to optical system design and helps to explain many critical factors such as modulation transfer function (MTF), beam shaping, beam smearing, and spatial interferometry.

We start here with a basic introduction and assume that the reader is familiar with the Fourier Transform (FT) pair for a regular function $f(x)$:

$$F(k) = \int_{-\infty}^{\infty} f(k)e^{ikx}\, dx \quad f(x) = \frac{1}{2\pi}\int_{-\infty}^{\infty} F(k)e^{ikx}\, dk. \tag{4.33}$$

The variable x can be a time-domain or space-domain coordinate (e.g. measured in seconds or metres) in which case k is then a temporal frequency or spatial frequency coordinate (cycles per second or cycles per metre).

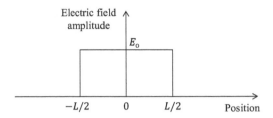

FIGURE 4.27 Plane-wave aperture field represented by a rectangular function.

To see how this relates to a telescope system, consider the electric field vector of a plane wave incident at the primary aperture. For simplicity, we consider a 1-D aperture of width L, equivalent to the case of diffraction due to a slit considered above. An incident plane wave is only accepted between $\pm L/2$, so the electric field is truncated to a top-hat function as shown in Figure 4.27, with $f(x)$ everywhere zero accept across the aperture where it is uniform:

$$f(x) = E_o \quad \text{for} - L/2 < x < L/2$$

$$= 0 \quad \text{otherwise.} \tag{4.34}$$

The FT of this field can be calculated using the FT relationship in equation (4.33), resulting in the set of spatial frequencies that define the field. Letting $e^{ix} = \cos x + i \sin x$, we can use the fact that $f(x)$ is an even function (symmetrical about the origin) and so ignore the sinusoidal terms in the exponential function to write

$$F(k) = \int_{-L/2}^{L/2} E_o \cos(kx)\,dx, \tag{4.35}$$

giving

$$F(k) = \frac{E_o}{k}\left[\sin(kx)\right]_{-L/2}^{L/2} = \frac{2E_o}{k}\sin\left(\frac{kL}{2}\right) = E_o L \frac{\sin\left(\dfrac{kL}{2}\right)}{\left(\dfrac{kL}{2}\right)} = E_o L \operatorname{sinc}(\beta), \tag{4.36}$$

where $\beta = kL/2$.

The spatial frequency composition of the electric field is a thus a sinc function. Each spatial frequency can be considered as a plane wave propagating from the aperture to a distant screen in the direction of the wave vector $\mathbf{k}$. Therefore, the amplitude distribution for a plane wave propagating through the aperture looks like a sinc function at a distant screen. The function $\operatorname{sinc}(\beta)$ has a FWHM of 3.79, corresponding to $k = \pm 3.79/L$, so there is a reciprocal relationship between the aperture size and the spread in spatial frequencies – a wider aperture corresponds to a narrower spread.

We now note the close similarity with the single-slit physical optics diffraction case considered in Section 4.6 above, in which we considered a uniform plane wave at a slit aperture and determined that the field at a point on a distant screen was equal to the sum of the contributions from all the point-like radiators across the aperture. Calculating the intensity using the Fraunhofer approach also gave a $\operatorname{sinc}^2(\beta)$, dependence, equivalent to the $\operatorname{sinc}(\beta)$ result derived here for the electric field. This leads to the important conclusion that the FT of the aperture field distribution is the field distribution in the Fraunhofer diffraction pattern. This can also be shown to hold for an arbitrary 2-D aperture. For a circular aperture, which is the most relevant for astronomical telescopes, the electric field is represented by a cylinder, for which the FT in spherical polar coordinates gives the same Bessel function result as equation (4.29).

When a telescope is used to map an extended source, structures smaller than the telescope beam are smeared by the beam profile. The main lobe of the beam profile, defined by the core of the Airy function can be reasonably well approximated by a Gaussian profile, $g(\theta)$ where we consider the 1-D case for simplicity. If this beam is scanned across an extended astronomical source represented by $f(\theta)$, then for each position in the scan the intensity detected, $h(\theta)$, is determined by to the overlap at that point of the telescope beam function with the source distribution. $h(\theta)$ is thus given by the convolution of the source intensity distribution with the beam profile:

$$h(\theta) = f(\theta) * g(\theta) = \int_{-\infty}^{\infty} f(\phi) g(\theta - \phi) \mathrm{d}\phi. \qquad (4.37)$$

The convolution theorem states that the FT of the convolution of two functions is the product of their FTs:

$$H(k) = F(k)G(k), \qquad (4.38)$$

where $H(k)$ is the FT of $h(\theta)$, $G(k)$ is the FT of $g(\theta)$, and $F(k)$ is the FT of $f(\theta)$, with k representing the spatial frequencies on the sky.

The resulting smeared image retains large-scale features but loses details on size scales smaller than the beam. Figure 4.28 shows a 1-D representation of this smoothing. The brightness distribution $f(\theta)$ is imaged by a system with a Gaussian beam, $g(\theta)$, resulting in the convolved (smoothed) image $h(\theta)$. The FTs of $f(\theta)$ and $g(\theta)$ are also shown, illustrating how the higher spatial frequency components of the object are attenuated by the FT of the beam profile. The finite spatial resolving power of the optical system thus corresponds to a low-pass spatial filter in the Fourier domain.

An extended source mapped by a telescope with a certain beam pattern thus results in a smeared image such as that shown in the left-hand panel of Figure 4.29, which shows a map of the Monoceros R-2 star formation region made with the *Herschel*-SPIRE camera at a wavelength of 500 μm, and with a native angular resolution of ~ 36″. Spatial details smaller than the beam have been lost. A common technique is to restore partly this lost detail using a process termed deconvolution. The measured source map, $h(\theta)$, can be Fourier transformed to obtain $H(k)$. Likewise, the known telescope beam pattern, $g(\theta)$, can be Fourier transformed to get $G(k)$. The convolution theorem gives

$$F(k) = \frac{H(k)}{G(k)}, \qquad (4.39)$$

and hence, by Fourier inversion, $f(\theta)$.

A problem in practice is that to obtain this inverse transform it is necessary to know the functions $h(\theta)$ and $g(\theta)$ accurately all the way to infinity. The presence of noise makes this impossible since both are determined experimentally. The deconvolution process thus requires high signal-to-noise data and an accurate knowledge of the beam pattern and will fail completely if at any point $G(k) = 0$. In practice, deconvolution-based data processing techniques can produce an effective improvement in angular resolution by about a factor of 2. The right-hand panel of Figure 4.29 was derived from the same data as used for the left-hand image, but post-processed using a deconvolution routine.

The FT of the PSF, $G(k)$, is known as the optical transfer function (OTF). It is a complex function that needs to be specified in terms of its amplitude and phase. The MTF is the modulus of the OTF (MTF=|OTF|) and characterises the extent to which the recorded image differs from the

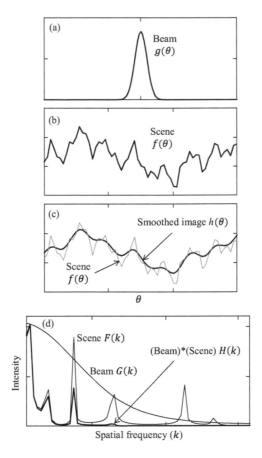

FIGURE 4.28 Convolution of a Gaussian beam (a) with a 1-D scene (b) to produce a smoothed image (c), with the FTs of the beam $G(k)$, the scene $F(k)$, and their product $H(k)$ (d).

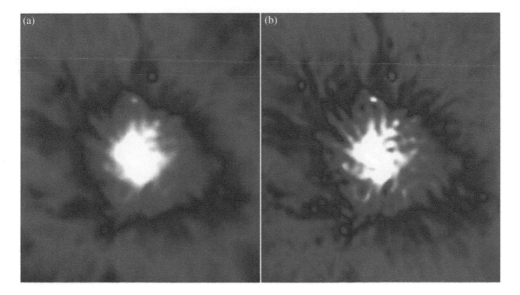

FIGURE 4.29 (a) False-colour image of the Monoceros R-2 star formation region made with the *Herschel*-SPIRE instrument at 500-μμm wavelength. (b) The same data post-processed using a deconvolution routine to enhance the angular resolution. (Credit: SPIRE Consortium; HOBYS Team; Tom Rayner.)

ideal image by representing how the system processes each spatial frequency. Such an analysis is important for any optical system and is used as a measure of its effectiveness for making specific observations.

4.8 SPATIAL INTERFEROMETRY

The telescope diameter needed to achieve a particular spatial resolution increases linearly with wavelength. To achieve the same angular resolution at 1 m wavelength as is achieved with a 1-m optical telescope would require a telescope 2 million times bigger – 2000 km in diameter, which is clearly impossible. The world's largest single-aperture radio telescope, the Five-hundred-metre Aperture Spherical Telescope (FAST) in Guizhou, China, uses a natural geographical bowl to support a 500-m diameter spherical dish. Astronomical sources are tracked by a moving secondary mirror, which can be placed at the appropriate position above the dish, and illuminate a 300-m diameter portion of the primary aperture.

Another approach is available that provides the high spatial resolution that is often needed, namely interferometry. The theory of the two-element interferometer will be described in the context of radio astronomy in Chapter 8, but the same technique can be, and is, used at shorter wavelengths. A simple two-element interferometer with two small and manoeuvrable telescopes is equivalent in spatial resolution (along the direction of its baseline) to using a single large dish of diameter equal to the baseline between the antennas, albeit with most of the dish collecting area missing so that the sensitivity is much reduced compared to a full aperture of the same size.

The basic properties of a two-element interferometer can be appreciated using a Fourier approach applied to the 1-D case of two slits (equivalent to a Young's slits experiment), as shown in Figure 4.30. We first consider two very narrow slits. A plane wave is incident on both slits and diffracted waves from the two slits interfere at point P on a distant screen, with distance y from the centre of the screen and making an angle θ with the optical axis. The path difference between the two waves is $d\sin\theta$, and the corresponding phase difference is $d\phi = 2\pi d\sin\theta/\lambda = kd\sin\theta$.

The waves from the two apertures at P can be written as

$$E_1(\theta) = Ee^{i(\omega t - kr)}; \; E_2 = Ee^{i(\omega t - kr + d\phi)}. \tag{4.40}$$

The resultant is therefore

$$E_{\mathrm{Res}}(\theta) = E_1(\theta) + E_2(t) = E\left(1 + e^{id\phi}\right) e^{i(\omega t - kr)}, \tag{4.41}$$

which has intensity

$$I(\theta) = E^2\left(1 + e^{id\phi}\right)\left(1 + e^{-id\phi}\right) = 2E^2\left(1 + \frac{e^{id\phi} + e^{-id\phi}}{2}\right) = 2E^2\left(1 + \cos(d\phi)\right). \tag{4.42}$$

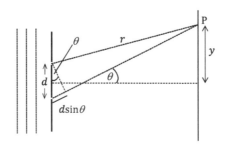

FIGURE 4.30   Two-slit interferometer.

Using the trigonometric identity $\cos^2(A/2) = (1 + \cos(A))/2$ and substituting $d\phi = kd\sin\theta$, we get

$$I(\theta) = 4E^2\cos^2\left(\frac{kd\sin\theta}{2}\right). \tag{4.43}$$

The intensity at the screen is thus a $\cos^2$ pattern extending indefinitely in both directions, which is equivalent to the FT of two δ-functions representing the slits. If the slits have finite widths, they can be modelled as the convolution of two δ-functions representing the slit positions with a top-hat function representing the uniform electric field across each slit.

The electric field in the focal plane of this double-slit interferometer is simply the FT of the aperture field. We have already shown that the FT of the top-hat is a sinc function. Since convolution in the spatial domain is equivalent to multiplication in the Fourier (spatial frequency) domain, we can conclude that the response in the focal plane is a cosine function with an envelope defined by a sinc function. The relationships are summarised in Figure 4.31 for two slits of width b separated by distance d.

The electric field disturbance, as seen at a position in the focal plane defined by a small angle θ, is

$$E(\theta) \propto \left(\frac{\sin\beta}{\beta}\right)\cos\alpha, \tag{4.44}$$

where

$$\beta = \frac{kb\sin\theta}{2} \quad \text{and} \quad \alpha = \frac{kd\sin\theta}{2}. \tag{4.45}$$

The intensity is thus given by

$$EI(\theta) = I(0)\left(\frac{\sin\beta}{\beta}\right)^2\cos^2\alpha. \tag{4.46}$$

Figure 4.32 shows the form of the $\cos^2$ fringes resulting from the interference between the waves from the two slits, with the broad sinc function envelope resulting from the diffraction at a single slit. If the slits are infinitely narrow, the sinc^2 function broadens into a uniform flat envelope. The broader they are, the narrower the envelope. Changing the baseline (distance between slits) changes

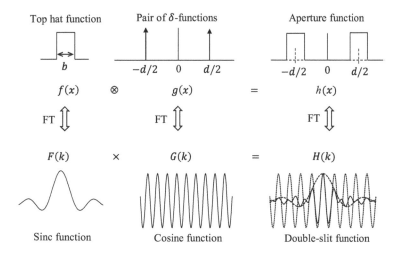

FIGURE 4.31 Fourier approach to two-slit diffraction or two-beam interferometry.

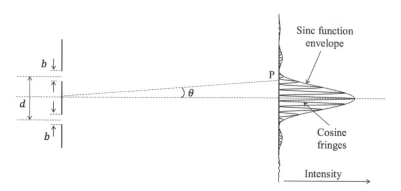

FIGURE 4.32   Interferometric fringes for a two-slit interferometer.

the width and number of fringes observable within the single-slit diffraction pattern. The longer the baseline, the narrower the fringes become.

The behaviour of a two-telescope interferometer is essentially the same. If a point-like source is observed and its intensity is undiminished with increasing baseline, it must be on-axis and within the central fringe – thus a spatial resolution comparable to the fringe width is achieved. By measuring the fringe patterns for all possible baselines, the exact location of a point-like source can be found and/or its extended structure can be mapped. Extending this model to three or more telescopes in a plane is easily achieved mathematically by using the 2-D FT adding in more δ-functions. In practice, several telescopes are usually used to provide good angular resolution in many directions as well as to provide more collecting area.

Establishing an interference pattern relies on being sensitive to the phase differences between the signals arriving at the different apertures. At radio wavelengths, the electromagnetic field amplitude and phase are measured directly with heterodyne receivers. This is why interferometric imaging at radio frequencies (aperture synthesis) has been successfully used in radio astronomy since the 1950s and has become a standard technique. Radio interferometers are described in Chapter 8.

In contrast, at optical and IR wavelengths, direct detectors are used, which respond to the radiation intensity and not its field amplitude. Indeed, it is inconceivable at present to consider an electronic readout system that could respond to optical wave (~10^{14} Hz) interference. Thus, a problem with optical/IR astronomical interferometry is that accurate optical delay lines and atmospheric wavefront aberration correction are required to bring the images from the separate apertures together so that the interference can be observed. This very demanding technology has only become possible since the 1990s. The European Southern Observatory's Very Large Telescope (VLT) facility in Chile includes an interferometric capability, the VLTI, which operates in the near-infrared. The signals from up to four of eight telescopes (the four 8.2-m VLT telescopes plus four additional smaller 1.8-m telescopes) can be combined, with a maximum baseline up to 130 m.

5 Key Concepts in Astronomical Measurement

5.1 INTRODUCTION

In this chapter, we briefly review some key general concepts and detector properties, which will be relevant to the design and performance of astronomical detection systems across the electromagnetic spectrum. Many of these concepts will arise again and will be covered in more depth in subsequent chapters.

5.2 TRANSDUCTION

The measurement of a physical parameter involves the conversion of the input quantity, electromagnetic power, for instance, into some easily measurable output quantity, such as voltage or current, by a transducer (which, for our purposes, is synonymous with sensor or detector). Examples of transducers in the measurement of electromagnetic power are:

i. a photoconductive detector, which produces a current proportional to the photon rate incident on the detector;
ii. a photovoltaic detector, which produces a voltage proportional to the photon rate;
iii. a radio antenna, which produces a current proportional to the amplitude of the incident electric field;
iv. a bolometric detector, which produces a change in temperature proportional to the electromagnetic power absorbed by the detector (the temperature change is usually then converted to a change in voltage);
v. a Charge Coupled Device (CCD) pixel, which produces an amount of charge proportional to the number of photons incident during the exposure.

Most of these will be dealt with in more detail in later chapters.

5.3 THE DECIBEL SCALE

A convention often adopted in measurement science is to represent ratios of physical quantities logarithmically. One of the most commonly adopted schemes is the decibel scale. If P_1 and P_2 are two values of power, the ratio of the two corresponds to a number of decibels (dB) given by

$$\text{Number of dB} = 10\log_{10}\left(P_1/P_2\right). \tag{5.1}$$

Zero dB corresponds to equal power levels, 10 dB corresponds to a factor of 10, 20 dB to a factor of 100, etc. A factor of 2 in power is equivalent to $10\log_{10}(2) \approx 3$ dB. Often, values of voltage or electric or magnetic field are expressed in dB, in which case the quadratic dependence of power on these quantities has to be taken into account:

$$\text{Number of dB} = 20\log_{10}\left(V_1/V_2\right) \quad \text{or} \quad 20\log_{10}\left(E_1/E_2\right). \tag{5.2}$$

A factor of 2 in power (3 dB) is equivalent to a factor of $1/\sqrt{2}$ in voltage or field amplitude. A voltage amplifier with a gain of, say 1000, can be described as having a gain of $20\log_{10}(1000) = 60$ dB.

5.4 RESPONSIVITY

The responsivity, S, is the relation between the output and input quantities:

$$S = \frac{\text{Change in output quantity}}{\text{Change in input quantity}}. \tag{5.3}$$

The definitions and typical units for the responsivities of some common detector types (to be described in later chapters) are given in Table 5.1. A high value of responsivity is generally a good feature but does not guarantee a sensitive detector.

5.5 RESPONSE TIME (SPEED OF RESPONSE)

For any detector, the output will not change instantaneously in response to an instantaneous change in the input. The detector has a certain response time, which is a measure of the delay between a change in the input and the corresponding change in the output. Sometimes, the response time is so short that it has no influence on the operation of the detector, and sometimes it is an important limitation. A finite response time has two practical implications for a detector: (i) some time must be allowed for the detector output to settle down after a sudden change in the input and (ii) if the input quantity varies at some frequency, the responsivity is less than the "DC" value defined above. The terminology used here comes from electrical circuit theory: the direct current (DC) responsivity is the value for 0-Hz input, corresponding to the change observed after waiting for an infinite time.

Many detectors have a response that is well described by that of a first-order linear system, for which the RC circuit is the best-known example. Many practical detectors are actually resistors, and any self-capacitance or capacitance of the read-out circuit effectively means that the detector response is literally that of an RC circuit. In other cases, the physical mechanisms are different, but the form of the response is still often dictated by a first-order differential equation and has the same essential features.

The response of a first-order system can be characterised by a single time constant, τ. The time constant represents the characteristic time delay between something happening at the input and the corresponding change occurring at the output. Here, we consider the transient response and the frequency response of an RC circuit as an example. Many detectors behave similarly. For a step

TABLE 5.1
Definitions and Typical Units for Responsivities of Some Types of Detector

Detector Type	Responsivity		Typical Units
	Definition	Equation	
Photoconductor	Output current I/Incident EM power P	$S = \dfrac{dI}{dP}$	A W^{-1}
Photovoltaic or bolometric detector	Output voltage V/Incident EM power P	$S = \dfrac{dV}{dP}$	V W^{-1}
CCD pixel	Charge produced Q/incident EM energy, W	$S = \dfrac{dQ}{dW}$	C J^{-1}

increase in the input voltage from 0 to V_{in} at $t = 0$, the output voltage of an RC circuit (the voltage across the capacitor) is given by

$$v_{out}(t) = V_{in}\left(1 - e^{-t/\tau}\right),\tag{5.4}$$

with $\tau = RC$.

For a step decrease from V_{in} to zero at $t = 0$, the corresponding change in the output voltage as a function of time is

$$v_{out}(t) = V_{in}e^{-t/\tau}.\tag{5.5}$$

These transients are illustrated in Figure 5.1. After a time τ, the response to a step increase has risen to a fraction $\left(1 - e^{-1}\right) = 67\%$ of the way towards its final value. The output only approaches the final value asymptotically, but after about 5τ, it has settled to within 1%. The response to a step decrease has fallen to a fraction $e^{-1} = 33\%$ of the total change after a time τ. Therefore, in either case, τ is roughly the time needed for two-thirds of the change to occur.

If the input voltage varies sinusoidally with some angular frequency ω (i.e. period $T = 2\pi/\omega$), such that

$$v_{in}(t) = V_{in}\sin(\omega t),\tag{5.6}$$

then the output voltage is given by

$$v_{out}(t) = V_{out}(\omega)\sin(\omega t + \phi),\tag{5.7}$$

where ϕ represents a phase lag between the output and the input.

The output amplitude, $V_{out}(\omega)$, is determined by the voltage divider constituted by the capacitive reactance, with complex impedance $1/(j\omega C)$, where $j = \sqrt{-1}$, and the resistance, R:

$$V_{out}(\omega) = \left|\frac{1/(j\omega C)}{R + 1/(j\omega C)}\right|V_{in} = \frac{1/(\omega C)}{\left(R^2 + 1/(\omega C)^2\right)^{1/2}}V_{in} = \frac{V_{in}}{\left[1 + (\omega\tau)^2\right]^{1/2}}.\tag{5.8}$$

The magnitude of the transfer function, $H(\omega)$, the ratio of the output and input amplitudes, is thus given by

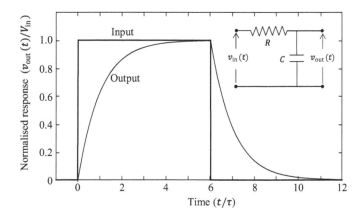

FIGURE 5.1 Normalised transient response of an RC circuit (inset), an example of a first-order system, to a step increase in the input voltage from zero to V_{in} at $t = 0$ followed by a step decrease back to zero at $t = 6\tau$.

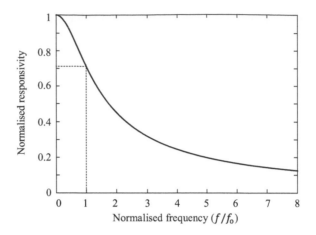

FIGURE 5.2 Responsivity vs signal modulation frequency for a detector with a first-order response (normalised to the 3-dB frequency, f_0).

$$H(\omega) = \frac{V_{\text{out}}(\omega)}{V_{\text{in}}} = \frac{1}{\left[1+(\omega\tau)^2\right]^{1/2}}. \tag{5.9}$$

For a detector characterised by a single time constant, the responsivity thus decreases with sinusoidal modulation frequency according to

$$S(\omega) = \frac{S(0)}{\left[1+(\omega\tau)^2\right]^{1/2}}, \tag{5.10}$$

as shown in Figure 5.2. A first-order system attenuates the input amplitude by a factor that depends on how its period compares to the time constant of the system. If the frequency is low enough that the period is long compared to the time constant, then the system has plenty of time to respond to the changing input and the attenuation is small. However, if the frequency is high, such that the input period is very short compared to the time constant, then the system cannot keep up with the changing input (as illustrated for a step change in Figure 5.1), and the output is strongly attenuated. A first-order system such as an RC circuit thus acts as a low-pass filter.

When $\omega\tau = 1$, i.e., the input frequency, f, is

$$f_0 = \frac{1}{2\pi\tau}, \tag{5.11}$$

the output voltage amplitude has decreased by a factor of $1/\sqrt{2} = 0.707$, corresponding to a reduction of approximately 3 dB in the output. Frequency f_0 is therefore called the 3-dB, or half-power frequency, and is a measure of how high the signal frequency can be before the detector response has faded away too much. If the input signal is modulated, as often needs to be the case, the modulation frequency must be low enough that the detector can respond to the changing power.

5.6   BACKGROUND RADIATION

In many cases, the electromagnetic radiation to be measured (the signal) is accompanied by a constant, or slowly varying, amount of unwanted additional radiation (background). For example, an infrared detector used to measure power from an astronomical source (the signal) must do so against a background of thermal radiation from the atmosphere and the telescope. Often, this background is much bigger than the signal. A standard way of subtracting the background is to "chop" between the source position (detector output proportional to source+background) and an adjacent patch of sky (detector output proportional to background only) and subtract the two measurements. Because the sky background can vary with time, this chopping must be done on a timescale faster than the timescale of sky background variation; otherwise, the background will have changed between the two measurements. This imposes a requirement on the detector time constant.

In addition to the inconvenience of having to measure and subtract it, the presence of background radiation often degrades the quality of measurement through contributing noise, as described in the next section. The effects of background radiation are discussed in detail in Chapter 6.

5.7   NOISE AND SIGNAL-TO-NOISE RATIO

Noise is any *fluctuating* unwanted contribution to the output signal that might affect the measurement by introducing uncertainty in the recorded value. It can be present in the signal before it reaches the detector, and can also arise in the detector or its electronics. An important figure of merit for any measurement is the signal-to-noise ratio (SNR), which characterises the sensitivity of the system: SNR = 10 corresponds to 10% rms uncertainty in the result, while the corresponding uncertainty for SNR = 100 is 1%.

Important noise sources include

  i. random fluctuation of the incident signal power – this cannot be eliminated even in principle;
 ii. random fluctuation of the incident background power – this can only be minimised by making the background as small as possible;
iii. random noise introduced by the detector and electronic circuits used to read out and process the detector signal (e.g. amplifiers), which can be reduced by careful design, use of low-noise components, cooling, etc.;
 iv. interference (e.g. due to radio signals) – this can be minimised by careful design, grounding, shielding, etc.

For any real detector system, the ultimate goal is to make the total noise dominated by the contributions from the signal and/or the unavoidable background. Chapter 6 will discuss in detail the most important noise mechanisms and the techniques used to minimise them.

5.8   ELECTRICAL FILTERING AND INTEGRATION

Making a measurement usually involves averaging the quantity to be measured over a certain period of time, often referred to as the integration time. The term is equivalent to "averaging time" since the average of a function over a certain time interval is just the integral divided by the time interval. The presence of noise means that a finite time is necessary to make a measurement, exploiting the fact that while the signal is steady, the noise fluctuates. The signal will always be present and will produce a positive average, but the noise tends to average out to zero, and the average will get closer and closer to zero as we increase the integration time. Thus, with signal and noise present, averaging for a sufficiently long time can allow the presence of the signal to be discerned. As we shall see in Chapter 6, the SNR tends to increase as the square root of the integration time. However, if the signal itself varies with time (for instance, in the case of short-duration radio pulses from a pulsar), then the integration time is naturally limited to the characteristic time over which the signal changes.

It is useful and important to consider the relationship between integration and filtering. A low-pass filter, such as an RC circuit, can be regarded as an integrator. The initial part of the step response (Figure 5.1; equation 5.4), when $t \ll \tau$ is approximately linear, representing the integral of the step function:

$$v_{\text{out}}(t) = V_{\text{in}}\left(1 - e^{-t/\tau}\right) \approx V_{\text{in}}\left(1 - \left(1 - \frac{t}{\tau}\right)\right) \approx \left(\frac{V_{\text{in}}}{\tau}\right)t. \qquad (5.12)$$

The RC circuit thus acts as an integrator of the input signal for timescales much shorter than τ.

In detector signal processing, an important parameter for the electrical circuit placed after the detector is its post-detection bandwidth. There is a reciprocal relationship between post-detection bandwidth and time constant – a narrow post-detection bandwidth corresponds to a long time constant. The constant of proportionality between the two depends on the shape of the filter passband.

It can be shown (Bracewell 1999; Kraus 1986) that for an ideal integrator which simply averages a signal over an integration time $t_{\mathrm{int-ideal}}$, the corresponding post-detection bandwidth is

$$\Delta f_{\mathrm{ideal}} = \frac{1}{2t_{\mathrm{int-ideal}}}. \tag{5.13}$$

For an RC circuit with time constant $\tau = RC$, the integration time and post-detection bandwidth are

$$t_{\mathrm{int-RC}} = 2\tau \quad \text{and} \quad \Delta f_{\mathrm{RC}} = \frac{1}{4\tau}. \tag{5.14}$$

5.9 NYQUIST SAMPLING

In making measurements of a continuous variable such as an electrical time series or the spatial distribution of intensity, the signal is normally digitised through samples taken at regular intervals. To ensure that no information is lost in this sampling, the samples must be sufficiently closely spaced. The necessary sampling frequency depends on the highest frequency of interest contained in the continuous signal, f_{max}; if f_{max} can be reconstructed from the sampled signal, then all lower frequencies can also be reconstructed. According to the Nyquist sampling theorem, the sample frequency must be at least $2 f_{\mathrm{max}}$ so that there are at least two samples for each period. The top panel of Figure 5.3 illustrates how if the sampling frequency is too small (in this case only 1.5 times the frequency of the signal), then the measured samples (circles) are consistent with either the signal (solid black) waveform at frequency f_{S} or the dashed one at frequency $f_{\mathrm{S}}/2$. In the bottom panel, the sampling frequency is $2 f_{\mathrm{S}}$, and in this case, no frequency lower than f_{S} is consistent with the measurements.

Applying the same principle to the case of astronomical imaging, this means that there must be at least two samples per beam FWHM to ensure that details in the image that the beam is capable of discerning are not lost. For example, Figure 5.4 shows the signal from two point sources that are just

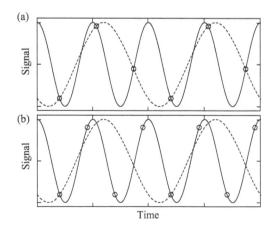

FIGURE 5.3 (a) A signal of frequency f_{S} (solid line) sampled (circles) at $1.5 f_{\mathrm{S}}$. The measured samples are also consistent with the lower-frequency waveform shown by the dotted line. (b) The same signal sampled at $2 f_{\mathrm{S}}$, resulting in an unambiguous reconstruction from the measured samples.

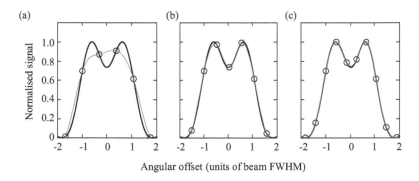

FIGURE 5.4 Double source signal profile for two point sources just resolved according to the Rayleigh criterion, with sample points (circles) for sampling rates of 1.5 (a), 2.0 (b), and 2.5 (c) samples per beam. The grey lines indicate cubic spline fits to the data points.

resolved according to the Rayleigh criterion (Section 4.6.1), sampled at 1.5, 2, and 2.5 samples per beam FWHM. The grey lines indicate spline fits that could be applied to the measured data points. It is clear that with only 1.5 samples per beam, the measurements can fail to show that there are two sources present, whereas with 2, or better still, 2.5, samples per FWHM, the double nature of the source is clearly seen, and good fits can be made to recover the detailed profiles.

5.10 LINEARITY

An ideal amplifier will have the same gain irrespective of the magnitude of the input voltage. Likewise, an ideal detector will have a constant responsivity regardless of the amount of power incident on it. Real detectors tend to depart from this ideal behaviour at high signal levels, resulting in a variation of responsivity with signal. Figure 5.5 shows how the response of a real detector can change depending on the input level. Over a certain range of input power, the detector behaves linearly, but for higher input, the responsivity decreases and can even flatten off for very high input levels.

In many experiments, it is necessary to make comparative measurements, for instance of an unknown source against a standard source, which requires taking ratios of signals to determine brightness ratios. This is a straightforward procedure in the linear regime, where equating V_1/V_2 with P_1/P_2 gives the correct result. Outside the linear region, a simple ratio gives an incorrect result and becomes more and more erroneous as the signal power goes up. Therefore, it is important that either the detector is operated only in the linear region or that the non-linear response curve is accurately known so that it can be corrected for in the analysis.

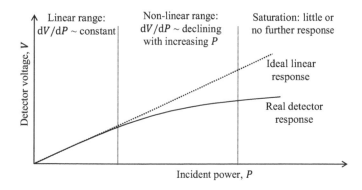

FIGURE 5.5 Schematic of output voltage vs input power for a non-linear detector.

5.11 DYNAMIC RANGE

The term "dynamic range" is usually taken to refer to the ratio of the largest to the smallest signal levels that a detection system can be used to measure. It is often limited at the lower end of the range by the noise level and can be limited at the upper end either by the detector ceasing to respond adequately or by the signal level exceeding the maximum value that can be handled by the electronics used to read the detector signal. For example, in detectors which operate by accumulating charge in proportion to the total number of photons incident during the integration time, a maximum "well capacity" is often specified, equal to the amount of charge that can be stored before the detector saturates. When observing bright sources or under a high background level, the integration time has to be kept short enough that the well capacity is not exceeded. The dynamic range with respect to the instantaneous signal level then depends on the integration time and can be large for short integration times. Dynamic range can also be defined as the well capacity divided by the noise level, which corresponds to the total energy collected in the integration rather than the instantaneous power on the detector.

5.12 TYPES OF MEASUREMENT: PHOTOMETRY, SPECTROSCOPY, SPECTRO-PHOTOMETRY

A particular detector of EM radiation will respond to a certain range of wavelengths. The human eye, for example, responds to wavelengths between about 400 and 700 nm. Sometimes, it is required to make a measurement over a narrower range of wavelengths, in which case the spectrum of the radiation that is being detected can be filtered to provide a well-defined spectral passband. This can be done by placing a filter or a spectrometer in front of the detector to allow only the desired range of wavelengths to be detected. Astronomical spectrometers will be described in Chapter 7.

The spectral resolving power of a measurement is defined as $\lambda_0/\Delta\lambda$ where λ_0 is the central wavelength and frequency and $\Delta\lambda$ is spectral resolution (the range of wavelengths detected together). An equivalent definition is $v_0/\Delta v$ where v is frequency. Ideally, the accepted range would be a perfect "top-hat" spectral response function (SRF), as shown in Figure 5.6, but in practice, it is usually rounded and may have some structure arising from the filters or other components used. Depending on the purpose of the measurement, the required passband can be very wide (photometry), narrow (spectroscopy), or in between (spectro-photometry).

Black body, free–free, and synchrotron radiation (described in Chapter 2) are examples of continuum radiation, for which the spectrum varies slowly over a broad range of wavelengths. In the case of photometry, the aim is to characterise the overall shape of the continuum (sometimes referred to as the spectral energy distribution, or SED) and/or to get an estimate of the total amount of power being emitted across the whole spectrum. For example, in determining the shape of a black body-like spectrum, observations in a few suitable bands with $\lambda_0/\Delta\lambda \sim 3$ are enough to determine the shape by fitting an appropriate function to the data points (or "joining the dots") as shown in

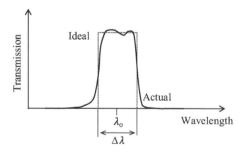

FIGURE 5.6   Ideal and realistic SRFs.

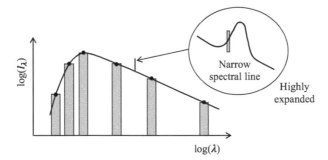

FIGURE 5.7 Photometry (defining the broad shape of the spectrum) and spectroscopy (characterising spectral lines or features).

Figure 5.7, allowing the temperature and the total brightness to be found. However, if the aim is to do spectroscopy – to measure spectral features superimposed on the continuum – then a much narrower passband is needed, with $\lambda_o/\Delta\lambda$ anything between ~100 and 10^7 depending on the width of the features.

A spectro-photometric measurement with $\lambda_o/\Delta\lambda$ ~ a few $\times$ 10, intermediate between the low resolution used for basic photometry and the higher resolution needed for spectroscopy, is useful when the objective is to gain more detailed information on the shape of the continuum or to study the broad spectral features that are characteristics of solid materials.

5.13 CALIBRATION

Calibration is the process by which the measured signal (e.g. voltage) is converted into astronomically meaningful units. The two most important kinds of calibration in astronomy are flux calibration (the assignment of a reliable brightness estimate) and spectroscopic calibration (knowledge of the wavelength or frequencies being measured). Absolute calibration through detailed knowledge of all the relevant properties of the system (detector, optical system, telescope, atmosphere, etc.) is often extremely difficult, and many astronomical calibration schemes get around this problem through the use of suitable standard astronomical sources, the properties of which are known sufficiently accurately through modelling or observation. Direct comparison of the signals generated by the standard and an unknown source then allows the brightness of the unknown object to be calculated without having to know various characteristics of the observing system (optical transmission, detector responsivity, electronics gain, etc.) because such parameters cancel out when the ratio of the signals is taken.

5.13.1 THE STELLAR MAGNITUDE SYSTEM

The Greek mathematician and astronomer Hipparchus first devised a scale in which the brightness of a star was characterised by its apparent magnitude. The brightest star in the sky had apparent magnitude 1, and the faintest visible star had apparent magnitude 6. In the 19th century, it was known that the eye's response is logarithmic so that a certain apparent magnitude difference was taken to correspond to a certain ratio in observed brightness, with 5 magnitudes corresponding to a factor of 100.

A difference of 1 magnitude corresponds to $\quad 100^{1/5}$ = a factor of 2.512.
A difference of 2 magnitudes corresponds to $\quad 100^{2/5}$ = a factor of $2.512^2 = 6.31$.

For two stars with fluxes (power per unit area) F_1 and F_2 at the Earth, we have

$$\frac{F_1}{F_2} = 100^{(m_2 - m_1)/5},$$ (5.15)

giving

$$m_2 - m_1 = -2.5 \log\left(\frac{F_1}{F_2}\right).$$ (5.16)

The magnitude of a star can be quoted in a number of ways: it can be determined by measuring the flux, F (W m^{-2}), within a certain standard passband, as above; it can be specified at a particular wavelength (monochromatic magnitude) by measuring the flux density, S_v (W m^{-2}Hz^{-1}) at that wavelength; or it can be based on the luminosity, L (W) – the total power emitted over the whole of the spectrum (bolometric magnitude).

The apparent magnitude of a star depends on its absolute brightness and its distance. The absolute magnitude, M, is a measure of its actual brightness and is defined as the apparent magnitude that the star would have if it were at a distance of 10 pc.

Let a star be at distance d pc and have flux at the Earth, F.

If it were at distance 10 pc, it would have flux $F_{10} = F(d/10)^2$. Therefore, the difference between the apparent and absolute magnitudes of a star is

$$m - M = -2.5 \log\left(\frac{F}{F\left(\frac{d}{10}\right)^2}\right) = 5 \log(d) - 5.$$ (5.17)

$m - M$ depends only on the distance and is called the distance modulus.

For two stars of different brightnesses at the same distance,

$$m_1 - m_2 = M_1 - M_2 = -2.5 \log\left(\frac{F_1}{F_2}\right).$$ (5.18)

Specifying the absolute brightness requires a standard value of brightness which is assigned magnitude zero. For instance, if the zero-magnitude reference has monochromatic flux density S_{v0}, then the flux density of a source of apparent magnitude M is given by

$$m = -2.5 \log\left(\frac{S_v}{S_{v0}}\right).$$ (5.19)

The star Vega (α Lyrae), which lies at a distance of 7.68 pc and has an effective (photospheric) temperature of around 9500 K, has, for many years, been adopted as the standard zero-magnitude star. However, because Vega is slightly variable, other schemes are sometimes adopted.

5.13.2 STANDARD FILTER BANDS

Usually, the brightnesses of stars are measured in particular wavelength bands defined by standard filters placed in front of the detector to define photometric bands. In the 1950s, to enable astronomers to inter-compare their observational data in the optical region Harold Johnson and William Morgan defined three photometric bands known as the U (Ultraviolet), B (Blue), and V (Visual) system. These were added to by later workers extending the system into the FIR, with the standard bands being designated UBVRIJHKLMNQ. Ground-based observatories use filter sets that

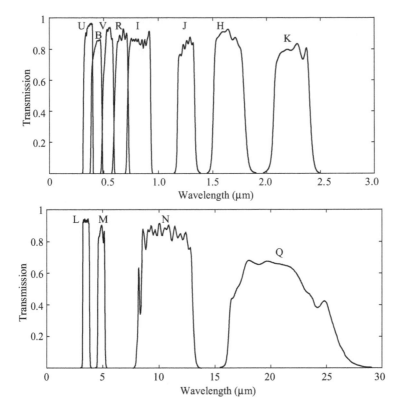

FIGURE 5.8 Typical photometric filter transmission profiles used by major ground-based observatories in the visible and infrared, all located on Maunakea in Hawai'i. The UBVRI filters are used on the Canada France Hawai'i (CFHT) telescope; the JHK bands are used in the United Kingdom Infrared Telescope (UKIRT) WFCAM instrument; the LMN bands are for the NSFcam instrument on the NASA Infrared Telescope Facility (IRTF) on Maunakea; and the Q band is used in the Gemini Observatory Michelle instrument. (Data are from the Spanish Virtual Observatory Filter Profile Service (Roderigo & Solano 2012; 2013) supported from the Spanish MINECO through grant AYA2017–84089; data available at http://svo2.cab.inta-csic.es/svo/theory/fps3/.)

generally adhere to these conventions, although the precise filter profiles vary significantly. Some typical filter bands from instruments at major observatories are shown in Figure 5.8.

The conversion between observed magnitude and flux density depends on the particular filter profile and calibration system adopted. In the Vega calibration system, Vega is defined to have a V-band magnitude of 0.03 and a magnitude of zero in all other bands. The absolute flux density of Vega at 556 nm is 3500 Jy. A comprehensive review of standard astronomical photometric systems is given by Bessell (2005).

The ideal astronomical "standard candle" calibration source is (i) an object whose absolute brightness is known with a high level of confidence through modelling or measurement or both; (ii) bright enough to provide a good SNR in a short time, but not so bright as to cause complications with detector saturation; (iii) point-like with respect to the telescope beamwidth, to enable simple characterisation of how it couples to the detector or direct comparison of signal levels when an unknown point source is observed; and (iv) non-variable or with well-understood variability (e.g. due to the varying distance to the Earth for a solar system calibrator).

Standard stars are the most convenient calibration sources for UV, optical and infrared observations. At visible and near-infrared wavelengths, brightness values are often quoted in magnitudes, whereas for longer wavelengths, they are more often quoted in terms of flux (W m^{-2}) or flux density

(Jy, equivalent to W m⁻²Hz⁻¹) – knowing the zero point of the magnitude system, it is possible to convert from one to the other.

At longer wavelengths, stars become fainter, and there can be additional uncertainties due to emission by circumstellar dust discs, making them unsuitable as calibration standards. Planets (Mars, Uranus, and Neptune) and large asteroids are commonly used as standards at the far infrared to millimetre wavelengths. At radio wavelengths, as will be discussed in Chapter 8, brightness is usually quoted in terms of a quantity known as antenna temperature. In the X-ray and γ-ray regions, in which individual photons can readily be detected, source brightness is often characterised in terms of counts per second per unit area (usually quoted in counts $cm^{-2}s^{-1}$) or in terms of power per unit area, usually quotes in ergs $s^{-1}cm^{-2}$ – although it is really an archaic unit of energy, the erg (1 erg ≡ 10^{-7}J) remains popular amongst astronomers. White dwarf stars can be used as calibrators below about 1 keV, but at higher energies, there are no suitable non-variable point source X-ray calibrators, so well-understood extended sources such as supernova remnants are often used. For many astronomical observations, absolute calibration accuracy is often determined by the knowledge of the calibration standard. Ten to twenty percent accuracy is often adequate, although accuracies of a few % are now achieved in many parts of the spectrum. Relative accuracy (i.e. brightness ratio with respect to some standard calibrator) can often be even better with careful high-SNR measurements.

5.13.3 Colour Correction

As noted above, the passbands of photometric instruments are often quite broad in order to maximise sensitivity (typically, $v_o/\Delta v \sim 3$). This means that the power incident on the detector can have a significant dependence on the shape of the source spectrum within the passband. If the standard calibration source and the unknown source to be measured have different spectral shapes, then a simple ratio of the measured signals will not yield a true value for the brightness of the source. A correction, known as the colour correction, needs to be applied in order to take the different spectral shapes of the calibrator and the source into account.

Consider an instrument with a SRF defined by $R(v)$, i.e., the instrument transmission as a function of frequency, v, taking into account both the variation of optical transmission across the passband and also any possible variation with frequency of the aperture efficiency (the fraction of the total power from an on-axis point source that is coupled to the detector). For an on-axis observation of an unknown source with spectral flux density $S_S(v)$ at the telescope aperture, the source power absorbed by the detector is directly proportional to the integral over the passband of the flux density weighted by $R(v)$, as shown in Figure 5.9.

Assuming that the detector voltage, V, is directly proportional to the absorbed power, the measured voltage is proportional to the SRF-weighted flux density, $\bar{S}_S$:

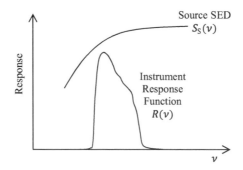

FIGURE 5.9 Dependence of measured signal on the integral over the passband of the product of the source SED and the instrument response function.

$$V_S = K\bar{S}_S = K \frac{\int_{Passband} S_S(v)R(v)dv}{\int_{Passband} R(v)dv},$$

(5.20)

where K is a constant.

The corresponding value for the standard calibration source, with spectral flux density $S_C(v)$, is

$$V_S = K\bar{S}_C = K \frac{\int_{Passband} S_C(v)R(v)dv}{\int_{Passband} R(v)dv}.$$

(5.21)

It is convenient and conventional to implement a standard data reduction procedure that produces monochromatic flux densities ascribed to a particular frequency within the band (e.g. the band centre, v_0). To do this, it is necessary to know the instrument response function (but because $R(v)$ appears in both the numerator and the denominator, it is only necessary to know the *relative* SRF rather than the absolute variation). It is also necessary to make some assumption about the spectral characteristics of the source:

$$S_S(v) = S_S(v_0)f_S(v,v_0),$$

(5.22)

where $f_S(v,v_0)$ characterises the shape of the spectrum. Letting the calibrator spectral shape be given by

$$S_C(v) = S_C(v_0)f_C(v,v_0),$$

(5.23)

we can then calculate the flux density of the source at frequency v_0 as

$$\frac{S_S(v_0)}{S_C(v_0)} = \frac{V_S}{V_C}\left[\frac{\int_{Passband} f_C(v,v_0)R(v)dv}{\int_{Passband} f_S(v,v_0)R(v)dv}\right].$$

(5.24)

The factor in brackets takes the different spectral shapes of the source and calibrator into account and is called the colour correction factor. Only in the case where the two spectra have the same shape is it legitimate simply to multiply the ratio of the measured signals by the calibrator flux density.

6 Sensitivity and Noise in Electromagnetic Detection

6.1 INTRODUCTION

In general, there are three categories of electromagnetic radiation detectors:

 i. coherent detectors, which measure both the amplitude and the phase of the incident electric field;
 ii. photon detectors, which produce a signal proportional to the incident photon rate; and
 iii. bolometric detectors, which measure the incident power (photon rate multiplied by photon energy).

Coherent detection is most relevant to the radio regime (wave picture of light) and is described in Chapter 8. The other two (incoherent) detection methods, which do not preserve phase information, are best understood using the photon picture of light and are considered in this chapter. We will consider the influence of noise – any fluctuating unwanted output – on the sensitivity of the detection system. We start with an ideal photon detector and derive an expression for the best achievable sensitivity, limited only by the properties of the incident radiation. Real detectors are subject to imperfections and additional noise contributions, and we will look at ways of characterising and minimising their effects on the overall sensitivity of the measurement.

6.2 AN IDEAL PHOTON DETECTOR

Consider an experiment in which we can regard the radiation to be measured as a stream of photons. For a measurement of electromagnetic power, we need to know the photon arrival rate and the energy (frequency) of the photons. Let us assume that the frequency is known by some other means, and that all the detector must do is count the photons within a narrow frequency band, Δv, centred on frequency v. For measuring the photon arrival rate, an ideal detector then has two essential properties:

 i. it registers an output for every photon that arrives (it is 100% efficient);
 ii. it never registers an output unless a photon has arrived (it is noiseless).

Let n_v be the average number of photons per unit time per unit radiation bandwidth incident on the detector so that the average photon arrival rate is $n = n_v \Delta v$. As noted in Chapter 5, astronomical signals often have to be measured in the presence of unwanted background radiation. To allow for this situation, let

$$n = (n_{vS} + n_{vB})\Delta v = n_S + n_B \left[\text{photons s}^{-1}\right], \qquad (6.1)$$

where subscript S indicates signal photons (from the source to be measured) and B denotes background photons (from any source of unwanted photons).

If the background is non-zero, we need to determine n_S by making two separate measurements:

 1. We observe the source plus background and count the number of photons, N_{S+B}, that have arrived in a given length of time, t_{obs}:

$$N_{S+B} = (n_{vS} + n_{vB})\Delta v t_{obs}. \tag{6.2}$$

2. We observe the background alone and count the number of photons, N_B, that have arrived in the same time interval:

$$N_B = n_{vB}\Delta v t_{obs}. \tag{6.3}$$

n_{vS} can then be found by subtracting these and dividing by $\Delta v t_{obs}$.

These two measurements could be executed sequentially if there is only one detector, or simultaneously if there are two detectors. Let the total time available for the measurement be t_{int} (the total integration time). Then for sequential measurement $t_{obs} = t_{int}/2$, and for simultaneous measurement $t_{obs} = t_{int}$.

In each of the measurements, we count the number of photons that arrive in a certain time interval. If this same measurement is repeated many times, the result will not be exactly the same every time because there will be statistical fluctuations in the arrival of photons at the detector, constituting a form of noise (as described in Section 5.7). Here we make an important assumption: the arrival of photons at the detector is random. The arrival of one photon at a certain instant provides no information about the arrival of the previous photon or the next photon – the photon arrival events are said to be completely uncorrelated. An analogy is the measurement of the number of raindrops per second hitting a certain patch of the ground during a shower. Because of these statistical variations, the number of photons arriving in a certain interval will not produce the same result every time – it is a random variable. The multiple measurements will produce a frequency (probability) distribution, which, in the case of such uncorrelated events, is the Poisson distribution. If the quantity being measured has a Poisson distribution with a mean value of M, and we make a single measurement, then the probability of the result being some value, N, is given by

$$p(N) = \frac{M^N}{N!}e^{-M}. \tag{6.4}$$

Figure 6.1 shows $p(N)$ for $M = 5$, 10, 20, 50, and 100. The distribution is not quite symmetrical, although the asymmetry is very small except for small M. While M is the most probable result, the chances of measuring exactly M are actually small (e.g. 4% for $M = 100$).

A single measurement is likely to give a result close to, but not equal to, M. The uncertainty or error in the measurement is characterised by the width, or standard deviation, of the Poisson distribution, σ, defined as the root-mean-square (rms) deviation of the individual measurements from the mean value. If a total of K measurements are made, then,

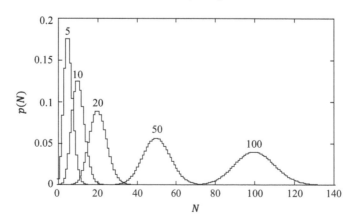

FIGURE 6.1   Poisson distributions for $M = 5$, 10, 20, 50, and 100.

$$\sigma = \left[\frac{1}{K} \sum_{i=1}^{K} (N_i - M)^2 \right]^{1/2}.$$

(6.5)

The Poisson distribution has the particular property that

$$\sigma = M^{1/2},$$

(6.6)

i.e., the standard deviation is equal to the square root of the mean. This is a very important general rule for counting things which are generated or arrive randomly: the rms uncertainty in one measurement is the square root of the number of things being counted.

The signal-to-noise ratio for the measurement is

$$\text{SNR} = \frac{\text{Mean}}{\text{Standard deviation}} = \frac{M}{M^{1/2}} = M^{1/2}.$$

(6.7)

Hence, the SNR is also equal to the square root of the number of things being counted. Therefore, if possible, it is a good idea to generate a large number of things to count.

In the case of our photon-counting measurements, the rms uncertainty in the number of photons collected, N, is given by

$$\Delta N = \sigma = (n_\nu t_{\text{obs}} \Delta \nu)^{1/2}.$$

(6.8)

For the measurement of source+background, we can therefore write

$$N_{\text{S+B}} = (n_{\nu \text{S}} + n_{\nu \text{B}}) t_{\text{obs}} \Delta \nu \pm \left[(n_{\nu \text{S}} + n_{\nu \text{B}}) t_{\text{obs}} \Delta \nu \right]^{1/2}.$$

(6.9)

Similarly, for the measurement of background alone,

$$N_{\text{B}} = n_{\nu \text{B}} t_{\text{obs}} \Delta \nu \pm \left[n_{\nu \text{B}} t_{\text{obs}} \Delta \nu \right]^{1/2}.$$

(6.10)

The resulting estimate of the source signal is then

$$N_{\text{S}} = n_{\nu \text{S}} t_{\text{obs}} \Delta \nu \pm \left[(2 n_{\nu \text{B}} + n_{\nu \text{S}}) t_{\text{obs}} \Delta \nu \right]^{1/2},$$

(6.11)

where, because the two measurements are independent and have uncorrelated uncertainties, the uncertainty in the difference is the quadrature sum (i.e. the square root of the sum of the squares) of the uncertainties in the two individual measurements.

For the combined measurement, the SNR is then given by

$$\text{SNR} = n_{\nu \text{S}} \left(\frac{t_{\text{obs}} \Delta \nu}{2 n_{\nu \text{B}} + n_{\nu \text{S}}} \right)^{1/2}.$$

(6.12)

Equation (6.12) is a very fundamental expression with some important implications:

1. It is impossible to get a better SNR than this value, which is dictated by so-called "photon noise", the random fluctuations in the photon arrival rate.
2. SNR increases as the square root of the integration time: the longer the time interval, the greater the number of photons being counted, so the greater the SNR.

3. SNR also increases as the square root of the accepted passband – broadening the passband results in more photons to count, but this is at the expense of spectral resolution: with a wide passband, we cannot ascribe a precise wavelength to the measurement.

4. SNR is degraded by the background because n_{vB} does not contribute to the signal, but it does contribute to the noise. If the background dominates $(n_{vB} \gg n_{vs})$, then SNR $\propto 1/\sqrt{n_{vB}}$. So background must be minimised.

5. If there is only one detector, the total integration time has to be divided between the two measurements and $t_{obs} = t_{int}/2$.

6. If two detectors can measure simultaneously the source plus background and the background alone, then $t_{obs} = t_{int}$, resulting an $\sqrt{2}$ improvement in SNR over the one-detector case.

7. If there is no background, then SNR $= (n_{vs}t_{obs}\Delta v)^{1/2}$ and no subtraction is needed so $t_{obs} = t_{int}$.

6.3   NOISE EQUIVALENT POWER

The sensitivity of a detector, or detector system, is often expressed in terms of the noise equivalent power (NEP), which is defined as the signal power which gives an SNR of unity for a total integration time of $t_{int} = 0.5$ second. It might seem more natural for NEP to be defined for an integration time of 1 second, but the convention is to define it with respect to a 1-Hz post-detection bandwidth, which corresponds to a 0.5-second integration time for an ideal integrator (equation 5.13).

We want the NEP to be as small as possible as low NEP means high SNR. Consider the SNR achieved in a measurement of power P in an integration time t_{int}. It will have the following dependencies on P, t_{int}, and NEP:

$$\text{SNR} \propto P; \quad \text{SNR} \propto t_{int}^{1/2}; \quad \text{SNR} \propto 1/\text{NEP}. \tag{6.13}$$

Therefore,

$$\text{SNR} = C\left(\frac{Pt_{int}^{1/2}}{\text{NEP}}\right), \tag{6.14}$$

where C is a constant. Invoking the definition of NEP, by putting $P = \text{NEP}$ when $t_{int} = 0.5$ second, gives $C = \sqrt{2}$. Therefore,

$$\text{SNR} = \frac{P\sqrt{2t_{int}}}{\text{NEP}} \quad \text{or} \quad \text{NEP} = \frac{P\sqrt{2t_{int}}}{\text{SNR}} \quad \text{or}$$

$$t_{int} = \left[\frac{(\text{SNR})(\text{NEP})}{\sqrt{2}P}\right]^2 \quad \text{or} \quad P = \left[\frac{(\text{SNR})(\text{NEP})}{\sqrt{2t_{int}}}\right]. \tag{6.15}$$

The first version of this equation can be used for predicting the SNR that will be achieved for a certain power level given a particular detector NEP and integration time (how good will my measurement be?); the second version can be used to estimate NEP from the result of a measurement of a known power (how good is my detector?); the third version can be used to determine the necessary integration time to measure a certain power level with a particular SNR (how long will my measurement take?); the fourth version quantifies the power detectable to a given SNR in a given integration time (what is the smallest signal that I can measure?).

Based on the above equation, the minimum detectable power (discernible with unit signal-to-noise) in an integration time t_{int} is defined as

$$\Delta P_{\text{min}} = \frac{\text{NEP}}{\sqrt{2t_{\text{int}}}}. \tag{6.16}$$

Since SNR is a dimensionless quantity, the units of NEP are (power) (time)$^{1/2}$ or W s$^{1/2}$, but it is more conventional to quote it in units of W Hz$^{-1/2}$.

6.3.1 BACKGROUND-LIMITED NEP

Ideally, the detector and other components will contribute a negligible amount of additional noise compared to the photon noise. Photon noise thus imposes an ultimate limit on the sensitivity for the measurement, and this can be expressed in terms of the photon noise-limited NEP, NEP_{ph}.

Consider the case in which the background dominates ($n_{\nu B} \gg n_{\nu S}$, which is indeed often true in practice for astronomical detectors). The fundamental expression for the photon noise-limited SNR, equation (6.12), then gives

$$\text{SNR} = n_{\nu S} \left(\frac{t_{\text{obs}} \Delta \nu}{2 n_{\nu B}} \right)^{1/2}. \tag{6.17}$$

Expressing this in terms of the signal and background powers rather than the photon rates using

$$P_S = n_{\nu S} h \nu \Delta \nu \quad \text{and} \quad P_B = n_{\nu B} h \nu \Delta \nu, \tag{6.18}$$

we get

$$\text{SNR} = \frac{P_S t_{\text{obs}}^{1/2}}{\left(2 P_B h \nu \right)^{1/2}}. \tag{6.19}$$

If a SNR of unity is achieved in a time of $t_{\text{int}} = 0.5$ second, then by definition $P_S = \text{NEP}_{\text{ph}}$.

We will distinguish between three different cases:

Case 1: There is only one detector, and it measures S+B and B sequentially. As noted above, the 0.5 second must be divided equally between the two measurements. Therefore, $t_{\text{obs}} = t_{\text{int}}/2 = 0.25$ second giving

$$\text{NEP}_{\text{ph}} = 2 \left(2 P_B h \nu \right)^{1/2}. \tag{6.20}$$

Case 2: There are two detectors, and S+B and B are measured simultaneously. Therefore, $t_{\text{obs}} = t_{\text{int}} = 0.5$ second so that

$$\text{NEP}_{\text{ph}} = 2 \left(P_B h \nu \right)^{1/2}. \tag{6.21}$$

Case 3: One detector measures S+B, and B is measured simultaneously by a large number of detectors, with their signals being averaged to reduce the noise on the B-alone measurement to a negligible level. This is an idealised equivalent of the case in astronomy in which a large array of detectors is used to observe a point source, with the signal from the source being concentrated in a small number of pixels (taken as one here) while many surrounding pixels in the array are averaged to make an independent measurement of the background. Here $t_{\text{obs}} = t_{\text{int}} = 0.5$ second, and the factor of 2 in the denominator of equation (6.15) also disappears (because the background only contributes noise to the S+B measurement). This gives

$$\text{NEP}_{\text{ph}} = \left(2 P_B h \nu \right)^{1/2}. \tag{6.22}$$

The difference between Cases 1 and 3 involves two factors of $\sqrt{2}$ – one due to having to split the time between the two measurements, and one due to the elimination of background noise in the B-alone measurement in Case 3.

Note that in all cases, the photon noise-limited NEP is proportional to the square root of the background power.

6.4 EFFICIENCY OF PHOTON DETECTORS

Our treatment of noise so far has assumed that the detector is perfect. The photon noise-limited SNR may be approached, but not achieved in practice because a real detector might not respond to every photon, and the detector and its electronics might produce additional noise. The following definitions are used to characterise detectors for infrared and shorter wavelengths (photon rather than wave description of light).

The responsive quantum efficiency (RQE, or η) is defined as the fraction of incident photons which contribute to the output signal. Clearly, RQE $\leq$ 1. It is easy to show that in the case of a detector with RQE η_d, the fundamental SNR equation (6.12) becomes

$$\text{SNR} = n_{\nu S}\left(\frac{\eta_d t_{obs} \Delta \nu}{2n_{\nu B} + n_{\nu S}} \right)^{1/2}. \tag{6.23}$$

The degradation in sensitivity is slower than η_d because although the signal level is reduced by imperfect absorption, so is the background.

The RQE accounts for imperfect absorption, but not for any additional noise. A figure of merit that takes both into account is the detective quantum efficiency (DQE), which is defined as follows:

$$\text{DQE} = \left[\frac{\text{SNR at detector output}}{\text{Photon noise limited SNR}} \right]^2 = \left[\frac{\text{Photon noise limited NEP}}{\text{Actual NEP}} \right]^2. \tag{6.24}$$

The DQE is the best overall figure of merit: it compares the actual sensitivity with the best achievable in principle and can be used to compare different types of detectors.

6.4.1 DQE AND NEP OF AN IMPERFECTLY ABSORBING BUT NOISELESS DETECTOR

Sometimes the absorption efficiency of the detector is not perfect, but it adds negligible noise. What is the DQE in this case? The photon noise-limited SNR is given by equation (6.12) and the actual SNR by equation (6.23), giving DQE = η_d, as expected. The corresponding photon noise-limited NEP is

$$\text{NEP}_{ph} = 2\left(\frac{P_B h\nu}{\eta_d} \right)^{1/2}. \tag{6.25}$$

The NEP is degraded in proportion to the square root of the RQE because although the signal decreases linearly with RQE, the photon noise declines as the square root of the RQE.

6.5 PHOTON SHOT NOISE AND WAVE NOISE

The assumption of random photon arrival involves regarding electromagnetic radiation as a stream of discrete, non-interacting particles (photons) with no correlation between the arrival of one photon and any other. This picture is usually justifiable at high frequencies. The random fluctuations in the photon arrival rate, as discussed above, are called photon shot noise. However, we know that at low frequencies it is often more appropriate to regard the incident light not as a stream of photons, but as EM waves. How is our photon noise theory modified if the wave description is more appropriate? Crudely, what we previously viewed as the arrival of a photon now corresponds to the arrival of the peak of wave so that there is a strong correlation with the arrival of the next (or previous) peak, and the previous assumption of random arrival is not valid. Photons are bosons, and so they can occupy identical quantum states. A beam of photons can be regarded as a stream of wave packets of finite length. It is therefore possible that they can interfere with each other – an effect which is neglected in the shot noise picture. When two photons/wave packets interfere it can be constructively or destructively so that the result could be equivalent to anything between zero (equivalent to no photons) and doubling of the amplitude (quadrupling the intensity – equivalent to four photons). So this wave packet interference corresponds to additional fluctuations (wave noise) in the numbers of photons, over and above the fluctuations in the arrival rate.

If the power incident on the detector originates from a thermal background (i.e. from the black body process) as is often the case in practice then a full treatment, using Bose–Einstein statistics (Landau and Lifshitz 2013), shows that an additional term must be included in equation (6.8) for the rms uncertainty in the number of photons incident:

$$\Delta N = \left[n_\nu t_{obs} \Delta \nu \left(1 + \varepsilon \eta_o b \right) \right]^{1/2}, \tag{6.26}$$

where ε is the emissivity of the background, η_o is the overall transmission efficiency between the background and the detector, and the Bose factor, or photon occupation number, b, is given by

$$b = \frac{1}{e^{h\nu/k_B T} - 1}. \tag{6.27}$$

We now have the shot noise term, as derived above, plus an additional wave noise term that depends on the Bose factor and on the overall factor by which perfect black body emission at the background temperature is attenuated before it reaches the detector. Proceeding as before, the background photon noise-limited SNR is found to be given by

$$\text{SNR} = n_{\nu S} \left[\frac{t_{obs} \Delta \nu}{2 n_{\nu B} \left(\dfrac{1}{\eta_d} + \varepsilon \eta_o b \right)} \right]^{1/2}, \tag{6.28}$$

and the photon noise-limited NEP (for Case 2 – two detectors; simultaneous S and S+B detection) by

$$\text{NEP}_{ph} = 2 \left[P_B h \nu \left(\frac{1}{\eta_d} + \varepsilon \eta_o b \right) \right]^{1/2}. \tag{6.29}$$

We can distinguish two regimes: one in which the noise is dominated by the shot noise component, and one in which the wave noise is most important.

i. Wien region ($h\nu \gg k_B T$; the photon picture of light is more appropriate): $e^{h\nu/k_B T}$ is large, so $b \approx 0$ and we are left with just the original shot noise term.

ii. Rayleigh–Jeans region ($h\nu \ll k_B T$; the wave picture of light is more appropriate):

$$b \approx \frac{1}{1 + \dfrac{h\nu}{k_B T} - 1} \approx \frac{k_B T}{h\nu}, \tag{6.30}$$

giving

$$\mathrm{NEP_{ph}} = 2\left[P_B \left(\frac{h\nu}{\eta_d} + \varepsilon\eta_o k_B T \right) \right]^{1/2}.$$ (6.31)

Provided that the background emissivity is not very low and that the overall efficiency with which background photons are collected is not too low, the second term dominates and

$$\mathrm{NEP_{ph}} = 2\left[P_B \varepsilon\eta_o k_B T \right]^{1/2}.$$ (6.32)

For thermal emission in the Rayleigh–Jeans regime,

$$P_B = \varepsilon\eta_o A\Omega B_\nu(T)\Delta\nu = \varepsilon\eta_o A\Omega \frac{2\nu^2 k_B T}{c^2}\Delta\nu,$$ (6.33)

where $A\Omega$ is the throughput of the detector, giving

$$\mathrm{NEP_{ph}} = \left[\frac{2\varepsilon\eta_o k_B \nu(2A\Omega\Delta\nu)^{1/2}}{c} \right]T = \left[2\varepsilon\eta_o k_B \left(2\frac{A\Omega}{\lambda^2}\Delta\nu \right)^{1/2} \right]T.$$ (6.34)

The factor in square brackets is a constant for the observation, so when viewing a thermal source in the wave-noise-dominated regime, $\mathrm{NEP_{ph}} \propto T$: the sensitivity no longer depends on the brightness of the background, only on its temperature. For a diffraction-limited antenna, with $A\Omega = \lambda^2$ (see Section 3.1.2), we have

$$\mathrm{NEP_{ph}} = \left(2\sqrt{2}\varepsilon\eta_o \right) k_B T (\Delta\nu)^{1/2}.$$ (6.35)

If the overall efficiency factor $\varepsilon\eta_o$ in equation (6.26) is small (e.g. if the optical transmission to the detector is low), then, even in the Rayleigh–Jeans regime, the wave noise term can be small due to the probabilistic removal of photons from the radiation stream reducing the correlations between arrival events. Under these conditions, the wave noise term is said to be diluted, and the overall photon noise is closer to the shot noise term.

6.6 BACKGROUND PHOTON NOISE-LIMITED NEP OF A BROADBAND DETECTOR

The treatment so far relates to a detector that receives only a narrow band of frequencies. We can extend it to cover a detector that responds over a broad range of frequencies, with the results being slightly different depending on whether it responds to the photon rate or to the incident power.

6.6.1 NEPPH FOR A PHOTON-COUNTING DETECTOR

Consider a photon-counting detector with RQE η_d, viewing a black body background of temperature T and emissivity ε. Let the throughput of the system be $A\Omega$. The background power absorbed in frequency interval $d\nu$ centred on some frequency ν is

$$dP_{BA} = A\Omega\eta_o\eta_d\varepsilon B_\nu(T)d\nu = A\Omega\eta_o\varepsilon \frac{2h\nu^3}{c^2\left(e^{h\nu/k_B T}-1\right)}d\nu.$$ (6.36)

Considering a 1-second measurement, and letting the number of absorbed photons in this frequency interval be dN_{BA},

$$dN_{BA} = \frac{dP_{BA}}{h\nu} = \frac{2A\Omega\varepsilon\eta_o\eta_d\nu^2 d\nu}{c^2\left(e^{h\nu/k_BT} - 1\right)} \text{ photons.} \tag{6.37}$$

The uncertainty (rms fluctuation) in dN_{BA} is given by equation (6.26) with $t_{obs} = 1$ second, $N = dN_{BA}$, and allowing for η_d:

$$(\Delta dN_{BA})^2 = dN_{BA}\left(1 + \frac{\varepsilon\eta_o\eta_d}{\left(e^{h\nu/k_BT} - 1\right)}\right) = \frac{2A\Omega\varepsilon\eta_o\eta_d\nu^2}{c^2\left(e^{h\nu/k_BT} - 1\right)}\left(1 + \frac{\varepsilon\eta_o\eta_d}{\left(e^{h\nu/k_BT} - 1\right)}\right)d\nu. \tag{6.38}$$

The uncertainty in the total detected background photon rate is obtained by integrating this over all frequencies (in practice, filters in the system often restrict the range of frequencies over which the optical efficiency, η_o, is significant):

$$\Delta N_{BA}^2 = \frac{2}{c^2}\int_0^\infty \frac{A\Omega\varepsilon\eta_o\eta_d\nu^2}{\left(e^{h\nu/k_BT} - 1\right)}\left(1 + \frac{\varepsilon\eta_o\eta_d}{\left(e^{h\nu/k_BT} - 1\right)}\right)d\nu. \tag{6.39}$$

Letting $x = h\nu/(k_BT)$, this can be written as

$$\Delta N_{BA}^2 = \frac{2}{c^2}\left(\frac{k_BT}{h}\right)^3\int_0^\infty \frac{A\Omega\varepsilon\eta_o\eta_d x^2}{\left(e^x - 1\right)}\left(1 + \frac{\varepsilon\eta_o\eta_d}{\left(e^x - 1\right)}\right)dx. \tag{6.40}$$

For simplicity, we consider a monochromatic signal power at some frequency ν_S, within the passband of the system. Let the incident signal power be NEP_{ph}, at frequency ν_S. The number of absorbed signal photons in 1 second is therefore

$$N_{SA} = \frac{(\eta_d)_{\nu_S} NEP_{ph}}{h\nu_S}. \tag{6.41}$$

From the definition of NEP, for an incident power equal to NEP_{ph} and an integration time of 1 second, the SNR must be $\sqrt{2}$ so that

$$\frac{N_{SA}}{\Delta N_{BA}} = \sqrt{2} \implies N_{SA}^2 = 2\Delta N_{BA}^2. \tag{6.42}$$

Therefore,

$$\left[\frac{(\eta_d)_{\nu_S} NEP_{ph}}{h\nu_S}\right]^2 = \frac{4}{c^2}\left(\frac{k_BT}{h}\right)^3\int_0^\infty \frac{A\Omega\varepsilon\eta_o\eta_d x^2}{\left(e^x - 1\right)}\left(1 + \frac{\varepsilon\eta_o\eta_d}{\left(e^x - 1\right)}\right)dx, \tag{6.43}$$

giving

$$NEP_{ph}^2 = \frac{4(k_BT)^3\nu_S^2}{(\eta_d)_{\nu_S}^2 c^2 h}\int_0^\infty \frac{A\Omega\varepsilon\eta_o\eta_d x^2}{\left(e^x - 1\right)}\left(1 + \frac{\varepsilon\eta_o\eta_d}{\left(e^x - 1\right)}\right)dx. \tag{6.44}$$

If η_d is independent of frequency, then

$$\text{NEP}_{\text{ph}}^2 = \frac{4(k_B T)^3 v_S^2}{c^2 h} \int_0^\infty \frac{A\Omega\varepsilon\eta_0 x^2}{(e^x - 1)} \left(\frac{1}{\eta_d} + \frac{\varepsilon\eta_0}{(e^x - 1)} \right) dx. \tag{6.45}$$

If the spectral content of the signal is more complex, then the procedure for evaluating NEP_{ph} is the same except that the signal photo-event rate must be calculated by integrating the source spectrum in the same way as for the background.

6.6.2 NEP$_{\text{PH}}$ for Broadband Power Detector

Some detectors, such as bolometric detectors (described in Chapter 9), respond to the absorbed power rather than the photon rate, and this leads to a different equation for the photon noise-limited NEP.

The uncertainty in the absorbed background photon rate within the frequency range dv is given, as before, by equation (6.38). The corresponding uncertainty in the background power absorbed within dv is

$$(\Delta dP_{\text{BA}})^2 = (hv\Delta N_{\text{BA}})^2 = \frac{2A\Omega\varepsilon\eta_0\eta_d h^2 v^4}{c^2(e^{hv/k_B T} - 1)} \left(1 + \frac{\varepsilon\eta_0\eta_d}{(e^{hv/k_B T} - 1)} \right) dv. \tag{6.46}$$

The uncertainty in the total absorbed background power is therefore

$$\Delta P_{\text{BA}}^2 = \frac{2h^2}{c^2} \int_0^\infty \frac{A\Omega\varepsilon\eta_0\eta_d v^4}{(e^{hv/k_B T} - 1)} \left(1 + \frac{\varepsilon\eta_0\eta_d}{(e^{hv/k_B T} - 1)} \right) dv. \tag{6.47}$$

Letting $x = hv/(k_B T)$ as before, we have

$$\Delta P_{\text{BA}}^2 = \frac{2(k_B T)^5}{c^2 h^3} \int_0^\infty \frac{A\Omega\varepsilon\eta_0\eta_d x^4}{(e^x - 1)} \left(1 + \frac{\varepsilon\eta_0\eta_d}{(e^x - 1)} \right) dx. \tag{6.48}$$

For a monochromatic signal power P_S at frequency v_S, the absorbed signal power is $P_{\text{SA}} = P_S(\eta_d)_{v_S}$. Therefore, the SNR is given by

$$\text{SNR} = \frac{P_S(\eta_d)_{v_S}}{\left[\frac{2(k_B T)^5}{c^2 h^3} \int_0^\infty \frac{A\Omega\varepsilon\eta_0\eta_d x^4}{(e^x - 1)} \left(1 + \frac{\varepsilon\eta_0\eta_d}{(e^x - 1)} \right) dx \right]^{1/2}}. \tag{6.49}$$

Putting $P_S = \text{NEP}_{\text{ph}}$, and SNR = 1, the integration time is 0.5 second, for which the power fluctuation is a factor of $\sqrt{2}$ larger, giving

$$\text{NEP}_{\text{ph}}^2 = \frac{4(k_B T)^5}{c^2 h^3 (\eta_d)_{v_S}} \int_0^\infty \frac{A\Omega\varepsilon\eta_0\eta_d x^4}{(e^x - 1)} \left(1 + \frac{\varepsilon\eta_0\eta_d}{(e^x - 1)} \right) dx. \tag{6.50}$$

If the detector RQE is independent of frequency, then

$$\mathrm{NEP}_{\mathrm{ph}}^2 = \frac{4(k_\mathrm{B}T)^5}{c^2 h^3} \int\limits_0^\infty \frac{A\Omega\varepsilon\eta_\mathrm{o}x^4}{(e^x-1)} \left(\frac{1}{\eta_\mathrm{d}} + \frac{\varepsilon\eta_\mathrm{o}}{(e^x-1)}\right) \mathrm{d}x. \tag{6.51}$$

In contrast to the corresponding expression for a photon detector, this does not depend on the signal frequency because the bolometer responds equivalently to signal power at any frequency.

6.7 ADDITIONAL SOURCES OF NOISE

A noiseless detector would only be subject to the unavoidable contribution from fluctuations in the incoming radiation (photon noise and wave noise). In practice, the detector and its associated electrical circuit generate additional noise through various mechanisms, some of the most important of which are outlined below.

6.7.1 THERMAL NOISE FROM A RESISTOR

One form of noise that is impossible to avoid at some level is thermal noise due to the random motion of charge carriers inside a resistive element. In practice, the detector and other components in the system will have finite resistances so some thermal noise will occur.

Inside a piece of resistive material, there will be random thermal motions of the electrons, as illustrated in Figure 6.2. This will cause the current passing through any cross section of the resistor to fluctuate randomly. A voltmeter connected across the resistor will register a fluctuating voltage, v_n, (where subscript n stands for noise) that fluctuates randomly about zero. This noise is called Johnson noise or Nyquist noise. Its average value is zero because positive and negative fluctuations are equally likely. However, the average noise power is not zero – it is proportional to v_n^2, and so is always positive.

The noise power produced by a resistor at temperature T, within an electrical bandwidth Δf, is given by

$$P = k_\mathrm{B}T\Delta f. \tag{6.52}$$

A resistor at temperature T is thus capable of delivering a power of $k_\mathrm{B}T$ per unit electrical bandwidth. This important equation can be proved by means of the following thermodynamic argument developed by Nyquist in 1928. Consider two equal resistances, R, connected together at each end of a lossless transmission line of length L, as shown in Figure 6.3, which has characteristic impedance equal to R. Let everything be in thermal equilibrium at temperature T.

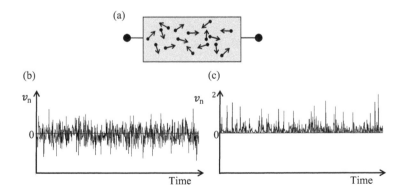

FIGURE 6.2 Motion of charge carriers in a resistor at finite temperature (a) produces a thermal noise voltage with zero mean (b), and finite thermal noise power (c).

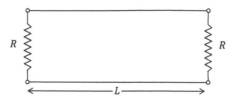

FIGURE 6.3 Equal resistors connected to each end of a lossless transmission line.

Let P be the average noise power produced by each resistor. This power is emitted by each resistor and travels as EM waves to the opposite end of the line, where it is absorbed by the other one. Now imagine that the line is suddenly short-circuited at both ends. Power will now be reflected instead of being absorbed, and the energy that was present on the line at the instant when it was short-circuited is now contained in the system in the form of standing waves. The boundary conditions that the electric field must be zero at both ends (short circuit) mean that the wavelength of a standing wave and the length of the line must be related by

$$(n\lambda_n)/2 = L \Rightarrow \lambda_n = 2L/n, \tag{6.53}$$

where each value of the mode number n (= 1, 2, 3, etc.) corresponds to an allowed mode of oscillation.

Let s be the speed of propagation of waves on the line. The standing waves are therefore at frequencies given by

$$f_n = s/\lambda = ns/2L. \tag{6.54}$$

The frequency interval between adjacent modes is given by

$$\Delta f = f_{n+1} - f_n = \frac{(n+1)s - ns}{2L} = \frac{s}{2L}. \tag{6.55}$$

The density of modes (i.e. the number of modes per unit bandwidth) is thus

$$\frac{\Delta n}{\Delta f} = \frac{2L}{s}. \tag{6.56}$$

By the equipartition theorem, each allowed mode contains energy $k_B T$. Therefore, the energy density (energy per unit bandwidth) on the line is

$$u = \frac{\Delta n}{\Delta f} k_B T = \frac{2L k_B T}{s}. \tag{6.57}$$

The total energy on the line, within bandwidth Δf, is

$$U = \frac{2L k_B T \Delta f}{s}. \tag{6.58}$$

This amount of energy originates from the two resistors. The energy due to each resistor is therefore half this amount:

$$U = \frac{L k_B T \Delta f}{s}. \tag{6.59}$$

Now, the time for energy to propagate from one of the resistors to the other end of the line is

$$t = L/s. \tag{6.60}$$

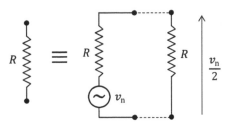

FIGURE 6.4 A real noisy resistor, represented by a hypothetical noiseless resistor in series with a noise voltage generator, is connected to a load resistance of the same value, resulting in a noise current i_n and a noise voltage across the load resistor of $v_n/2$.

Therefore, the corresponding power (energy per unit time) emitted by each resistor in bandwidth Δf is

$$P = k_B T \Delta f. \tag{6.61}$$

The noise of a resistor is usually represented by a noise voltage source, v_n, in series with a hypothetical noiseless resistor, as illustrated in Figure 6.4.

Under matched conditions (i.e. we connect the resistor up to another equal resistor), the noise power delivered by the noise source, in a bandwidth Δf, is $P = kT\Delta f$, which is fully absorbed by the load resistor and is therefore given by

$$P = k_B T \Delta f = i_n^2 R = \left(v_n/2R \right)^2 R = \frac{v_n^2}{4R}, \tag{6.62}$$

where i_n is the noise current.

Thus, we can represent a real resistor, R, at temperature T, by a noiseless resistor in series with a noise voltage generator

$$v_n = \left(4 k_B T R \Delta f \right)^{1/2}. \tag{6.63}$$

This result points at two common methods of reducing noise: cooling components to low temperature (making T small) and using a small bandwidth for the measurement of the electrical signal from the detector (making Δf small). Note that this *post-detection bandwidth* (see Chapter 5) is not the same as the bandwidth of electromagnetic frequencies being detected – rather it is the bandwidth of the electrical circuit used to read out the detector signal. For example, the noise voltage measured across the terminals of a 100-kΩ resistor at 300 K is 407 nV for $\Delta f = 100\,$Hz, or 40.7 nV for $\Delta f = 1\,$Hz. As we discussed in Chapter 5, making the post-detection bandwidth small is equivalent to making the integration time long.

The noise voltage measured across a resistor depends on the post-detection bandwidth used, which is not a property of the component itself. The noise voltage spectral density, which is a property only of the component, not the measurement system, is defined as

$$e_n = \frac{v_n}{\Delta f^{1/2}} = \left(4 k_B T R \right)^{1/2}. \tag{6.64}$$

Therefore, in the case of thermal noise, a plot of e_n vs. f is flat: the same amount of noise power ($kT\Delta f$) will be measured in any bandwidth Δf regardless of what frequency it is centred on. Noise which has a flat spectrum is called *white* noise. However, it must decrease at high frequency, just as $B_v(T)$ does, as shown in Figure 6.5.

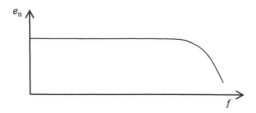

FIGURE 6.5 Typical white noise spectrum.

6.7.2 ELECTRON SHOT NOISE

Johnson noise, as described in the previous section, occurs at any finite temperature and whether or not any current is flowing. When charge flows as a current, an additional noise contribution arises because the charge flowing is quantised – it is made up of individual electrons. Just as in the case of photon noise, there will be random fluctuations in the flow rate, resulting in a shot noise current superimposed on the average current.

Consider a current constituted by a total of n_e electrons flowing during a time interval t. The current is therefore given by

$$i = \frac{n_e e}{t}. \tag{6.65}$$

The rms fluctuation in the number of electrons flowing in time t is

$$\Delta n_e = n_e^{1/2} = \left(\frac{it}{e}\right)^{1/2}. \tag{6.66}$$

and the corresponding current fluctuation is

$$\Delta i = \frac{n_e^{1/2} e}{t} = \left(\frac{ei}{t}\right)^{1/2}. \tag{6.67}$$

The noise current spectral density is obtained by putting $t = 0.5$ second, corresponding to a 1-Hz bandwidth, giving

$$i_{n-s} = (2ei)^{1/2}. \tag{6.68}$$

Like Johnson noise, electron shot noise also has a white (frequency-independent) spectrum.

6.7.3 GENERATION-RECOMBINATION NOISE

Generation-recombination (G-R) noise occurs in semiconductors and is due to the random thermal generation and recombination of electron/hole pairs giving rise to fluctuations in carrier concentration and so leading to a noise current if there is an applied voltage. The expression for G-R noise is similar to that for electron shot noise except for two additional factors: (i) a factor of $\sqrt{2}$ due to the fact that the random generation and recombination processes are uncorrelated and both affect the carrier concentration to the same extent and (ii) an additional factor, g, called the photoconductive gain, which represents the lifetime of the electrons compared with the time needed to traverse the sample (see Section 9.6.1 for more details). The G-R noise current spectral density is thus given by

$$i_{n-s} = (4egi)^{1/2}. \tag{6.69}$$

6.7.4 PHONON NOISE

Bolometric detectors involve a flow of heat from the bolometer to a heat sink. This flow of heat is quantised in the form of phonons (quantised lattice vibrations), leading to random fluctuations in the temperature of the bolometer. Bolometric detectors and their noise contributions are described in Chapter 9.

6.7.5 1/f - NOISE

In most devices, excess noise is observed at very low frequencies, sometimes for reasons that are not well understood, and is known as $1/f$ noise or pink noise. This is a very important noise source in practice and is usually present to some degree. In practice, the frequency dependence is usually of the form $1/f^\beta$, with $\beta \sim 1$–2. The spectrum is often characterised in terms of the "knee frequency", f_0, at which the noise level has increased by a factor of $\sqrt{2}$ compared to the white noise level. Figure 6.6 illustrates $1/f$ spectra for knee frequencies of 2 and 6 Hz.

6.7.6 κTC NOISE

When a capacitance, C, is charged or discharged, the charge flows in or out through a resistance. The noise voltage of that resistance leads to an uncertainty in the amount of charge left on the capacitor at the end of the operation.

The noise voltage of the resistance is given by equation (6.63), and the bandwidth Δf is here determined by that of the RC circuit (equation 5.14):

$$\Delta f = \frac{1}{4RC}, \tag{6.70}$$

giving

$$v_n^2 = \frac{k_B T}{C}, \tag{6.71}$$

which is independent of the resistance. The corresponding charge uncertainty is given by

$$\Delta q = C v_n = \left(k_B TC\right)^{1/2}. \tag{6.72}$$

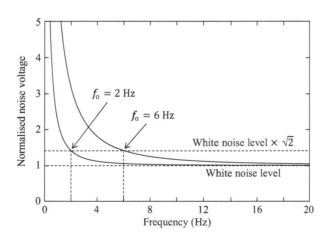

FIGURE 6.6 Typical $1/f$ noise spectra.

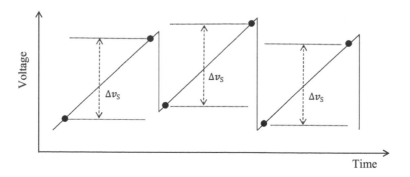

FIGURE 6.7   Cancellation of kTC noise by correlated double sampling.

Reading out a detector for which the signal is in the form of a stored electric charge (e.g. a CCD – see Chapter 10) introduces a random uncertainty of this magnitude, which can be significant in terms of the signal levels to be measured and the inherent noise of the detector. For example, with $T = 77$ K (liquid nitrogen temperature) and $C = 0.1$ pF, $\Delta q \equiv 65$ electrons.

Fortunately, when working with detectors which operate by generating a charge proportional to the number of photons collected during the integration time, the effects of kTC noise can be eliminated using a technique known as correlated double sampling. In such detectors, the resistance of the RC circuit is extremely large during the integration period. The capacitor is charged up by the signal current flowing into it, accumulating a signal charge q_S, and the integration time is such that operation is on the initial linear part of the charging transient. At the end of an integration period, the output voltage is measured and the capacitor discharged in order to commence a new integration. During the discharge, the resistance is very low and the time constant is very short. The discharge leaves a random value of charge, Δq, on the capacitor when the resistance is returned to its large value. However, the timescale for Δq to change is given by the RC time constant of the circuit, which is now very long compared to the integration time. The noise charge is thus "frozen" during the integration. Non-destructive readout with differencing of samples at the beginning and end of the integration results in both readings having the same frozen value of kTC noise charge, Δq, which therefore cancels out in the subtraction, as shown in Figure 6.7: the change in voltage during the integration due to the detector current, $\Delta v_S = q_S/C$, is the same regardless of the value of Δq.

6.7.7 INTERFERENCE

Wires and components in electrical circuits tend to act as antennas, even when they are not designed as such. They can thus "pick up" strong man-made signals (e.g. radio waves, 50-Hz mains, spikes due to switching, etc.). Interference can usually be minimised by careful design and shielding, but sometimes it is impossible to eliminate completely, in which case signal frequencies must be kept away from frequencies at which interference is significant.

6.7.8 MICROPHONIC NOISE

In a sensitive detector system, even very low-level mechanical vibrations can cause voltages to be induced due to the physical movement of conducting elements in electric or magnetic fields, as illustrated in Figure 6.8. For example, in the case of a signal propagating in the presence of stray capacitance C, mechanical movements can cause C to change. From the definition of capacitance, we have

$$\frac{dq}{dC} = C\frac{dV}{dt} + V\frac{dC}{dt}. \tag{6.73}$$

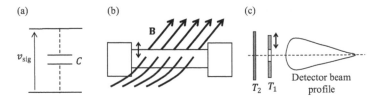

FIGURE 6.8 Microphonic noise mechanisms: (a) capacitive; (b) inductive; and (c) optical.

The second term constitutes microphonic noise induced by changes in the capacitance and shows that one way of reducing microphonic noise is to ensure that the DC voltage level on the signal line is kept close to zero.

Vibrations can also lead to magnetically induced voltages arising from changes in loop area or motion of wires in a magnetic field. A third way in which mechanical vibrations can introduce spurious voltages is through optical modulation – the relative movement of emitting elements seen by the detector. In the example shown in Figure 6.8, the detector views a background at temperature T_2 and movement of the stop at temperature T_1 that defines the detector field of view causes the coupling of the detector both to the scene and the stop to be modulated.

6.8 COMBINATION OF NOISE FROM SEVERAL SOURCES

For a practical measurement, the total noise spectral density will be the combination of all of the individual noise sources present (e.g. photon noise, Johnson noise, shot noise, etc.). Usually, the various noise contributions are uncorrelated, arising in different physical components and/or by different mechanisms. In that case, the instantaneous value of any one of them is completely independent of that of any of the others. To combine them, we do not add them directly – to be added directly, they would all have to be in phase such that all of the noise voltages are at their peak values simultaneously. As there are no phase correlations, the overall rms value will be less because the different noise sources will sometimes tend to cancel each other out. Uncorrelated noise voltages or spectral densities are added in quadrature:

$$e_{\text{n-tot}} = \left(e_{\text{n1}}^2 + e_{\text{n2}}^2 + e_{\text{n3}}^2 + \cdots \right)^{1/2}.$$ (6.74)

6.8.1 OVERALL NOISE AND NEP

Recall the definition of NEP as the signal power which gives SNR of 1 in an integration time of 0.5 seconds, or, equivalently, the signal power which gives an SNR of 1 in a post-detection bandwidth of 1 Hz. Let the detector responsivity be S (SI units: V W^{-1}), the radiant power on the detector be P (SI units: W), and the total noise voltage be v_n (SI units: V).

By definition for responsivity, the signal voltage is

$$v_{\text{sig}} = SP,$$ (6.75)

and with $\Delta f = 1$ Hz, the noise voltage is

$$v_n = e_n.$$ (6.76)

By definition of the NEP, if $P = \text{NEP}$, then the signal voltage is equal to the noise voltage, so $SP = e_n$, giving

$$\text{NEP} = \frac{e_n}{S} = \frac{\text{Noise spectral density}}{\text{Responsivity}} \left[\text{SI Units W Hz}^{-1/2} \right].$$ (6.77)

Therefore, the NEP of a detector or a complete system can be straightforwardly determined from the responsivity and noise spectral density, both of which are relatively straightforward to measure.

6.9 OPTIMISING A SYSTEM FOR BEST SENSITIVITY

Bearing in mind the various noise mechanisms that can affect a measurement, it is clear that we can keep the overall noise to a minimum if we

  i. keep the signal frequency (or frequency band) well away from interference at discrete frequencies (e.g. mains and its harmonics);
 ii. make the signal frequency (or frequency band) high enough that the $1/f$ noise is small;
iii. make the post-detection bandwidth as small as possible;
 iv. ensure that the detector and its readout and the signal-processing electronics introduce as little additional noise as possible.

6.9.1 Choosing the Signal Frequency or Frequency Band

Because of $1/f$ noise, it is often not possible to "stare" at the source for very long continuous integration times since that would involve working at very low frequencies where the $1/f$ noise would be too severe. To avoid this, the signal is often modulated in some way. For instance, the telescope beam can be switched between the source and an adjacent patch of sky producing a signal at the switching frequency or it can be scanned across the source producing a range of signal frequencies that depends on the scan rate and the spatial distribution of the sky brightness.

As we have already seen, another advantage of modulation is that it can be used to subtract out the background by switching between source and background. The modulation frequency should be high enough to get above the worst of the $1/f$ noise, preferably well above the knee frequency, but not so high that the detector frequency response results in too much loss of signal. In addition, if there are any unavoidable sources of interference at particular frequencies, then those frequencies should be avoided.

Figure 6.9 shows an example with a zero-frequency responsivity of 10^8 V W^{-1}, a time constant of 10 ms, a $1/f$ knee frequency of 25 Hz, and a mains pickup feature at 50 Hz. In this particular case, the best frequency range for signal modulation is ~10–20 Hz.

6.9.2 Choosing the Post-Detection Bandwidth

With the signal at a known frequency or band, it would clearly be good to restrict the post-detection bandwidth to match this. That way, the system is still sensitive to noise at or near the signal frequencies (which is unavoidable) but not by noise at any other frequencies. One of the best ways of doing this is to use a technique known as phase-sensitive detection (the terms synchronous demodulation, synchronous rectification, and lock-in amplification are sometimes used to refer to the same thing). This is often done in the optical, infrared, and radio regions to measure small signals in the presence of noise.

Figure 6.10 shows the essential features of a phase-sensitive detection (PSD) system. The signal from the source is modulated at frequency ω (e.g. by a rotating blade chopper). After amplification, it is fed into a multiplier circuit, together with a reference signal from the modulator. Both signal and reference are at frequency ω, but they are not necessarily in phase. Finally, the output of the multiplier is passed through a low-pass filter.

Sometimes the reference signal is a square wave and sometimes a sine wave – it does not make much difference to the final outcome. Here, to illustrate the essential features, we consider the case of a sine-wave reference.

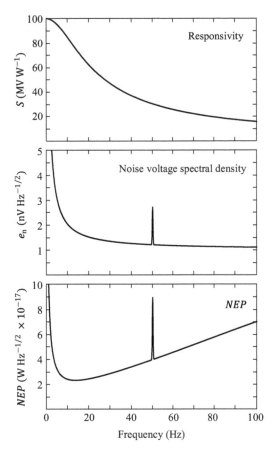

FIGURE 6.9 Example of optimisation of signal modulation frequency for the best NEP.

Let

$$v_S(t) = V_S \sin(\omega t) \tag{6.78}$$

be the signal voltage, and

$$v_R(t) = V_R \sin(\omega t + \phi) \tag{6.79}$$

be the reference voltage. The multiplier output is

$$v_M(t) = V_S V_R \sin(\omega t)\sin(\omega t + \phi). \tag{6.80}$$

From the trigonometric identity $\sin(A)\sin(B) = \left[\cos(A - B) - \cos(A + B)\right]/2$, we have

$$v_M(t) = \frac{1}{2}V_S V_R \left[\cos(\phi) - \cos(2\omega t + \phi)\right]. \tag{6.81}$$

Therefore, the multiplier output has a constant term and an oscillating term at frequency 2ω. The low-pass filter is designed to attenuate all frequencies except those very close to DC, so the oscillating term is suppressed and the output signal is

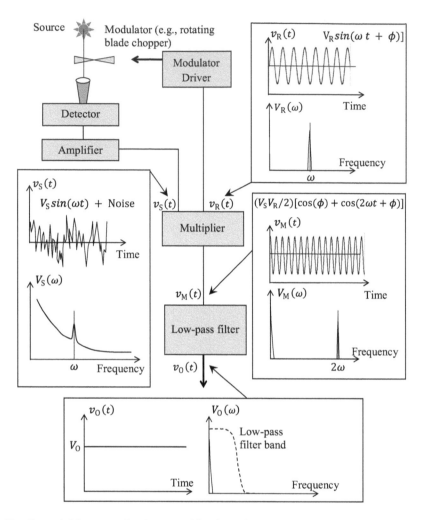

FIGURE 6.10   Essential features of a phase-sensitive detection system.

$$v_O(t) = V_O = \frac{1}{2} V_S V_R \cos(\phi). \tag{6.82}$$

The output is thus a DC voltage, and provided the reference amplitude and phase are constant, it is directly proportional to V_S, the amplitude of the signal to be measured. The proportionality to the phase difference between the signal and the reference voltages is what leads to the term phase-sensitive detection.

   In a practical PSD system, the phase of the reference can be adjusted at will so that ϕ can be made equal to zero at the multiplier input, giving maximum output signal. Some noise will also be present at the output, depending on how much noise is at the input. Clearly, any noise at the signal frequency will unavoidably affect the output, but the PSD has the very desirable feature that it rejects noise at all frequencies except those within in a narrow band around the signal frequency. To see why, let a noise voltage at the input, occurring at some arbitrary frequency ω_1, be given by

$$v_n(t) = V_n \sin(\omega_1 t) \tag{6.83}$$

Proceeding as before, we find that the multiplier output due to v_n is given by

$$v_M(t) = \frac{1}{2} V_S V_R \left[\cos\left[(\omega_1 - \omega)t\right] - \cos\left[(\omega_1 + \omega)t\right] + \phi \right]. \tag{6.84}$$

The output corresponding to the noise at ω_1 has two oscillating terms: one at $\omega' = |\omega_1 - \omega|$, which is passed by the low-pass filter, and one at $\omega_1 + \omega$, which is rejected by the filter. If the passband of the low-pass filter is $\Delta\omega_{LPF}$, then any noise which is more than $\Delta\omega_{LPF}$ away from the modulation frequency will be rejected. Therefore, by making $\Delta\omega_{LPF}$ very small, the output can be made insensitive to any noise contributions that are not very close to the signal frequency.

For example, let the modulation frequency be 100 Hz. A low-pass filter time constant of $\tau_{LPF} = 1$ second gives $\Delta f_{LPF} = 1/(2\tau_{LPF}) = 0.5$ Hz. Therefore, only frequencies between 99.5 and 100.5 Hz have any significant influence on the output. With phase-sensitive detection, we can make the post-detection bandwidth very narrow indeed. Increasing the filter time constant to 10 seconds will make the post-detection bandwidth ten times smaller again (0.05 Hz), reducing the noise by a factor of $\sqrt{10}$. However, there is a price to pay: the overall response time of the whole system will be ten times longer, making it impossible to measure any changes in the source power on time-scales shorter than this. Effectively, by reducing the post-detection bandwidth, we are increasing the integration time (the low-pass filter is an integrator).

6.9.3 MINIMISING NOISE FROM THE DETECTOR AND ITS READOUT AND SIGNAL-PROCESSING ELECTRONICS

All practical electronic circuits used to read out and amplify detector signals will introduce additional noise (due, for instance, to Johnson noise, G-R noise, shot noise, $1/f$ noise, etc. in their electronic components). Ideally, with the input short-circuited, the output would be zero, but in reality, due to noise generated by the amplifier components, the output noise voltage is non-zero. A real amplifier is often represented by an ideal noiseless amplifier in combination with a voltage source, v_{n-out}, representing its noise, connected to the input, as shown in Figure 6.11. The value of the input short noise is v_{n-out}/G, the output noise divided by the amplifier gain to replicate the measured output noise voltage of the amplifier.

To achieve the best possible sensitivity, the noise contribution of the signal amplification and subsequent processing stages should be made as small as possible. The best way to do this is to put a lot of effort into the design of the first stage of amplification so that it has high gain and low noise. Then, subsequent amplifier stages need not be of such high quality.

Consider a case in which signal power P_S is incident on a detector with responsivity S and overall noise voltage v_{n-det} followed by two amplifier stages in series with gains G_1 and G_2 and input short noise voltages v_{nA1} and v_{nA2}, as shown in Figure 6.12.

The signal level at the output is

$$v_{S-c} = G_1 G_2 S P_S, \tag{6.85}$$

and the noise level for a 1-Hz post-detection bandwidth is

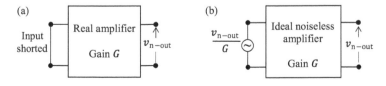

FIGURE 6.11 (a) Real amplifier of gain G with input short noise v_{n-out}; (b) Equivalent noiseless amplifier with noise source v_{n-out}/G at the input.

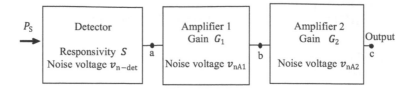

FIGURE 6.12   A detector followed by two stages of amplification.

$$v_{\text{n-c}} = \left(G_2^2 \left(G_1 \left(\text{NEP}_{\text{ph}} S \right)^2 + \left(G_1 v_{\text{n-det}} \right)^2 + v_{\text{nA1}}^2 \right) + v_{\text{nA2}}^2 \right)^{1/2}. \tag{6.86}$$

Therefore, the overall SNR is given by

$$\text{SNR} = \frac{G_1 G_2 S P_{\text{S}}}{\left(G_2^2 \left(G_1 \left(\text{NEP}_{\text{ph}} S \right)^2 + \left(G_1 v_{\text{n-det}} \right)^2 + v_{\text{nA1}}^2 \right) + v_{\text{nA2}}^2 \right)^{1/2}}. \tag{6.87}$$

The overall SNR can be maximised by making the detector responsivity, S, high and by having a low-noise, high-gain first-stage amplifier. A high detector responsivity boosts the photon noise voltage to a level which makes the contribution of the detector noise negligible. If the noise voltage of the first stage, $v_{\bar{\text{n}}\text{A1}}$, can be made much lower than the combination of detector and photon noise, then it too becomes negligible. Likewise, a high first-stage gain, G_1, means that the contribution of $v_{\bar{\text{n}}\text{A2}}$, the noise voltage of the second stage, can be neglected. Then the noise performance of the system can approach the ideal situation (for unit post-detection bandwidth) in which

$$\text{SNR} = \frac{P_{\text{S}}}{\text{NEP}_{\text{ph}}}. \tag{6.88}$$

6.10   NOISE EQUIVALENT FLUX DENSITY

The overall sensitivity of an astronomical detection system depends on many other factors as well as the detector sensitivity, such as the transmission of the atmosphere, the size of the telescope, the accepted passband, the transmission efficiency of the optics, etc. The noise equivalent flux density (NEFD) combines all of the relevant parameters to characterise the overall sensitivity. It is defined, in a way similar to the NEP, as the source flux density that can be detected to a signal-to-noise ratio of unity in and integration time of 0.5 second.

As illustrated schematically in Figure 6.13, the detector views the astronomical source through the instrument, the telescope, and the atmosphere (except for a space-borne system). We can relate the NEFD and the overall NEP of the detector as follows.

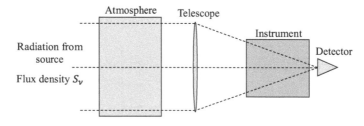

FIGURE 6.13   Coupling of a detector to the radiation from an astronomical source through the instrument, telescope, and atmosphere.

Let

t_a be the transmission of the atmosphere (= 1 for a space observatory);

η_0 be the efficiency of the instrument optics, filters, etc., representing the losses between the telescope and detector;

η_{Ap} be the aperture efficiency of the system – the fraction of the power from a point source that is coupled to the detector;

A_{tel} be the collecting area of the telescope; and

$\Delta\nu$ be the band of frequencies accepted by the system.

For a point source of flux density S_ν (power per unit area per unit bandwidth) above the atmosphere, the signal power incident on the detector is

$$P_S = S_\nu A_{tel} \Delta\nu t_a \eta_0 \eta_{Ap}. \tag{6.89}$$

The case where $P_S = \text{NEP}$ corresponds to $S_\nu = \text{NEFD}$, so

$$\text{NEFD} = \frac{\text{NEP}}{A_{tel}\Delta\nu t_a \eta_0 \eta_{Ap}}. \tag{6.90}$$

The SNR achieved after integration time t_{int} is

$$\text{SNR} = \frac{\sqrt{2}S_\nu t_{int}^{1/2}}{\text{NEFD}}, \tag{6.91}$$

which is equivalent to equation (6.15) for NEP, but now in a form that characterises the astronomical sensitivity of the complete system.

For an integration time t_{int}, the uncertainty in the measured flux density is

$$\Delta S_\nu = \frac{\text{NEFD}}{\sqrt{2}\, t_{int}^{1/2}}. \tag{6.92}$$

7 Astronomical Spectroscopy

7.1 INTRODUCTION

Photometric and spectroscopic measurements together, as described in Chapter 5, provide information on the broad shape of the spectrum (the SED) and on particular spectral lines or features. At radio frequencies, information on the spectral nature of the signal can be preserved by the detector and analysed in the post-detection electronics using techniques that will be discussed in Chapter 8. However, most detectors of UV, optical, and infrared wavelengths are sensitive to the intensity of the incident radiation across a wide spectral band. Therefore, for spectroscopic observations, we need to select the wavelength band or discriminate the spectral nature of the radiation before it reaches the detector. In this chapter, we describe several types of spectrometers that are used to select and measure the astronomical spectral information in the UV to infrared range and the characteristics that make them suitable for particular kinds of observations.

The spectral resolving power $\left(R = \lambda/\Delta\lambda = v/\Delta v\right)$ required depends on the features to be detected, what physical information is to be determined, and to what extent the lines are broadened by any of the various phenomena described in Chapter 2. Resolving atomic or molecular lines usually requires $R \sim 10^5$, whereas dust features in the infrared region only require $R \sim 20–100$. If it is only desired to measure the total line flux (power per unit area at the telescope aperture) integrated over the whole of the line profile, then $\Delta\lambda$ can be significantly larger than the line width. A resolving power of a few hundred is often adequate for this purpose.

The spectrometer is installed in the light path between the telescope and the detector(s). Some collimation and re-imaging optics are usually necessary to ensure optimal performance. Figure 7.1 shows a schematic of a suitable optics scheme. Here the telescope beam is brought to a focus at an aperture or slit, which controls the field of view of the spectrometer and, by limiting the range of angles admitted, influences the achieved spectral resolution. The beam is collimated before entering the spectrometer and then condensed to re-image the radiation at the detector(s).

7.2 SPECTROMETER TYPES

Spectrometers can be characterised by the physical optics principles of their operation. For the UV to FIR region, there are two basic types of instrument which can be further subdivided depending on the spectral resolution requirements:

i. Dispersive devices contrive to direct different wavelengths along different directions. Prism spectrometers use the wavelength-dependent refractive index of the prism material to disperse the light, producing a low-resolution spectrum. Grating spectrometers use a diffraction grating, either in reflection or in transmission.

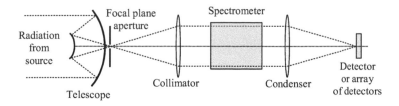

FIGURE 7.1   Schematic of optics coupling a spectrometer to a telescope.

ii. Interference-based spectrometers involve splitting the beam by partial reflection and allow-ing the corresponding beams to interfere, with certain wavelengths interfering construc-tively or destructively depending on the wavelength and the path difference. The Fourier transform spectrometer (FTS) uses a Michelson-type dual-beam interferometer, and the Fabry-Perot (FP) spectrometer uses multiple reflections between two partially reflecting parallel surfaces.

In the following sections, we will describe each of these before discussing the applicability of each to specific kinds of observations.

7.3 PRISM SPECTROMETERS

To examine the spectroscopic resolving power of a prism, we consider an isosceles prism of apex angle α, base length b, and slant length L, and examine the deviation of an incoming ray as it is refracted twice, once on entry and once on exit, as shown in Figure 7.2. As a further simplification, we assume the simplest case in which the overall deviation, δ, is a minimum, which corresponds to the path of the ray inside the prism being parallel to the base, and the angle of incidence on entry, θ, being at a particular value such that the angle of refraction on exit is the same.

Taking the refractive index of the air or vacuum to be unity, and the refractive index of the prism to be n, we have from Snell's law,

$$\sin\theta = n\sin\beta. \tag{7.1}$$

From the geometry of the prism and the first refraction, we can write

$$90° - \alpha/2 + \beta = 90° \Rightarrow \beta = \alpha/2, \tag{7.2}$$

and the total deviation after the second refraction is

$$\delta = 2(\theta - \beta) = 2\theta - \alpha/2. \tag{7.3}$$

From equation (7.1), the variation of the angle of incidence, θ, with n is given by

$$\frac{d\sin\theta}{dn} = \frac{d\sin\theta}{d\theta}\frac{d\theta}{dn} = \cos\theta\frac{d\theta}{dn} = \sin\beta = \sin(\alpha/2) = b/(2L), \tag{7.4}$$

so

$$\frac{d\theta}{dn} = \frac{b}{2L\cos\theta}. \tag{7.5}$$

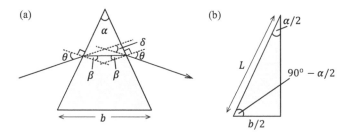

FIGURE 7.2 (a) Minimum deflection of a light ray by an isosceles prism; (b) prism geometry.

Equation (7.3) implies that the variation of δ with the prism refractive index is

$$\frac{\mathrm{d}\delta}{\mathrm{d}n} = 2\frac{\mathrm{d}\theta}{\mathrm{d}n}. \tag{7.6}$$

Therefore,

$$\frac{\mathrm{d}\delta}{\mathrm{d}n} = \frac{b}{L\cos\theta}. \tag{7.7}$$

To determine how the deviation varies with wavelength, we can write

$$\frac{\mathrm{d}\delta}{\mathrm{d}\lambda} = \frac{\mathrm{d}\delta}{\mathrm{d}n}\frac{\mathrm{d}n}{\mathrm{d}\lambda} = \frac{b}{L\cos\theta}\frac{\mathrm{d}n}{\mathrm{d}\lambda}, \tag{7.8}$$

where $\mathrm{d}n/\mathrm{d}\lambda$, a property of the prism material, indicates how the refractive index changes with wavelength. The resolving power depends not only on the properties of the prism but also on the diameter of the incident beam. Consider a prism spectrometer setup as shown in Figure 7.3. The narrow slit source and collimator define a beam of width W which illuminates the full extent of the prism at angle of incidence θ so that $\cos\theta = W/L$. A condensing optical element focuses the beam onto the focal plane, with different wavelengths λ and $\lambda + \mathrm{d}\lambda$ being imaged to slightly separated positions.

From equation (7.7), we can write the angular dispersion formula for the prism:

$$\frac{\mathrm{d}\delta}{\mathrm{d}\lambda} = \frac{b}{W}\frac{\mathrm{d}n}{\mathrm{d}\lambda}. \tag{7.9}$$

The resolving power can now be determined by applying the Rayleigh criterion (Section 4.6.1) to the images for λ and $\lambda + \mathrm{d}\lambda$ in Figure 7.3. Assume that the images are just resolved – i.e., the peak focal plane intensity for $\lambda + \mathrm{d}\lambda$ coincides with the first minimum of the intensity distribution for λ. Applying equation (4.28), the first minimum of the diffraction pattern is at an angular separation of

$$\mathrm{d}\delta = \lambda/W. \tag{7.10}$$

Substituting this in equation (7.9) gives the resolving power:

$$R = \frac{\lambda}{\mathrm{d}\lambda} = b\left|\frac{\mathrm{d}n}{\mathrm{d}\lambda}\right|, \tag{7.11}$$

where we take the absolute value of $\mathrm{d}n/\mathrm{d}\lambda$ as it is usually negative (i.e. shorter wavelengths are refracted more strongly). The resolving power thus depends on the base length of the prism and the variation of refractive index with wavelength for the material from which it is made.

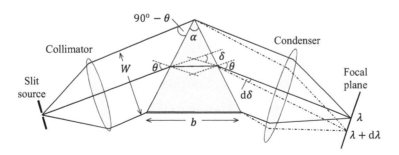

FIGURE 7.3 Prism spectrometer optics.

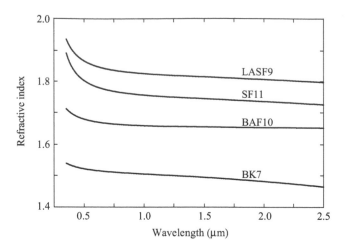

FIGURE 7.4 Refractive index vs wavelength in the visible and near-infrared region for four types of glass as indicated. (Data are from the Refractive Index Database: https://refractiveindex.info/; Mikhail Polyanskiy.)

The variation of refractive index with wavelength for a material is well described by the Sellmeier equation:

$$n(\lambda) = \left[1 + \frac{B_1\lambda^2}{\lambda^2 - C_1} + \frac{B_2\lambda^2}{\lambda^2 - C_2} + \frac{B_3\lambda^2}{\lambda^2 - C_3}\right]^{1/2}. \tag{7.12}$$

The Sellmeier coefficients, B_i and C_i, are tabulated for different materials in various optics databases. Figure 7.4 shows $n(\lambda)$ vs λ at visible and near-infrared wavelengths for some typical kinds of glass.

Prisms are suitable for low-resolution $(R <\sim 100)$ spectroscopy in the visible and near infrared.

7.4 GRATING SPECTROMETERS

A diffraction grating is a periodic structure that disperses radiation through diffraction. The periodic features are parallel grooves (called rulings or lines) or metallic facets. The dispersion can be in transmission or reflection, and it is orientated in a direction perpendicular to the grooves. Reflecting gratings with specially shaped parallel grooved surfaces are often preferred as this minimises the absorption losses which arise in dielectric transmission gratings. Like the prism, a particular wavelength exits in a particular direction. With a single-pixel detector, it is necessary either to move the detector or to rotate the grating to get complete spectral coverage. Alternatively, a linear detector array can detect all of the desired wavelengths simultaneously for a single beam on the sky.

Figure 7.5 illustrates a grating spectrometer using reflecting optics. The essential features are the same as for the prism spectrometer described above, except that the prism is replaced by the reflection grating.

7.4.1 THE GRATING EQUATION AND GRATING DISPERSION

Figure 7.6 shows the optical path delay for light diffracted by a reflection grating. The grooves are perpendicular to the plane of the page and have period d. GN is the normal to the grating. The grating facets are at an angle θ_B, called the blaze angle, to the grating. FN is the facet normal. Input beam rays are shown incident on adjacent facets and the corresponding reflected rays. The input and output beams are at angles α and β with respect to the grating normal, with the angle between the

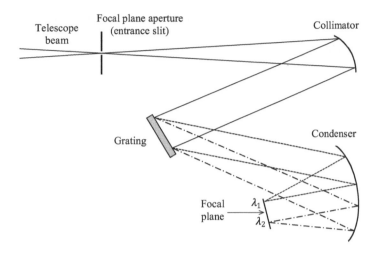

FIGURE 7.5 Grating spectrometer using reflecting optics.

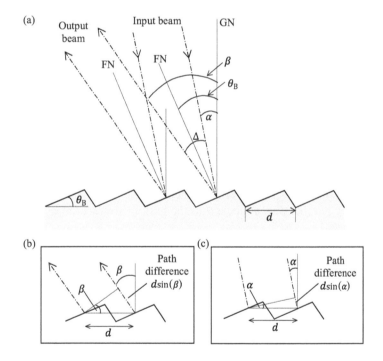

FIGURE 7.6 (a) Operation of a reflection grating showing input and output rays from two adjacent facets; (b) path difference between the two output rays; (c) path difference between the two input rays.

input and output beams being $\Delta = \beta - \alpha$. The path differences between the two input rays and the two output rays are also shown.

For monochromatic radiation of wavelength λ, constructive interference occurs when the total optical path difference between rays reflected from neighbouring facets is an integer, m, number of wavelengths:

$$d\left(\sin\alpha + \sin\beta\right) = m\lambda \quad \text{(the grating equation).} \tag{7.13}$$

With the input angle fixed, different monochromatic wavelengths will be observed at different diffraction angles, thus separating them spatially. The variation of β with wavelength is the angular dispersion, determined by differentiating the grating equation with respect to λ:

$$\frac{d\beta}{d\lambda} = \frac{m}{d\cos\beta} = \frac{\sin\alpha + \sin\beta}{\lambda\cos\beta}. \tag{7.14}$$

We see that the angular dispersion depends on the grating period, d, and the spectral order, m. High angular dispersion requires small d if the grating is operated in low order $(m = 1\,\text{or}\,2)$ or high order if d is large.

Another relevant parameter is the linear dispersion in the detector focal plane, which depends on the focal length of the condensing optics, f_C, and the angular dispersion. Letting x represent position in the focal plane,

$$\frac{dx}{d\lambda} = f_C \frac{d\beta}{d\lambda}. \tag{7.15}$$

It is common to quote the reciprocal of this, termed the plate factor, P, which has units of wavelength per unit distance:

$$P = \frac{d\lambda}{dx} = \frac{d\cos\beta}{mf_C}. \tag{7.16}$$

To examine the resolving power of the grating – how well it is possible to discriminate between neighbouring monochromatic wavelengths – we first need to determine the form of the dispersed input from a monochromatic source.

7.4.2   GRATING RESPONSE TO A MONOCHROMATIC SOURCE

We already met the two-slit diffraction case in Chapter 4, and we can generalise that treatment to a large number of slits. Figure 7.7 shows a diffraction grating in transmission, with N slits (equivalent to grooves) with separation d. The grating is illuminated by a collimated monochromatic beam with wavelength λ. Outgoing parallel rays at angle θ to the grating are focused by the condenser onto point P in the focal plane.

To begin with, let the slit width be negligible. Extending equation (4.40) for two slits to this situation with N illuminated slits, the resultant field amplitude at P is given by

$$E_P(\theta) = E\left(1 + e^{id\phi} + e^{i2d\phi} + e^{i3d\phi} + \cdots + e^{i(N-1)d\phi}\right) = E\sum_{n=0}^{N-1} e^{ind\phi}. \tag{7.17}$$

with $d\phi = 2\pi d\sin\theta/\lambda = kd\sin\theta$.

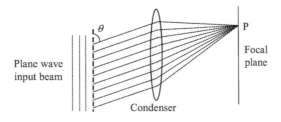

FIGURE 7.7   Parallel rays from a diffraction grating condensed to a particular point in a focal plane.

Using the series sum:

$$\sum_{n=0}^{N-1} x^N = \frac{x^N - 1}{x - 1},$$ (7.18)

we have

$$E_P(\theta) = E\frac{e^{iNd\phi} - 1}{e^{id\phi} - 1} = \frac{e^{\frac{iNd\phi}{2}}\left(e^{\frac{iNd\phi}{2}} - e^{-\frac{iNd\phi}{2}}\right)}{e^{\frac{id\phi}{2}}\left(e^{\frac{id\phi}{2}} - e^{-\frac{id\phi}{2}}\right)} = Ee^{\frac{i(N-1)d\phi}{2}}\left[\frac{\sin\left(\frac{Nd\phi}{2}\right)}{\sin\left(\frac{d\phi}{2}\right)}\right].$$ (7.19)

The intensity is obtained by multiplying this by its complex conjugate, giving

$$I(\theta) = E^2\left[\frac{\sin(Nd\phi/2)}{\sin(d\phi/2)}\right]^2 = I_0\left[\frac{\sin(N\gamma)}{\sin(\gamma)}\right]^2,$$ (7.20)

where $\gamma = (kd\sin\theta)/2$.

I_0 is the intensity in the direction $\theta = 0$ emitted by each of the slits individually. The intensity in the $\theta = 0$ direction is given by

$$I(0) = I_0 \lim_{\alpha\to 0}\left[\frac{\sin(N\gamma)}{\sin(\gamma)}\right] = I_0\left(\frac{N\gamma}{\gamma}\right)^2 = I_0 N^2.$$ (7.21)

Therefore, the focal plane intensity is given by

$$I(\theta) = \frac{I(0)}{N^2}\left[\frac{\sin(N\gamma)}{\sin(\gamma)}\right]^2.$$ (7.22)

This is illustrated in Figure 7.8 for $N = 3, 5, 10$ and 20. Peaks, called principal maxima, occur when $\sin\gamma \to 0$, so at

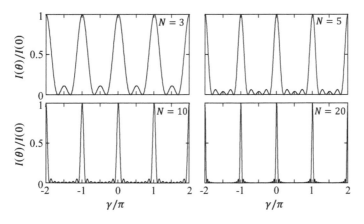

FIGURE 7.8 Focal plane intensity vs γ for a grating with N slits of negligible width, for different values of N.

$$\gamma = \pm m\pi \quad \text{where } m = 0, 1, 2 \ldots \Rightarrow d \sin\theta = m\lambda. \tag{7.23}$$

As the number of slits increases, the peaks become sharper. The half-width of a peak, $\Delta\gamma$, is determined by $\sin(N\gamma)$ changing from 1 to 0, so $\Delta\gamma = \pi/N$.

7.4.3 SPECTRAL RESOLVING POWER

The resolving power of the grating is determined by the Rayleigh criterion whereby the principal maximum of wavelength $\lambda + \Delta\lambda$ coincides with the first minimum of the response for wavelength λ.

For small θ, $\gamma = \pi d\theta/\lambda$, so,

$$\Delta\gamma = \frac{\pi d}{\lambda}\Delta\theta = \frac{\pi}{N}. \tag{7.24}$$

Therefore, the half-width of the principal maximum is

$$\Delta\theta = \frac{\lambda}{Nd}. \tag{7.25}$$

The peaks for wavelengths λ and $\lambda + \Delta\lambda$ are separated by

$$\Delta\theta = \frac{m\Delta\lambda}{d} \tag{7.26}$$

Equating these two expressions gives the resolving power of the grating:

$$R = \frac{\lambda}{\Delta\lambda} = mN. \tag{7.27}$$

Substituting for m from the grating equation (equation 7.13),

$$R = \frac{Nd(\sin\alpha + \sin\beta)}{\lambda} = \frac{L(\sin\alpha + \sin\beta)}{\lambda}, \tag{7.28}$$

where $L = nd$ is the illuminated length of the grating. The maximum possible resolving power achievable with a grating of a given length is thus

$$R_{\max} = \frac{2L}{\lambda}. \tag{7.29}$$

Achieving high resolving power means a large grating and/or operating the grating in high order, m, – but the latter solution comes with an increasing restriction of spectral coverage, as described in the next section.

7.4.4 FREE SPECTRAL RANGE AND ORDER-SORTING

The whole of the spectrum appears in each spectral order. As the order increases, the wavelengths spread more. Therefore, there comes a point at which we have spectral overlap, where monochromatic components originating from adjacent spectral orders will satisfy the diffraction equation for the same input and output beam angles for two different wavelengths:

$$d(\sin\alpha + \sin\beta) = m\lambda_1 = (m+1)\lambda_2. \tag{7.30}$$

For a certain diffraction order, the free spectral range (FSR), $\delta\lambda$, is defined as the range of wavelengths which is free from contamination from other orders. Letting $\delta\lambda = \lambda_1 - \lambda_2$, we have

$$\delta\lambda = \frac{\lambda_2}{m}. \tag{7.31}$$

When operating in low order, $\delta\lambda$ is large, so unwanted contamination from other orders can be easily eliminated by placing a suitable filter (order-sorting filter) in front of the detector. For high orders (required for high resolving power), the FSR is small, in which case an echelle spectrometer configuration is often used, as described in Section 7.4.8.

7.4.5 EFFECT OF FINITE SLIT WIDTH

We can incorporate the effect of finite slit width in the same way as we did in Chapter 4 (Section 4.6). If the slits have width b, then superimposed on the intensity pattern as above will be the sinc^2 function due to a single slit:

$$I(\theta) = \frac{I(0)}{N^2}\left[\frac{\sin(N\gamma)}{\sin(\gamma)}\right]^2 \text{sinc}^2(\rho), \tag{7.32}$$

with $\rho = (kb\sin\theta)/2$.

If γ is small and N is large, as is usually the case, then equation (7.32) becomes

$$\frac{I(\theta)}{I(0)} = \lim_{N\to\infty}\left[\frac{\sin(N\gamma)}{N\sin(\gamma)}\right]^2 \text{sinc}^2(\rho) = \lim_{N\to\infty}\left[\frac{\sin(N\gamma)}{N\gamma}\right]^2 \text{sinc}^2(\rho) = \text{sinc}^2(N\gamma)\text{sinc}^2(\rho). \tag{7.33}$$

Figure 7.9 shows an example of the intensity distribution as a function of diffraction angle for the case of $d = 30\ \mu\text{m}$, $b = 7.5\ \mu\text{m}$, and $N = 30$ (a lot smaller than would typically be used, but chosen to bring out the key features of the grating response). The principal maxima, and the sinc^2

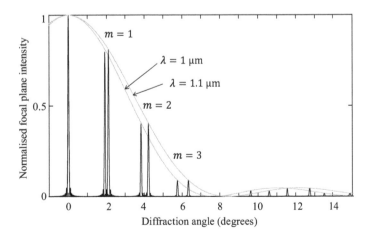

FIGURE 7.9 Intensity distribution as a function of diffraction angle for $N = 30$, $d = 20\ \mu\text{m}$, $b = 7.5\ \mu\text{m}$, and wavelengths, $\lambda_1 = 1\ \mu\text{m}$ and $\lambda_2 = 1.1\ \mu\text{m}$. The grey lines are the sinc^2 envelopes for the diffracted intensity from a single grating slit or facet.

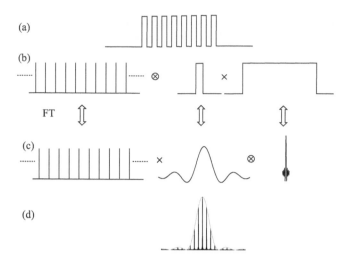

FIGURE 7.10 Fourier optics representation of the response of a diffraction grating. (a) The electric field pattern at the grating; (b) representation of the electric field pattern as the convolution of an infinite series of δ-functions with one grating slit and multiplied by a top-hat function representing the extent of the grating; (c) Fourier domain representation of (b), with an infinite series of δ-functions in the frequency domain multiplied by the FT of the slit profile, then convolved with the FT of the grating extent function; (d) the resulting intensity profile.

envelopes from a single grating slit or facet, are shown for two nearby wavelengths, $\lambda_1 = 1$ μm and $\lambda_2 = 1.1$ μm. The spectral orders are labelled.

It is useful to look at the operation of a grating in terms of Fourier optics, as shown in Figure 7.10. The electric field at the grating can be represented as the convolution of a single top-hat function of width b (a single grating facet or slit) with an infinite comb of δ-functions of separation d, then multiplied by a wider top-hat, width L, function representing the extent of the grating. Recall that the FT of this aperture field then represents the diffraction field. The FT of the infinite comb of δ-functions separated by d is also an infinite comb of δ-functions. This is multiplied by the (broad) FT of a single slit and then convolved with the (narrow) FT of the grating as a whole. The magnitude of the result then gives the overall response as a series of narrow principal maxima following the envelope of the broad single slit response.

7.4.6 BLAZED GRATINGS AND GRATING EFFICIENCY

From Figure 7.9, it is evident that significant power goes into the zero order $(m = 0)$ with no diffraction, and thus no wavelength discrimination (white light) and the remaining spectral power is spread over the other spectral orders. Without modification, the grating is therefore inefficient as a spectrometer. The purpose of the blaze angle (Figure 7.6) is to rectify that by angling the facets so that the grating will specularly reflect radiation into a specific order (only possible with gratings that operate in reflection rather than transmission). This moves the sinc^2 intensity envelope so that it coincides with the intended order of usage, as shown in Figure 7.11. Offsetting the angle β by the blaze angle does not affect the diffraction peak locations or their width.

From the grating equation for constructive interference (7.13) and the trigonometric identity

$$\sin A + \sin B = 2\sin\left(\frac{A+B}{2}\right)\cos\left(\frac{A-B}{2}\right), \tag{7.34}$$

we have

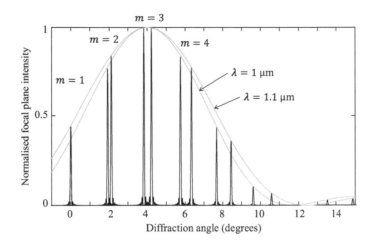

FIGURE 7.11 Equivalent of Figure 7.9, but with a blaze angle of 4°, chosen to maximise efficiency for third order.

$$2d \sin(\theta_B) \cos\left(\frac{\alpha - \beta}{2}\right) = m\lambda. \tag{7.35}$$

Inspection of Figure 7.6 shows that the blaze angle is related to the angles of incidence and diffraction by

$$\theta_B = \frac{\beta + \alpha}{2} \Rightarrow \frac{\alpha - \beta}{2} = \alpha - \theta_B. \tag{7.36}$$

Therefore,

$$2d \sin(\theta_B) \cos(\alpha - \theta_B) = m\lambda. \tag{7.37}$$

The corresponding wavelength is called the blaze wavelength, λ_B:

$$\lambda_B = \frac{2d \sin(\theta_B) \cos(\alpha - \theta_B)}{m}. \tag{7.38}$$

This is the wavelength at which the grating has maximum efficiency, defined as the fraction of the incident light that is diffracted into the particular order being measured. It depends on various properties of the grating including the grating geometry, especially the blaze angle, the material from which the grating is made, the surface quality, the wavelength and the order of diffraction, and the polarisation state of the incident light. Typically, the efficiency curve has a broad peak around the blaze wavelength. Figure 7.12 shows typical efficiency vs wavelength plots for gratings used in astronomy. The curves are for three reflection gratings used in the GMOS spectrometer at the Gemini Observatory.

7.4.7 OPTICAL MATCHING BETWEEN AN ASTRONOMICAL SOURCE AND A GRATING SPECTROMETER

Figure 7.13 shows an optical schematic of a grating spectrometer integrated with a telescope. In this system the collimator lens, focal length f_1, produces a parallel beam at the grating and the condensing lens, focal length f_2, re-images the diffracted rays at detector. Therefore, the focal plane slit aperture, W_s, which determines the field of view, is re-imaged at the detector aperture, W_d, with

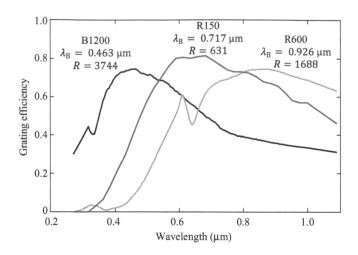

FIGURE 7.12 Grating efficiency for three gratings (designated as B1200, R150, and R600, where the numbers refer to number of grating rulings/mm), used in the GMOS instrument at the Gemini Observatory in Hawai'i. The blaze wavelength and resolving power (at λ_B) are indicated for each grating. (Data are from the Gemini web site: http://www.gemini.edu/sciops/instruments/gmos/spectroscopy-overview/gratings; credit Gemini Observatory/NOIRLab.)

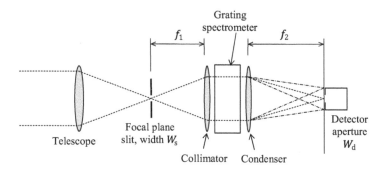

FIGURE 7.13 Schematic of telescope and grating spectrometer optics.

$$W_d = W_s \frac{f_2}{f_1}. \tag{7.39}$$

If the system is diffraction-limited, W_s is matched to the width of the Airy pattern which defines the spatial resolution. To avoid degrading the spectral resolution, the detector aperture should be no larger than the width of the principal maximum:

$$W_d \leq \frac{\Delta\lambda}{\Delta\theta}. \tag{7.40}$$

If the detector aperture is larger than this, then its size will dictate the instrument spectral resolution, $\Delta\lambda_{inst}$, with

$$\Delta\lambda_{inst} = \Delta\theta W_d = \Delta\theta W_s \frac{f_2}{f_1}. \tag{7.41}$$

There is a subtle interplay between the spatial and spectral resolutions. Ideally, the instrument resolving power, $R_{inst} = \lambda/\Delta\lambda_{inst}$, should equal the chromatic resolving power, $R = mN$.

However, achieving the high spectral resolution afforded by a grating requires small apertures and consequently low throughput ($A\Omega$ product). To enhance the spectral SNR ratio, R_{inst} is sometimes made much less than R, for instance by increasing the slit width W_S – sacrificing spectral resolution for better SNR, and also enlarging the field of view of the spectrometer. Thus, SNR, spectral resolution, and spatial resolution are all interconnected, and an optimum compromise between them needs to be arrived at, based on the scientific objectives of the instrument.

7.4.8 THE ECHELLE SPECTROGRAPH

A grating spectrometer has two major drawbacks when used in high-resolution mode. First, a single sky pixel produces a linear output of diffracted radiation at the detector and thus needs a linear detector array, (which can be inconveniently long) or else a grating rotation mechanism can be used so that the complete spectrum can be observed sequentially with a limited number of detectors (which is inefficient). Most modern detector arrays are two-dimensional, allowing the non-dispersion axis to be used to get additional spatial coverage by observing a strip of sky, thus regaining some efficiency. Second, achieving high resolution means operating in high order, with restricted FSR, requiring disentangling radiation from contaminating lower or higher orders. A solution to both of these problems is provided by an echelle configuration.

The echelle grating structure, illustrated in Figure 7.14, is designed to have a large blaze angle (typically ~70°) for high efficiency when operated in high order (typically $m \sim 100$). The input and output beam directions are close to the facet normal, and the three angles α, β, and θ_B are nearly equal. The grating equation can be then approximated by

$$2d \sin \beta = m\lambda. \tag{7.42}$$

With the FSR given by λ/m, the width of each order is very small – for instance, with $m = 100$ and $\lambda = 1\,\mu m$, wavelengths of 1 and 1.01 μm share the same diffraction angle.

To order-sort the output of the echelle grating, a second lower-resolution disperser, either another grating or a prism, is used to separate the orders spatially in the orthogonal plane. It can be placed either before or after the echelle grating. The principle is illustrated in Figure 7.15, which shows a prism being used to pre-disperse the beam before it is incident on the echelle grating.

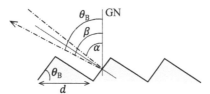

FIGURE 7.14 Detail of echelle grating.

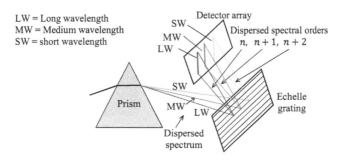

FIGURE 7.15 Principle of operation of an echelle spectrometer.

A certain wavelength range is represented by contiguous short (SW), medium (MW), and long wavelength (LW) parts. A low-resolution full spectrum is produced by the prism and is incident on the echelle grating which disperses the radiation into the orthogonal direction. High-resolution spectra for the three sub-bands are projected onto the detector array in successive orders. The result is a 2-D spectrum from a single sky pixel, with simultaneous observation of all of the orders and thus of the complete spectrum. This is ideal for CCD and other imaging cameras, as it makes full use of the detection capability. For high spectral resolution, astronomical observations in the optical and near-infrared spectral regions, the echelle spectrometer is thus the favoured design.

7.4.9 GRISMS

A grating directs the beam at specific angles away from the optical axis of the input beam depending on the spectral order of operation. For a narrow band of frequencies, a prism simply refracts the beam by an amount that depends on the refractive index of its material. Therefore, by combining a transmission diffraction grating, designed to operate in a specific order over a narrow band of frequencies, with an appropriate prism, the refracted beam can be steered back to be parallel to the input beam. Such a combination is called a grism. Figure 7.16 shows the basic configuration. The advantage of a grism is that it converts an imaging camera into a spectroscopic imaging camera by simply inserting it into any collimated beam section in the instrument. Thus, observations of an astronomical field of stars or galaxies can be quickly changed from photometric to spectroscopic mode with the insertion of a grism, resulting in a short low-resolution spectrum appearing around the position of each object. Grisms are useful for wide field surveys with a modest spectral resolution ($R \sim 1000 - 3000$), searching for objects with particular spectral characteristics (e.g. redshift surveys). A grism has to be designed for a particular spectral band so we cannot adjust the central wavelength, although several selectable grisms can be mounted in an interchange wheel. A single grism spectral image is vulnerable to contamination due to spectra of different sources overlapping. To counteract this and enable uncontaminated spectra to be measured, exposures can be taken with different rotation angles of the grism so that spectra that overlap in one image do not overlap in another.

7.5 FABRY-PEROT (FP) SPECTROMETERS

The FP spectrometer relies on multiple beam interference when the beam encounters two parallel surfaces that partially reflect and transmit. Depending on the relationship between the wavelength and the optical path difference between the two surfaces, constructive or destructive interference can occur.

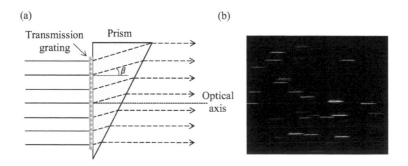

FIGURE 7.16   (a) A basic grism configuration showing counteraction of dispersion and refraction to keep the beam parallel with the optical axis; (b) example focal plane image of a field of objects with a spectrum formed for each one. (Credit NASA and ESA.)

7.5.1 The Fabry-Perot Interferometer

Consider a plane parallel beam, propagating in a medium of refractive index unity (e.g. air or vacuum), incident on a slab of material of refractive index n and thickness d. Such a component is often referred to as an FP etalon. The beam will undergo multiple back-and-forth reflections, with part of the radiation transmitted and part reflected at each encounter with a surface, and interference between the reflected beams means that the component operates as a wavelength-selective interferometer. The situation is illustrated in Figure 7.17, in which the angle of incidence is much exaggerated to enable the first few reflections to be shown. For simplicity, assume no loss due to absorption in the slab.

We define the following as the field amplitude reflectances and transmittances of the surfaces to incident waves:

r, t = reflectance and transmittance of a surface to an externally incident wave;
r_1, t_1 = reflectance and transmittance of a surface to an internally incident wave.

As discussed in standard optics textbooks (Hecht 2002; Born & Wolf 1980), when a wave encounters an interface between two media, coming from the side with higher refractive index, the reflected component experiences a 180° phase change. However, when a wave meets the interface propagating from the side with lower refractive index, there is no phase change on reflection. This means that $r_1 = -r$. We can also derive a relationship for the transmittances by considering power conservation at the first two interactions:

$$E_0^2 = E_0 r^2 + E_0 t^2 \Rightarrow r^2 + t^2 = 1, \tag{7.43}$$

$$\left(E_0 t\right)^2 = \left(E_0 t r_1\right)^2 + \left(E_0 t t_1\right)^2 \Rightarrow r_1^2 + t_1^2 = 1. \tag{7.44}$$

Therefore,

$$\left(t t_1\right)^2 = \left(1 - r^2\right)\left(1 - r_1^2\right) = \left(1 - r^2\right)^2 \Rightarrow t t_1 = 1 - r^2. \tag{7.45}$$

Let δ be the phase change with respect to the incoming wave as a result of each double-pass through the slab so that each successive reflected component experiences an additional phase change of δ.

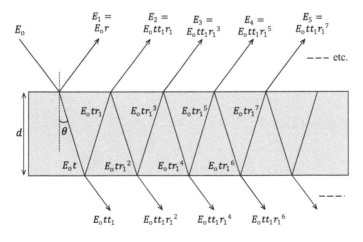

FIGURE 7.17 Multiple reflection and transmission of a wave of amplitude E_0 incident on a plane parallel slab of material with properties as given above. The amplitudes of the various transmitted and reflected components are indicated.

Letting θ be the angle with respect to the normal of the internal rays, the path difference on each double-pass is $2d/\cos\theta$, so the optical path difference is $2nd/\cos\theta$. The corresponding phase difference is thus

$$\delta = \frac{2nd}{\cos\theta}\frac{2\pi}{\lambda}. \tag{7.46}$$

Fabry-Perot interferometers (FPIs) are almost always operated with normal incidence to prevent beam walk-off due to the multiple reflections. For normal incidence, $\cos\theta = 1$ and $\delta = 4\pi nd/\lambda$.

Letting the incident wave be given by $E_o(t) = E_o e^{i\omega t}$, we have

$$E_1(t) = E_o r e^{i\omega t}$$

$$E_2(t) = E_o tt_1 r_1 e^{i(\omega t - \delta)}$$

$$E_3(t) = E_o tt_1 r_1^3 e^{i(\omega t - 2\delta)} \tag{7.47}$$

$$E_4(t) = E_o tt_1 r_1^5 e^{i(\omega t - 3\delta)}$$

$$\text{etc.}$$

This pattern is represented for the Nth-reflected component by

$$E_N(t) = E_o tt_1 r_1^{(2N-3)} e^{i(\omega t - (N-1)\delta)}, \tag{7.48}$$

which can be rewritten as

$$E_N(t) = E_o e^{i\omega t} tt_1 r_1 e^{-i\delta} \left(r_1^2 e^{-i\delta} \right)^{N-2}. \tag{7.49}$$

The overall reflected field, $E_r(t)$, is the superposition of a very large number of these reflected components, which we can express as

$$E_r(t) = E_o r e^{i\omega t} + E_o e^{i\omega t} tt_1 r_1 e^{-i\delta} \left[1 + \left(r_1^2 e^{-i\delta} \right) + \left(r_1^2 e^{-i\delta} \right)^2 + \left(r_1^2 e^{-i\delta} \right)^3 + \cdots \right]. \tag{7.50}$$

Letting $x = r_1^2 e^{-i\delta}$, and assuming that its magnitude is less than unity, the term in square brackets is equivalent to the convergent summation

$$\sum_{N=1}^{\infty} x^N = \frac{1}{1-x}. \tag{7.51}$$

Therefore, the total reflected field is

$$E_r(t) = E_o e^{i\omega t} \left[r + tt_1 r_1 e^{-i\delta} \left(\frac{1}{1 - r_1^2 e^{-i\delta}} \right) \right]. \tag{7.52}$$

Substituting $r_1 = -r$ and $tt_1 = 1 - r^2$ gives

$$E_r(t) = E_o e^{i\omega t} r \left[1 - \left(\frac{\left(1-r^2\right)e^{-i\delta}}{1 - r^2 e^{-i\delta}} \right) \right] = E_o r \left(\frac{1 - e^{-i\delta}}{1 - r^2 e^{-i\delta}} \right) e^{i\omega t}. \tag{7.53}$$

The reflected intensity, I_r, is obtained by multiplying the amplitude by its complex conjugate (see Section 1.9):

$$I_r = I_0 r^2 \left(\frac{1-e^{-i\delta}}{1-r^2 e^{-i\delta}} \right) \left(\frac{1-e^{i\delta}}{1-r^2 e^{i\delta}} \right) = \left(\frac{2-\left(e^{i\delta}-e^{-i\delta}\right)}{1+r^4-r^2\left(e^{i\delta}-e^{-i\delta}\right)} \right). \tag{7.54}$$

Substituting $e^{i\delta} - e^{-i\delta} = 2\cos\delta$ gives

$$I_r = 2I_0 r^2 \left(\frac{1-\cos\delta}{1+r^4-2r^2\cos\delta} \right). \tag{7.55}$$

We can also derive the transmitted intensity, I_t, by noting that since there are no losses, $I_0 = I_r + I_t$. Therefore,

$$I_t = I_0 \left(1 - 2I_0 r^2 \left(\frac{1-\cos\delta}{1+r^4-2r^2\cos\delta} \right) \right), \tag{7.56}$$

which reduces to

$$I_t = I_0 \left(\frac{\left(1-r^2\right)^2}{1+r^4-2r^2\cos\delta} \right). \tag{7.57}$$

Applying the trigonometric identity $\cos\delta = 1 - \sin^2(\delta/2)$, we can reduce the two intensity expressions to:

$$\frac{I_t}{I_0} = \frac{1}{1+\left(\dfrac{2r}{1-r^2}\right)^2 \sin^2\left(\dfrac{\delta}{2}\right)} = \frac{1}{1+F\sin^2\left(\dfrac{\delta}{2}\right)}, \tag{7.58}$$

$$\frac{I_r}{I_0} = \frac{2\left(\dfrac{2r}{1-r^2}\right)^2 \sin^2\left(\dfrac{\delta}{2}\right)}{1+\left(\dfrac{2r}{1-r^2}\right)^2 \sin^2\left(\dfrac{\delta}{2}\right)} = \frac{2F\sin^2\left(\dfrac{\delta}{2}\right)}{1+F\sin^2\left(\dfrac{\delta}{2}\right)}. \tag{7.59}$$

where F, called the coefficient of finesse, is given by

$$F = \left(\frac{2r}{1-r^2} \right)^2 = \frac{4R}{(1-R)^2}, \tag{7.60}$$

with $R = r^2$ representing the intensity reflectivity. For highly reflective surfaces, F is large.

The fractional transmission is maximised, and is equal to unity, when $\sin^2(\delta/2) = 0$ (denominator is minimum), i.e., when $\delta = 2\pi m$, with the order of interference $m = 0, 1, 2,...$ The corresponding wavelengths are given by $\lambda = 2nd/m$, when the optical path difference between the surfaces is $m\lambda/2$. Transmission minima, with fractional magnitude $1/(1+F)$, when $\cos\delta = -1$ (denominator is minimum), i.e., when $\sin^2(\delta/2) = 0$, i.e., $\delta = 2\pi(2m+1)$, or when $\lambda = nd/(2m+1)$.

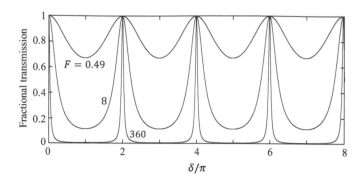

FIGURE 7.18 Transmission of a lossless FPI as a function of phase difference $\left(\delta/\pi = 4nd/\lambda\right)$ with plate reflectivities of 0.1 (upper curve), 0.5 (middle curve) and 0.9 (lower curve), corresponding to coefficient of finesse values of 0.49, 8 and 360, respectively. With no absorption, the reflected power will mirror the transmitted power, so when $R \sim 1$ the etalon reflects like a mirror at all wavelengths except for a narrow band at the transmission peak.

The contrast ratio between the peak and minimum transmission is

$$C = 1 + F = \left(\frac{1+R}{1-R}\right)^2. \tag{7.61}$$

It is remarkable that 100% transmission can occur through two highly reflecting surfaces. One way to visualise this is that a photon can tunnel through the potential barrier constituted by the device because in the space between the reflectors there is an allowed state with precisely the same energy as that of the photon.

Figure 7.18 shows how the fractional transmission depends on the phase difference $\left(\delta = 4\pi nd/\lambda\right)$ and the reflectivity. Transmission peaks occur when the phase difference is an even multiple of π, so at evenly spaced frequencies given by

$$v_m = \frac{cm}{2nd}. \tag{7.62}$$

7.5.2 FREE SPECTRAL RANGE AND RESOLVING POWER

The free spectral range of the FPI is the frequency separation between peaks and is a constant for a given material and thickness:

$$\Delta v_{\text{FSR}} = \frac{c}{2nd}. \tag{7.63}$$

As the reflectivity increases, the coefficient of finesse increases, and the peaks become sharper.

The spectral resolving power is determined by half width of the transmission peak. For a peak occurring at phase δ, the transmission will be 0.5 when the phase has changed by $\Delta\delta_{\text{HWHM}}$ such that

$$\frac{1}{1 + F\sin^2\left(\dfrac{\Delta\delta_{\text{HWHM}}}{2}\right)} = \frac{1}{2}. \tag{7.64}$$

Assuming high reflectivity so that the peaks are sharp and $\Delta\delta$ is small, this reduces to

$$\frac{\Delta\delta_{\text{HWHM}}}{2} = \frac{1}{\sqrt{F}} \Rightarrow \Delta\delta_{\text{FWHM}} = 2\Delta\delta_{\text{HWHM}} = \frac{(1-R)}{\sqrt{R}}. \tag{7.65}$$

The spectral resolution is therefore

$$\Delta v_{\text{FWHM}} = \frac{c(1-R)}{2\pi d \sqrt{R}}. \tag{7.66}$$

The finesse, $\mathcal{F}$, (as opposed to the coefficient of finesse, F) is defined as the ratio of the FSR to the spectral resolution:

$$\mathcal{F} = \frac{c}{2nd} \frac{2\pi d \sqrt{R}}{c(1-R)} = \frac{\pi \sqrt{R}}{n(1-R)}, \tag{7.67}$$

and the resolving power is

$$\frac{v_{\text{Peak}}}{\Delta v_{\text{FWHM}}} = \frac{cm}{2nd} \frac{2\pi d \sqrt{R}}{c(1-R)} = \frac{m\pi \sqrt{R}}{n(1-R)} = \frac{m\pi}{n} \mathcal{F}. \tag{7.68}$$

Therefore, high reflectivity (high finesse) and high order of operation give high resolving power.

7.5.3 FABRY-PEROT SPECTROMETERS

A practical FP spectrometer uses an FPI comprising a pair of parallel partially transmitting plates with air or vacuum separation and a mechanism for changing the separation between the plates. For optical or near-infrared applications, the reflectors are usually made from pairs of highly polished dielectric slabs with coatings on the cavity faces to give high reflectivity in the desired wavelength range. In the far infrared, patterned metallic grids are used to form the reflectors. One of the reflectors can be on a translation stage so that the gap, d, can be tuned for a particular wavelength, λ_o, and a chosen order. There is a comb of transmission peaks with $\lambda = 2d/m$, so to achieve spectral purity filters are used to remove the unwanted orders. The plate separation can be scanned to sequentially to detect the intensity of adjacent monochromatic components. An FP spectrometer consists of a collimator to produce a parallel beam, a pair of reflecting plates with an adjustable gap, an element and/or filters to reject unwanted spectral orders (as described in Section 7.5.5), a condenser to provide a focal plane image, and a detector or detector array. Because the FP spectrometer involves no angular dispersion, it is suitable for direct imaging spectroscopy in which every pixel in an imaging 2-D detector array provides a spectrum as the FP is scanned.

Figure 7.19 shows the spectral transmission as a function of wavelength for two FP spectrometers with $R = 0.9$ and two different plate separations of $d_1 = 2$ mm and $d_2 = 5$ mm. The peaks for the two separations coincide when $m_1/m_2 = 2$.

7.5.4 EFFECTS OF PLATE ABSORPTION AND OTHER IMPERFECTIONS

The transmission of a practical system will be less than one due to finite absorption by the plates. Because the device relies on a very large number of reflections, the peak transmission is very sensitive to the absorptivity and even a very small absorptivity will have a big effect.

The equivalent of equation (7.45) when the plate absorptivity, A, is taken into account is

$$tt_1 + r^2 + A = 1 \quad \text{or} \quad T + R + A = 1, \tag{7.69}$$

where T is the fractional transmission for the intensity.

A full analysis (Hecht 2002) results in a modified version of equation (7.58) for the fractional transmission:

$$\frac{I_t}{I_o} = \left(\frac{1-R-A}{1-R}\right)^2 \frac{1}{1+F\sin^2\left(\frac{\delta}{2}\right)} = \left(1 - \frac{A}{1-R}\right)^2 \frac{1}{1+F\sin^2\left(\frac{\delta}{2}\right)}, \tag{7.70}$$

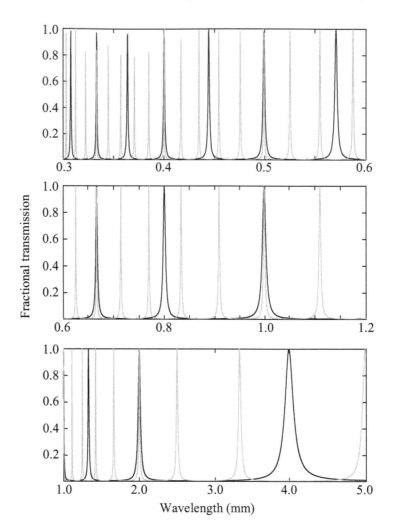

FIGURE 7.19 Transmission vs wavelength of a two lossless FP spectrometers with $R = 0.9$ and plate separations of 2 mm (FP-1; black lines) and 5 mm (FP-2; grey lines), with the three panels covering the wavelength range 0.3–5 mm.

which has a peak value of

$$\left[\frac{I_t}{I_o}\right]_{max} = \left(1 - \frac{A}{1-R}\right)^2. \tag{7.71}$$

Figure 7.20 shows an example with plate reflectivity of $R = 0.97$. Even 1% absorptivity results in the transmission efficiency decreasing by more than a factor of 2.

As well as plate absorption, other imperfections can affect the spectral performance of an FP, in particular departures from perfect parallelism of the reflectors, and from perfect collimation of the incident beam. These effects are often characterised as additional contributions to the overall finesse. The flatness finesse, $\mathcal{F}_F$ is given by

$$\mathcal{F}_F = \frac{\lambda}{2\Delta d}, \tag{7.72}$$

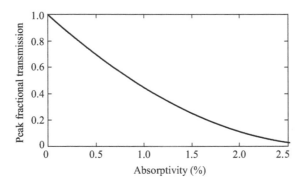

FIGURE 7.20 Peak transmission vs plate absorptivity for $R = 0.97$.

where Δd is the rms deviation from flatness due to either imperfect parallelism or surface roughness. For instance, a flatness finesse of 200 at 100 μm wavelength would require $\Delta d < 0.25$ μm.

The aperture finesse is determined by the spread of angles of the incident beam, which results in a range of phase differences between the interfering beams. It is given by

$$\mathcal{F}_A = \frac{8 f^2}{m}, \tag{7.73}$$

where f is the focal length of the collimator. Therefore, high finesse requires a high focal length (i.e. a close approximation to a parallel beam).

All imperfections result in degradation of the overall finesse, which is given by

$$\frac{1}{\mathcal{F}_{tot}^2} = \frac{1}{\mathcal{F}^2} + \frac{1}{\mathcal{F}_F^2} + \frac{1}{\mathcal{F}_A^2}. \tag{7.74}$$

7.5.5 Rejection of Unwanted Orders

When operated in high order (high resolving power), the FSR is small, so it is important to reject nearby unwanted spectral orders. One FP can be used to reject the unwanted orders from another one, in a configuration known as a tandem FP. For example, referring to Figure 7.19, if the required wavelength were 0.5 mm (20th order for FP-2), then by putting FP-1 in series, the adjacent for orders on either side could be rejected. It is then relatively easy to incorporate suitable filters in the optical train to remove any remaining unwanted orders. A grating can also be used for the same purpose, and some instruments employ a combination of two FPs or a grating and an FP to enable observations to be made in low-resolution mode (low-resolution FP or grating alone) or high-resolution mode (tandem FP or FP plus grating). Order-sorting with a tandem FP configuration rather than a grating has the advantage that no dispersion is involved, so the imaging capability is retained.

7.6 FOURIER TRANSFORM SPECTROMETERS

7.6.1 The Michelson Interferometer as a Fourier Transform Spectrometer

The Michelson interferometer, as used by Michelson and Morley in their famous experiment to investigate the existence of the aether, the postulated medium in which EM waves propagate, is a basic two-beam interferometer. In the FTS, such an interferometer is used to form an autocorrelation spectrometer – that is one that compares the beam with a delayed version of itself over a range of delays. The essential features of an FTS are shown in Figure 7.21. The beam from the source at Port 1 is collimated and its intensity is divided (ideally 50:50) by the beam splitter. Beam A is

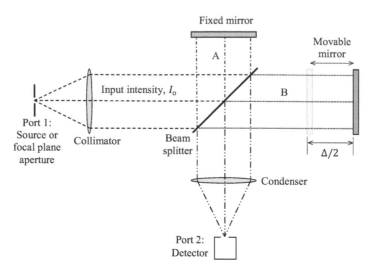

FIGURE 7.21 Schematic of a Fourier transform spectrometer. The beam reflected in path A (dash dot) is combined with the beam reflected in the variable path B (dotted lines) and condensed onto the detector (dash dot dot).

reflected back by the fixed mirror and the other (B) by the movable mirror. After reflection, the returning beams are again divided at the beam splitter where two recombined beams are created, one of which goes to the detector through the condensing optics with the other being sent back towards the source.

Before reaching the detector, each beam has undergone two reflections (one by the beam splitter and one by a mirror), producing a total phase change of 2π, which is equivalent to zero. However, there is a phase difference δ between the two beams due to the path difference. Considering the two beams that are directed back towards the source, beam A has had three reflections, and so experiences a total phase change of π, while beam B has undergone the phase change due to the path difference but only one reflection, and so experiences a total phase change of $\pi + \delta$. The outputs at the two ports are thus complementary, with one being at a maximum when the other is at a minimum. On average, half the source power goes to the detector and the other half returns to the source. Likewise, if there is any emission from the detector port, half will go towards the source and half will be returned.

Letting Δ be the optical path difference between the two beams (twice the physical mirror translation as the beam is reflected), the phase difference between waves travelling along path A and B is, $\delta = 2\pi\Delta/\lambda = 2\pi\sigma\Delta$, where

$$\sigma = \frac{1}{\lambda}, \tag{7.75}$$

and is known as the wavenumber, a term commonly adopted by Fourier transform spectroscopists. Note that various symbols are used for this parameter in different works, including ν, $\tilde{\nu}$, and k, and that it is conventionally quoted in cm^{-1}.

To explore the operation of the Michelson interferometer as a spectrometer, we start by considering an incident monochromatic electromagnetic wave with amplitude E_o:

$$E(t) = E_o e^{i\omega t}. \tag{7.76}$$

The beam divider splits the input into two waves, one reflected with amplitude $a_1 E_o$ and the other transmitted with amplitude $a_2 E_o$. Assuming no absorption loss, conservation of energy requires that $a_1^2 + a_2^2 = 1$. For instance, if the beam divider splits 50:50 in power, then $a_1 = a_2 = 1/\sqrt{2}$. After reflection, the beams arrive back at the beam splitter and are divided again, with the originally reflected beam now being transmitted towards the detector with amplitude $a_1 a_2 E_0$, and the originally transmitted beam being reflected towards the detector with amplitude also $a_1 a_2 E_0$.

Taking into account the phase difference of δ between the two beams, the wave at the detector is given by

$$E_{\text{det}}(t) = a_1 a_2 E_0 e^{i\omega t} + a_1 a_2 E_0 e^{i(\omega t + \delta)} = a_1 a_2 E_0 \left(1 + e^{i\delta}\right) e^{i\omega t}. \tag{7.77}$$

Similarly, the wave directed back towards the source is given by

$$E_{\text{source}}(t) = a_1 a_2 E_0 \left(1 + e^{i\delta}\right) e^{i(\omega t + \pi)}, \tag{7.78}$$

which is equal in amplitude but exactly out of phase with the wave at the detector.

The oscillations can be at very high frequency – far too high for the detector to be able to respond directly – it can only measure the resultant intensity, corresponding to the time average of the squared amplitude. This is given by multiplying the amplitude by its complex conjugate:

$$I_{\text{det}}(\delta) = (a_1 a_2 E_0)^2 \left(1 + e^{i\delta}\right)\left(1 + e^{-i\delta}\right) = (a_1 a_2 E_0)^2 \left(2 + 2\left(\frac{e^{i\delta} + e^{-i\delta}}{2}\right)\right). \tag{7.79}$$

Therefore,

$$I_{\text{det}}(\delta) = 2(a_1 a_2 E_0)^2 (1 + \cos\delta) = 2(a_1 a_2)^2 I_0 (1 + \cos\delta), \tag{7.80}$$

where the input intensity is $I_0 = E_0^2$.

If the beam divider splits 50:50 in intensity, then $(a_1 a_2)^2 = 1/4$ and

$$I_{\text{det}}(\delta) = \frac{I_0}{2}(1 + \cos\delta) \text{ or } I_{\text{det}}(\Delta) = \frac{I_0}{2}(1 + \cos(2\pi\sigma\Delta)). \tag{7.81}$$

The intensity at the detector has a constant term corresponding to the average intensity reaching the detector and an interference term which is the time averaged cross-correlation term between the two waves of relative phase difference. The maximum detector intensity is equal to I_0, the input intensity, and occurs when $\Delta = 0$ so that path A=path B, termed zero path difference, or zpd.

The average intensity at the detector or the source is $I_0/2$, as on average half of the input light finds its way to the detector and the other half goes back out the input port towards the source. At the zpd position, all frequencies in the recombined waves are in phase at the detector, whereas as seen at the source, they are all out of phase, so at zpd no power is directed back towards the source and all the power arrives at the detector. A plot of $I_{\text{det}}(\Delta)$, or the detector output proportional to it, vs Δ is termed the interferogram. The interferogram observed at the detector port and the one that would be observed at the input port are mirror images of each other about the average intensity level, $I_0/2$, as dictated by conservation of energy arguments. For a monochromatic input, the output is a constant level plus a cosine term so that the interferogram is symmetric about the zpd position. Figure 7.22 shows an example detector port interferogram for a monochromatic input with $\lambda = 10~\mu m$ ($\sigma = 1000~\text{cm}^{-1}$). The distance between two peaks (constructive interference) corresponds to an optical path difference of one wavelength, and the peak signal corresponds to the input wave intensity, so the recorded interferogram allows the measurement of both the wavelength and the intensity of the input.

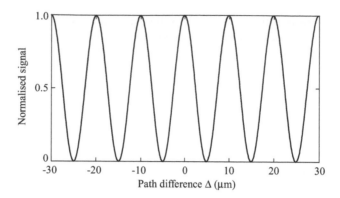

FIGURE 7.22   Interferogram for a monochromatic input with $\lambda = 10$ μm ($\sigma = 1000$ cm^{-1}).

Let the detector output (e.g. voltage) be proportional to the intensity at the detector:

$$v_{\text{det}}(\Delta) = KI_{\text{det}}(\Delta),\tag{7.82}$$

where K is a constant.

If instead of a monochromatic input with intensity I_{o}, there is an arbitrary input spectrum given by $I(\sigma)$, the detector output signal as a function of Δ is proportional to the sum of many cosine wave monochromatic components:

$$v_{\text{det}}(\Delta) = K\int_{0}^{\infty}\frac{I(\sigma)}{2}\bigl(1+\cos(2\pi\sigma\Delta)\bigr)\,\mathrm{d}\sigma.\tag{7.83}$$

At zpd, all waves are in phase and

$$v_{\text{det}}(0) = K\int_{0}^{\infty}I(\sigma)\,\mathrm{d}\sigma.\tag{7.84}$$

Splitting the integral in equation (7.83) into two separate integrals, we have

$$v_{\text{det}}(\Delta) = \frac{K}{2}\int_{0}^{\infty}I(\sigma)\,\mathrm{d}\sigma + \frac{K}{2}\int_{0}^{\infty}I(\sigma)\cos(2\pi\sigma\Delta)\,\mathrm{d}\sigma = \frac{v_{\text{det}}(0)}{2} + \frac{K}{2}\int_{0}^{\infty}I(\sigma)\cos(2\pi\sigma\Delta)\,\mathrm{d}\sigma,\tag{7.85}$$

and, since cosine is an even function,

$$v_{\text{det}}(\Delta) - \frac{v_{\text{det}}(0)}{2} = \frac{K}{4}\int_{-\infty}^{\infty}I(\sigma)\cos(2\pi\sigma\Delta)\,\mathrm{d}\sigma.\tag{7.86}$$

The right-hand side of this equation represents the real part of a Fourier transform (FT). Applying the Fourier integral theorem,

$$f(\Delta) = \int_{-\infty}^{\infty}F(\sigma)\cos(2\pi\sigma\Delta)\,\mathrm{d}\sigma \quad \text{and} \quad F(\sigma) = \int_{-\infty}^{\infty}f(\Delta)\cos(2\pi\sigma\Delta)\,\mathrm{d}\Delta,\tag{7.87}$$

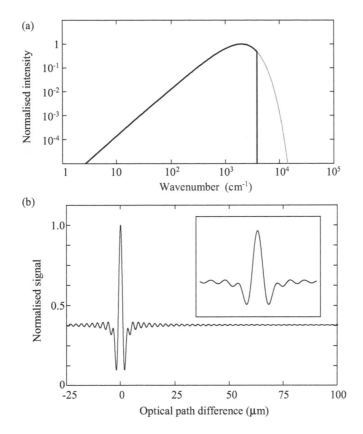

FIGURE 7.23 (a) Normalised input spectrum (black line) corresponding to a 1000-K black body filtered to reject frequencies greater than 4000 cm⁻¹ (wavelengths shorter than 2.5 μm) shown by the grey line; (b) the corresponding normalised interferogram. The inset is an expanded version of the portion near the central maximum around zpd.

gives

$$I_{det}(\sigma) = \frac{4}{K} \int_{-\infty}^{\infty} \left(v_{det}(\Delta) - \frac{v_{det}(0)}{2} \right) \cos(2\pi\sigma\Delta) d\Delta. \tag{7.88}$$

This is the fundamental principle in FT spectroscopy. The spectral content of the input (frequency domain) can be derived from the measured interferogram (spatial domain) by performing an FT. As a simple example, the FT of the cosine interferogram function in Figure 7.22 is a δ-function at 1000 cm⁻¹, representing the single-frequency (monochromatic) input.

Figure 7.23 shows an example interferogram for which the input spectrum is a 1000-K black body filtered to reject frequencies higher than 4000 cm⁻¹ (wavelengths shorter than 2.5 μm).

7.6.2 SPECTRAL SAMPLING

In a practical measurement, the interferogram is sampled at regular path difference steps dΔ from $\Delta = 0 - \Delta_{max}$. The maximum allowed step size is determined by the requirement that, according to the Nyquist sampling theorem (Section 5.9), there must be at least two samples per cycle for the highest frequency that we want to reconstruct, σ_{max}. Thus,

$$d\Delta \le \frac{1}{2\sigma_{max}}. \tag{7.89}$$

If higher frequencies are present in the signal, then the whole spectrum is folded, or "aliased", into the computed region $0-\sigma_{max}$ in the FT process, contributing to false intensities for wanted frequencies. In practice, it is customary to use optical or electrical filtering to roll off the response to higher frequencies and to over-sample the interferogram. This practice eliminates aliasing and has the added advantage that there are many samples around the zpd position to enable it to be located precisely. An error in determining the zpd position leads to baseline errors (phase errors) in the spectral data.

7.6.3 Spectral Resolution and the Instrument Response Function

A perfect recovery of the input spectrum would require an infinite optical path difference, but a real interferometer will have some maximum value, Δ_{max}, and this determines the achievable spectral resolution. Only frequencies for which at least one cycle occurs in the region $0 < \Delta < \Delta_{max}$ can be unambiguously identified. Therefore, the minimum determinable frequency interval is:

$$d\sigma_{min} \le \frac{1}{\Delta_{max}}. \tag{7.90}$$

For an interferogram measured over $\pm\Delta_{max}$, the interferogram is multiplied by a rectangular truncation function equal to unity between those limits and zero outside:

$$T(\Delta) = 1 \quad \text{for } -\Delta_{max} < \Delta < \Delta_{max};$$
$$= 0 \quad \text{otherwise.} \tag{7.91}$$

Since multiplication in the spatial domain is equivalent to convolution in the frequency domain, we can represent this schematically as shown in Figure 7.24.

The measured spectrum for a monochromatic input (i.e. a δ-function in the frequency domain) is known as the instrument response function (or instrument line-shape function) and is the FT of the truncation function:

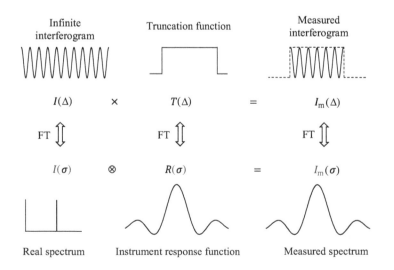

FIGURE 7.24 Spatial and frequency domain representation of FTS operation.

$$R(\sigma, \Delta_{max}) = \int_{-\Delta_{max}}^{\Delta_{max}} \cos(2\pi\sigma\Delta)d\Delta = \frac{\sin(2\pi\sigma\Delta_{max})2\Delta_{max}}{2\pi\sigma\Delta_{max}} = 2\Delta_{max}\mathrm{sinc}(2\pi\sigma\Delta_{max}). \quad (7.92)$$

The measured spectrum is given by

$$I_m(\sigma) = \int_0^{\infty} I_{det}(\sigma')R(\sigma - \sigma', \Delta_{max})d\sigma'. \quad (7.93)$$

Figure 7.25 shows the shape of the instrument response function. The spectral resolution is defined as its FWHM and is given by

$$\delta\sigma_{FWHM} = \frac{0.603}{\Delta_{max}}. \quad (7.94)$$

There are strong secondary features ("ringing") due to the shape truncation of the interferogram at $\pm\Delta_{max}$, which can be problematic when, for example, there are closely spaced lines in the spectrum. One way of mitigating this effect, but at the cost of some spectral resolution, is to taper the truncation function according to some recipe, known as the apodisation function, from the Greek, meaning to remove the feet. A simple case is linear apodisation, whereby the interferogram is tapered linearly from zpd to Δ_{max}, with the rectangular truncation function modified to

$$T(\Delta) = 1 - \frac{|\Delta|}{\Delta_{max}} \quad \text{for } -\Delta_{max} < \Delta < \Delta_{max};$$
$$= 0 \quad \quad \text{otherwise.} \quad (7.95)$$

The effective instrument response function then becomes the FT of this triangular apodisation function:

$$R(\sigma, \Delta_{max}) = \mathrm{sinc}^2(2\pi\sigma\Delta_{max})\Delta_{max}. \quad (7.96)$$

As shown in Figure 7.25, this results in a large reduction in the secondary features, but at the expense of the FWHM which is now rather wider (by 47%). Other apodisation functions can be adopted (Norton & Beer 1976), which involve different trade-offs between sidelobe suppression and the width of the instrument response function.

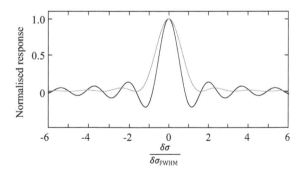

FIGURE 7.25 Shapes of the FTS instrument response function with no apodisation (black line) and linear apodisation (grey line). The wavenumber scale is normalised to the FWHM for no apodisation.

An important feature of the FTS is that the user can adapt the form of the response function post-measurement by choosing whether or how to apodise the measured interferogram to suit the type of spectral information being retrieved. As an alternative to apodisation, model-fitting of the unapodised interferogram is an effective method of distinguishing nearby spectral lines and has the advantage that there is no loss of native spectral resolution.

7.6.4 Effect of a Non-Parallel Beam on Spectral Resolving Power

There is no fundamental link between the input aperture size and the spectral resolving power – high resolving power only requires a large path difference. However, as for the FP spectrometer, the spectral performance is influenced by any departure from parallelism of the beam (which is inevitable when viewing an extended source). Non-parallelism results in phase shifts that affect the way in which the beams interfere. Figure 7.26 illustrates the ray geometry for a beam at off-axis angle θ.

The optical path difference between the rays reflected by the fixed and movable mirrors is

$$2y - z = \frac{\Delta}{\cos\theta} - \Delta\tan\theta\sin\theta = \Delta\left(\frac{1 - \sin^2\theta}{\cos\theta}\right) = \Delta\cos\theta, \qquad (7.97)$$

with y and z defined as shown in Figure 7.26.

Equation (7.81) is thus modified to

$$I_{\text{det}}(\Delta) = \frac{I_0}{2}\left(1 + \cos(2\pi\sigma\Delta\cos\theta)\right). \qquad (7.98)$$

The off-axis ray produces a spectral feature at $\sigma\cos\theta$ instead of at σ. Therefore, the corresponding resolution limit is (for small θ)

$$\delta\sigma = \sigma(1 - \cos\theta) \approx \sigma\theta^2/2, \qquad (7.99)$$

and the resolving power is thus limited to

$$\frac{\sigma}{\delta\sigma} \leq \frac{2}{\theta^2} = \frac{2\pi}{\Omega}, \qquad (7.100)$$

where $\Omega = \pi\theta^2$ is the solid angle of the beam. For final condensing optics with a focal length f and a detector aperture diameter d, the beam solid angle which the detector sees is

$$\Omega = \frac{\pi d^2}{4 f^2}, \qquad (7.101)$$

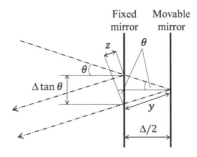

FIGURE 7.26 Path difference for off-axis rays in an FTS. For clarity, the paths through both arms of the FTS have been superimposed.

giving

$$\frac{\sigma}{\delta\sigma} \leq \frac{8f^2}{d^2}. \qquad (7.102)$$

This limitation is known as the Jacquinot criterion or the Jacquinot advantage. It represents an advantage in that in practice the FTS can accept a much larger range of solid angles for a given resolving power than can other spectrometers such as gratings or prisms, which require a narrow entrance slit to restrict the range of angles.

7.6.5 FTS Operation and Data Reduction

To measure the interferogram, the detector output must be sampled at various positions of the moving mirror. Two operating modes can be used: step-and-integrate or continuous scan. In the first method, the mirror is kept stationary at each position while an integration takes place and the sample is recorded, and is then moved to the next position for another data point, and so on. In the second method, the mirror is made to move at a constant speed. Recall that the interferogram for a monochromatic input is a cosine wave which is converted to an electrical frequency in the detector output that depends on the scan rate. Therefore, the incident radiation frequencies are converted to audio frequencies in the output by the motion of the scan mirror. The electrical frequency, f_{elec}, corresponding to the optical path scan speed v_s and wavenumber σ is given by

$$f_{elec} = v_s\sigma. \qquad (7.103)$$

One advantage of continuous scan mode is that the scanning can place frequencies of interest above the $1/f$ noise knee (Section 6.7.5) of the detector system.

The recorded interferogram must be Fourier transformed to retrieve the spectrum. The most efficient way of doing this is to use the computationally efficient fast Fourier transform (FFT) algorithm. It is important to identify the zpd position accurately otherwise so-called phase errors can result in significant distortion of the spectrum.

7.7 ADVANTAGES AND DISADVANTAGES OF DIFFERENT SPECTROMETER TYPES

Prism, grating, and FP spectrometers are monochromators: the detector sees only a narrow band of wavelengths at one time. The FTS is fundamentally different in that the detector views all wavelengths simultaneously. If there is only one detector, then to sample a monochromator's spectrum completely, we must contrive to scan the spectrum across the detector with each resolution element being observed sequentially. If a spectrum is resolved into M resolution elements, and if the required integration time is t_{int} for each, then the monochromator needs total time $t_{tot} = Mt_{int}$, but the FTS only needs t_{int}. Since SNR $\propto \sqrt{t_{tot}}$, if the same total time is spent for each case, the FTS has an SNR advantage of $\sqrt{M}$. This is called the multiplex advantage (or Fellgett advantage) and can be very significant for high spectral resolution. However, it only holds if the overall noise is dominated by detector noise. If the system is photon noise limited then, since the FTS observes a much broader instantaneous wavelength range taking in all M spectral elements at once, the photon noise is also increased by $\sqrt{M}$, resulting in an exact cancellation of this advantage. In the photon-noise-limited regime, a monochromator instrument with a detector array that observes the whole spectrum simultaneously has an advantage over an FTS with one detector because the photon noise for each detector is $\sqrt{M}$ lower.

Prism-based spectrometers have the advantages of simple optical design and no need for order-sorting. They are highly efficient in the optical and near infrared, and for this reason are sometimes preferred over grating designs, even though they can be more expensive. Their main disadvantages are that the wavelength dispersion is non-linear, and they cannot achieve high resolution.

The main advantages of the FP configuration are that it can achieve high resolution with a very compact design, and that it can provide imaging spectroscopy without the complications needed in the case of a grating configuration. The main disadvantages are that the broad wings of the instrument response function can pose problems in the case of feature-rich spectra; it is difficult to achieve efficiency as high as for grating or FTS spectrometers; it provides no possibility for wavelength multiplexing so that mechanical scanning is essential; order-sorting can be awkward; and it is difficult to achieve more than an octave (factor of two) in wavelength coverage.

The FTS is also well-suited to high-resolution imaging spectroscopy with a reasonably compact instrument and has broad instantaneous wavelength coverage suitable for survey spectroscopy. Wavelength calibration is automatic, and the instrument response function is well-defined. The FTS configuration also has ability to match the spectral resolution to the science programme by selecting the maximum path difference and/or the apodisation function. The main disadvantage is that in the photon-noise-limited regime, the additional photon noise from wide-band operation leads to lower sensitivity than for an optimised grating design.

The grating spectrometer is capable of achieving high resolution. When operated with sensitive large-format detector arrays such as modern optical and infrared arrays (see Chapters 9 and 10), it can achieve high SNR (being a monochromator) with no need to scan the grating. It can be implemented in a variety of configurations optimised in terms of sensitivity and efficiency for particular scientific observations – single object or long slit or echelle spectroscopy, integral field observations, or multi-object spectroscopy.

Case studies of different kinds of spectrometer instruments for UV to γ-ray wavelengths are also given in Chapters 9–11, and some of the most commonly used grating instrument configurations are described in the next section.

7.8   GRATING SPECTROMETER INSTRUMENTS

At UV, optical, and near-infrared wavelengths most astronomical spectrometers use gratings. The overall optimisation of a grating spectrometer involves a complex trade-off between spatial resolution, spectral resolution, throughput, and additional factors connected to the particular scientific observations (point source, imaging or multi-object spectroscopy, line-to-continuum ratio, etc.) and the capabilities of the telescope (e.g. pointing accuracy and stability). For a single observation of a point source, a short slit disperses the radiation. To observe another object, the telescope must be repointed and a separate integration carried out. One-dimensional imaging can be achieved using a longer slit so that spectroscopy of a narrow strip of the sky can be achieved simultaneously by projecting the dispersed image of the slit onto a 2-D detector array. To provide a 2-D image, the slit needs to be stepped in the orthogonal direction. Such long-slit spectrographs were the common method to getting high-SNR astronomical spectra in the past, but are inefficient for imaging spectroscopy and multi-object spectroscopy, both of which are important astronomical observations. Grating instrument configurations suitable for these applications are described below.

7.8.1   INTEGRAL FIELD UNITS (IFU) AND MULTI-OBJECT SPECTROMETERS (MOS)

An integral field spectrometer is capable of observing a spatially resolved 2-D field, broken up into an array of pixels, simultaneously producing a spectrum for each pixel (sometimes referred to as a "spaxel" meaning spectral pixel). To do this with a grating spectrometer, the focal plane image is manipulated optically to present it to the long slit of the spectrometer as a succession of linear image segments along the slit. The system then operates as a conventional long-slit spectrometer, and the 2-D image is reconstructed in the analysis. Two methods are commonly used to transfer the 2-D image onto the 1-D slit: lenslet/optical fibre coupling and image slicing. In the first method,

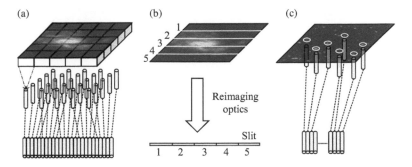

FIGURE 7.27 Principles of operation of a lenslet/fibre-coupled IFU (a), an image-slicer-coupled IFU (b), and a fibre-coupled MOS (c).

illustrated in Figure 7.27a, an array of micro-lenses (lenslets) is located in the image plane with each lens corresponding to one pixel. Each lens is coupled to a low-loss optical fibre, which acts as a flexible light-guide. The bundle of fibres is routed from the focal plane to the spectrometer, where they are arranged in a line that constitutes the entrance slit. In this way, a spectrum is generated for each pixel. Conveniently, the spectrometer, which is usually a physically large instrument, can be in a fixed position some distance away.

As an alternative to fibre-coupling, an image slicer uses a set of mirrors to divide the image at the focal plane into linear sections and then re-images them onto the spectrometer input slit (Figure 7.27b). The optical arrangement is complicated, but the efficiency is high given the polished mirror surfaces as opposed to the inevitably lossy optical fibres. Image slicing is useful for the UV wavelength range, for which optical fibres are not suitable, and for longer infrared wavelengths for which image slicing mirrors can be produced cheaply.

The multi-object spectrometer (MOS) (Figure 7.27c) is a configuration that enables simultaneous spectroscopic observations of large numbers of (often faint) point-like sources such as stars or galaxies, which would be very time-consuming if each source had to be observed sequentially. The field of view of a telescope is usually big enough to contain a large number of such objects, and it is far more efficient to observe many of them in one go. The light from a set of discrete positions is collected, usually via optical fibres, and directed to the input of a high-resolution spectrometer. The positions of the fibres are adjustable and configured robotically to match the positions of a set of known point-like sources in the field, which are the objects to be observed. As with the fibre-coupled IFU, the spectrometer can be in a fixed location some distance from the focal plane.

It is important that losses associated with fibre coupling are kept to a minimum. Although the optical fibre transmission is good in the optical, it is not perfect and deteriorates towards longer infrared wavelengths. In addition, the cone of light at the output of the fibre is larger than at the input (focal ratio degradation) due to imperfections such as small variations in fibre diameter. Efficient light collection at the output therefore requires that the collimator and grating size are larger than would otherwise be required for a direct spectrometer feed and leads to degradation in resolving power. However, the gains in observing efficiency made possible by IFU and MOS instruments far outweigh these drawbacks.

8 Radio Instrumentation

8.1 INTRODUCTION

In this chapter, we consider the techniques used for coherent detection in radio astronomy, which are used today for frequencies up to several THz (wavelengths down to ~ 100 μm), and review the key performance measures for radio telescopes and receivers.

Considering coherent detection necessarily means adopting the wave picture of electromagnetic radiation, and radio receivers are capable of measuring both the radio-frequency amplitude and phase. An astronomical radio receiver has to select the observing frequency and bandwidth, detect and amplify the signal, and measure the power arriving from the source in that band. At low frequencies, the receiver electronics can operate at the same frequency as the signal, but electronic circuits do not perform well at high radio frequencies (>100 GHz) because components and the links between them operate inefficiently (for instance, due to stray electrical capacitances tending to short the signals to ground). Many receivers, therefore, utilise a technique known as mixing, whereby the signal is combined with a locally generated reference at a slightly different frequency to produce a beat-frequency signal at the difference frequency. The lower-frequency signal still contains the desired information, which has been "down-converted" and can now be further processed by electronic circuits.

Radio techniques have developed enormously since their early development during and after the Second World War and have been applied to higher and higher frequencies. To achieve good angular resolution at radio wavelengths, very large telescopes are needed. Modern radio observatories include single-dish facilities and arrays of many antennas to form interferometers which can, through combining their signals, provide high angular resolution through a technique known as aperture synthesis.

8.2 ANTENNAS

8.2.1 ANTENNA BEAM PATTERN

The characteristics of an antenna are often described by considering power radiated outward from the antenna rather than the incoming power detected – according to the principle of reciprocity (Section 3.1.2), the key properties of the antenna are the same for transmission and reception of electromagnetic radiation. Figure 8.1 shows on a polar plot the form of a typical antenna beam pattern (note that this shows a 2-D cut through the 3-D beam beam). A large fraction of the power radiated (i.e. of the beam solid angle) is in the main beam of the antenna, but some proportion is in sidelobes, which extend out to more extreme angles. The sidelobes originate mainly from diffraction associated with the antenna geometry. For instance, the discontinuity at the edge of a single-dish antenna can result in significant sidelobes, as shown in Chapter 4. In an observation, the main beam will collect emission from the source while the sidelobes usually do not – but they may couple undesirably to unwanted off-axis radiation.

A useful parameter is the angular width of the main beam. There are a number of ways to define this, the two most common being the angle between first nulls in the radiation pattern and the angle measured between the half-maximum points in the power pattern, θ_{FWHM}. The latter is usually adopted as it relates more closely to the spatial resolution achievable. If we define the beam width in this way, we have

$$\frac{\pi \theta_{\mathrm{FWHM}}^2}{4} \approx \Omega_{\mathrm{MB}} \Rightarrow \theta_{\mathrm{FWHM}} \approx 2\sqrt{\frac{\Omega_{\mathrm{MB}}}{\pi}}. \tag{8.1}$$

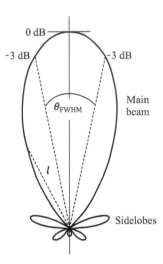

FIGURE 8.1 Slice through the 3-D polar beam pattern of an antenna showing main beam and sidelobe response. The radial length l in a certain direction represents, in dB, the magnitude of the response in that direction relative to the maximum response (0 dB). The dotted line represents the main beam full width at half maximum, θ_{FWHM}, corresponding to −3 dB.

8.2.2 GAUSSIAN TELESCOPE ILLUMINATION

At radio wavelengths, it is common to couple the radiation collected by the main antenna to the detector elements using a second antenna (called a feed antenna) at the focus. A conical horn feed is often used for this purpose, although there are many other types. The feed antenna guides the electric fields into a waveguide, and the detector element is placed either in the waveguide or in a cavity at the end of it. To understand the way in which the use of a feed antenna affects the propagation of radiation through an instrument, it is useful to consider the feed as an emitter and examine how the power that it radiates is coupled to the sky through the main antenna. For simplicity, we assume that the beam emerging from the feed has cylindrical symmetry with peak emission on-axis, and that the beam pattern of the feed can be closely approximated by a 2-D Gaussian. This is a good approximation for a conical horn feed, with the width of the Gaussian inversely proportional to the horn aperture size. It can be less accurate for other kinds of feed antenna.

We can no longer assume plane-wave propagation with a flat wavefront phase, but must consider a spherical wave with a curved wavefront emerging from the feed. Only at large distances from the feed will the curved wavefront approach a planar form.

The central core of a circular aperture diffraction pattern can also be reasonably well approximated by a Gaussian profile (as illustrated in Figure 4.23). To optimise the telescope coupling efficiency, it is necessary to match the feed beam profile to the central Gaussian part of the antenna diffraction pattern. It is important here not to confuse Gaussian optics (first-order paraxial optics), as discussed in Section 4.4, with the Gaussian beams that are referred to here – they are fundamentally different concepts.

For a feed radiating spherical waves over its aperture, solutions to the wave equation must have a variation perpendicular to the optical axis, reducing in amplitude with transverse distance from the axis, and have a form that changes smoothly as the wave propagates. Solutions that match these criteria have a Gaussian transverse profile, which changes as the beam emerges from the feed. A detailed derivation and discussion of this formalism can be found in Goldsmith (1998). Here we select some of the most useful results of this methodology to illustrate the differences between this approach and the geometric optics case.

Figure 8.2 illustrates a Gaussian beam near a beam waist (the equivalent of a focus) in an optical system and defines some of its characteristic parameters.

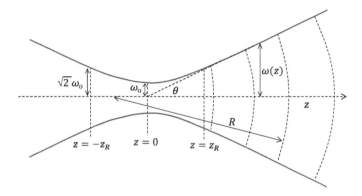

FIGURE 8.2 Parameterisation of Gaussian beam waist.

The beam, propagating in the z direction, has a minimum diameter at $z = 0$, with $\omega(z)$ representing a lateral distance from the beam axis to a point where the beam amplitude has decreased to $1/e$ times its on-axis value. The quantity ω_0, the minimum beam radius, is termed the beam waist radius. The radius of curvature of the phase-front, R, is given, as a function of distance from the beam waist, by the formula

$$R = z\left[1 + \left(\frac{\pi\omega_0^2}{\lambda z}\right)^2\right]. \tag{8.2}$$

Therefore, at the beam waist position, $z = 0$, the radius of the wavefront is infinite (a plane phase-front). As the beam propagates from $z = 0$, in either direction $(\pm z)$, the radius decreases and the width of the beam increases. The size of the beam waist depends on the horn aperture or a limiting stop in the optical system, and the beam spread depends on ω_0. The spread of the beam as it propagates is given by

$$\omega(z) = \omega_0\left[1 + \left(\frac{\lambda z}{\pi\omega_0^2}\right)^2\right]^{1/2}. \tag{8.3}$$

For $z = 0$, the $1/e$ width is ω_0, and for large z it approaches $\lambda z/(\pi\omega_0)$. As the beam propagates away from the waist the angular spread, defined by the angle θ, quickly reaches its far-field limit given, for small θ, by

$$\theta = \tan^{-1}\left(\frac{\lambda}{\pi\omega_0}\right) \approx \frac{\lambda}{\pi\omega_0}. \tag{8.4}$$

The beam spreads less rapidly than the radiation from a uniformly illuminated circular aperture as it is better controlled.

The combination of the wavelength and beam waist specifies the way the beam propagates, and the two can be combined into a single parameter, z_R, known as the Rayleigh range:

$$z_R = \left(\frac{\pi\omega_0^2}{\lambda}\right). \tag{8.5}$$

This is the distance over which the beam cross-sectional area doubles, which is equivalent to saying that it is the value of z for which $\omega(z) = \sqrt{2}\omega_0$ This is used as a practical measure of the beam divergence and conveniently separates the "near-field" and "far-field" regions of the system.

To achieve high optical coupling efficiency, it is necessary to match the fields set up by the feed with those of the incoming waves. This means matching the Gaussian core of the feed beam to that of the incoming Gaussian beam. Because the EM waves propagating through the feed structure depend on the boundary conditions particular to the feed geometry, it is possible by optimising the feed design (e.g. shaping the metal walls of a horn feed) to tune this interface between guided and free-space propagation to ensure a smooth transition, thus minimising the loss of beam power and the generation of sidelobes in the antenna pattern due to discontinuities in the fields. The calculation of this coupling efficiency is complex but tractable with reasonable computing power.

From this analysis, we can ensure that the feed generates the appropriate beam waist size and wavefront curvature for optimum efficiency. The fundamental Gaussian beam mode field amplitude and power are represented by

$$E(r,z) = E(0,z)e^{-\left(\frac{r}{\omega}\right)^2} \quad \text{and} \quad P(r,z) = P(0,z)e^{-2\left(\frac{r}{\omega}\right)^2}, \tag{8.6}$$

where r is the perpendicular distance from the propagating axis. This Gaussian beam emanating from the telescope focus will illuminate the telescope with a Gaussian profile. This is very different from the geometric optics approach, appropriate to uniformly absorbing detectors such as CCD arrays, in which the radiation incident on the outer parts of the telescope is collected with the same efficiency as that incident at the centre, corresponding to a top-hat telescope illumination profile (these differences are considered further in Chapter 9).

Gaussian illumination of the telescope is illustrated in Figure 8.3. To parameterise this illumination, we define the edge taper of the beam, T_e, as:

$$T_e = \frac{P(r_e)}{P(0)} = e^{-2\left(\frac{r_e}{\omega}\right)^2} = e^{-2f}, \tag{8.7}$$

where r_e is the radius of the aperture and the factor $f = (r_e/\omega)^2$.

The edge taper is often expressed in decibels (dB): $T_e[\text{dB}] = -10\log_{10}(T_e)$. Equation (8.7) then gives

$$T_e[\text{dB}] = (8.69)f. \tag{8.8}$$

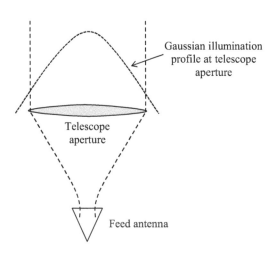

FIGURE 8.3 Illumination of telescope with a Gaussian beam.

Tapering the primary illumination in this way may appear to be wasteful as it involves reduced response to radiation collected by the outer parts of the telescope. However, the effect is not as severe as it might first appear. The fraction of the power in the Gaussian beam intercepted by the aperture is

$$\frac{P(r_e)}{P(\infty)} = \frac{\int_0^{r_e} |E(r)|^2 2\pi r\, dr}{\int_0^{\infty} |E(r)|^2 2\pi r\, dr} = \frac{\int_0^{r_e} e^{-2\left(\frac{r}{\omega}\right)^2} r\, dr}{\int_0^{\infty} e^{-2\left(\frac{r}{\omega}\right)^2} r\, dr} = 1 - e^{-2f} = 1 - T_e. \tag{8.9}$$

For example, with the $1/e$ width of the Gaussian beam matching the telescope aperture $(r_e = \omega)$, we have $T_e = 0.14$ ($\equiv 8.7\,\text{dB}$), which results in 86% of the power in the Gaussian beam coupling to the primary. The major advantage of tapering the illumination of the primary is that by reducing the discontinuity in the field at the edge of the antenna it reduces unwanted beam sidelobes, and this usually outweighs the small penalty in efficiency. This is quantified in terms of the taper efficiency, η_t, as discussed below.

The edge taper is controlled mainly by the aperture size of the feed – a large feed aperture produces a narrow feed beam which under-illuminates the outer parts of the antenna, corresponding to a large edge taper. Conversely, a small feed aperture produces a wider feed beam and a more uniform illumination of the antenna.

8.2.3 ANTENNA EFFICIENCIES

Various efficiency factors characterise the performance of a radio antenna. An important one is the overall antenna aperture efficiency, defined as the fraction of the total power, intercepted by the antenna from a point source, that is coupled to the detector. This incorporates all effects that reduce the ability of the antenna to couple to an incoming plane wave. For an antenna system made up of a main reflector and a feed antenna coupled to a receiver or transmitter, there are two main contributions to the aperture efficiency. The taper efficiency, η_t, represents imperfect coupling to the outer parts of the antenna. The spillover efficiency, η_s, arises because the Gaussian beam illumination of the primary aperture, although tapered towards the edge, does extend beyond it. The power in that part of the feed beam is not intercepted by the aperture and is termed spillover. The spillover efficiency is equal to the proportion of the total solid angle of the feed beam that the telescope intercepts. Spillover must be carefully controlled because the feed will pick up any spurious signals from local sources in the portion of its beam that is not intercepted by the antenna and couple them directly into the detection system. Spillover also means that some thermal radiation from the background around the telescope is seen by the receiver and can thus contribute to the overall noise level.

Optimum matching of the receiver to the telescope involves the best compromise between the overall coupling efficiency to the source and minimisation of the sidelobe response, secondary blockage, and spillover.

Equation (8.8) gives a relationship between the edge taper in dB and the ratio of the aperture radius, r_e, to the Gaussian beam waist radius ($1/e$ field radius). The calculation of the taper efficiency requires an overlap integral between the electric fields at the telescope focus generated by an incident plane wave and the Gaussian electric fields that would propagate from the feed placed at its focus. These integrals of the power contained within a given aperture radius can be used to generate analytical forms for the edge taper and spillover efficiencies as a function of edge taper (Goldsmith 1998):

$$\eta_s = 1 - e^{-2f}, \tag{8.10}$$

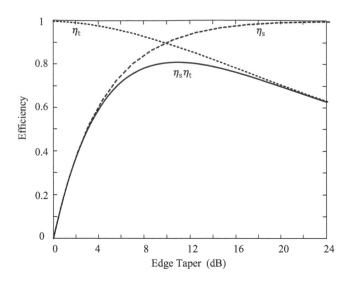

FIGURE 8.4 Taper and spillover efficiencies and the product of the two as a function of edge taper in dB.

and

$$\eta_t = \frac{2\left(1 - e^{-f}\right)^2}{f\eta_s}.$$ (8.11)

With low edge taper (small f) the telescope illumination is near-uniform giving maximum taper efficiency, but most of the solid angle of the feedhorn spills over the aperture giving low η_s. As the edge taper increases, we have reduced coupling to power near the telescope edge, the taper efficiency decreases and the spillover efficiency increases. This is depicted in Figure 8.4, showing that the overall efficiency, $\eta_s\eta_t$, is highest, at about 80%, for an edge taper between 8 and 14 dB. A value of ~14 dB is commonly used given the advantages of having low sidelobes.

Various other effects also contribute to the overall efficiency. Many single-dish radio telescopes use a Cassegrain configuration, which has a central blockage to incoming radiation in the form of the secondary mirror and its support struts. This leads to a loss, typically a few %, characterised by the blockage efficiency η_b, which depends on the fraction of the primary area that is obscured by the secondary and its supports. A further unwelcome consequence of the secondary mirror and its associated support structure is that they can introduce additional sidelobe structure through diffraction.

Surface roughness and other departures from the ideal shape of the antenna also cause some incoming signal power to be deflected at wrong angles and so fail to be coupled to the detection system, or to be subject to unwanted phase changes adversely affecting the field distribution in the focal plane. Likewise, unwanted background power from the surroundings can be deflected into the receiver. Gaussian distribution of surface error amplitudes with an rms value of σ leads to an efficiency at wavelength λ, known as the Ruze efficiency (Ruze 1966), given by

$$\eta_R = e^{-\left(\frac{4\pi\sigma}{\lambda}\right)^2}.$$ (8.12)

This is plotted as a function of σ/λ in Figure 8.5. As a rule of thumb, to avoid significant degradation of performance, surface imperfections need to be no more than ~$\lambda/20$.

There may also be some additional ohmic loss, η_O because the reflectors are not perfect conductors.

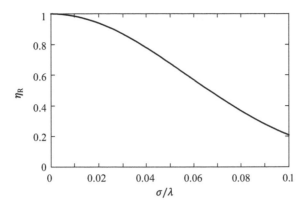

FIGURE 8.5 Ruze efficiency factor vs rms surface error as a fraction of the observing wavelength.

The overall efficiency, corresponding to the aperture efficiency – the fraction of the power from a point source arriving at the antenna that is coupled to the detector, is the product of all the contributing factors:

$$\eta_A = \eta_s \eta_t \eta_b \eta_R \eta_0. \tag{8.13}$$

The overall efficiency is typically ~60%.

Besides suppressing sidelobes, the edge taper also affects the width of the main lobe of the beam pattern: a strong edge taper broadens the beam by under-using the outer parts of the antenna. In the case of Gaussian illumination of the antenna, the following relationship applies (Goldsmith 1998):

$$\theta_{FWHM} = \left(1.02 + 0.0135 T_e \, [\mathrm{dB}]\right)\frac{\lambda}{D}. \tag{8.14}$$

For a 14-dB edge taper, $\theta_{FWHM} = 1.21\lambda/D$.

When an extended source is being observed, it is useful to define another efficiency parameter: the main beam efficiency, which is the fraction of power contained within the main beam compared to the total power integrated over the whole of the beam pattern, or, equivalently, the ratio of the main beam solid angle, Ω_{MB}, to the total solid angle of the antenna, Ω_A:

$$\eta_{MB} = \frac{\iint_{\mathrm{Main\ beam}} P_n(\theta,\phi)\,\mathrm{d}\theta\,\mathrm{d}\phi}{\iint_{4\pi} P_n(\theta,\phi)\,\mathrm{d}\theta\,\mathrm{d}\phi} = \frac{\Omega_{MB}}{\Omega_A}, \tag{8.15}$$

where θ and ϕ are angular coordinates defining the direction with respect to the direction of maximum response, and $P_n(\theta,\phi)$ is the power, normalised with respect to the peak, radiated per unit solid angle in that direction (equivalent to $B(\theta, \phi)$ as defined in Chapter 3).

8.3 RADIO RECEIVERS

The radio-frequency emission received from astronomical sources has the characteristics of random electromagnetic noise (Condon & Ransom 2016): the electric fields at the antenna fluctuate randomly about zero, and so do the currents and voltages induced in the antenna. However, as for a resistor, as discussed in Chapter 6, the noise power, which is proportional to field or voltage or current squared, is always positive. It is this noise power that is proportional to the brightness of the source.

8.3.1 POWER RECEIVED BY A RADIO ANTENNA

It is conventional to characterise the source brightness in terms of a temperature, as though the emission were by the black body process. Even if this is not the case, we can ascribe a brightness temperature to the source, equal to the temperature which a black body would need to have in order to radiate the same power as the source. The throughput of a radio antenna with effective area A_e and total beam solid angle Ω_{Beam} (see Chapter 3) is

$$A_e \Omega_{\text{Beam}} = \lambda^2. \tag{8.16}$$

At frequency ν, the intensity of a source of brightness temperature T_S is, by definition,

$$I_\nu = \frac{2\nu^2 k_B T_S}{c^2} = \frac{2k_B T_S}{\lambda^2}, \tag{8.17}$$

where we use the Rayleigh–Jeans approximation $(h\nu \ll k_B T_S)$, which is appropriate for radio wavelengths. The power received by the antenna in bandwidth $\Delta\nu$ is therefore

$$P = A_e \Omega_{\text{Beam}} \Delta\nu I_\nu = 2k_B T_S \Delta\nu. \tag{8.18}$$

Most radiometers are only sensitive to only one polarisation, which reduces the detected power by a factor of two, so

$$P = k_B T_S \Delta\nu. \tag{8.19}$$

Note that this is the same as the formula derived in Chapter 6 for the thermal noise power from a resistor at temperature T_S, which is not a coincidence. The output of the antenna is equivalent, from the point of view of the detection process to the thermal noise of a resistor within the same bandwidth.

8.3.2 THE TOTAL POWER RADIOMETER

A radiometer is a radio receiver system designed to measure the average noise power collected by an antenna in a particular band of frequencies. The simplest form of radiometer (Figure 8.6) uses a bandpass filter to select the frequency range to be detected. A low-noise amplifier is often placed at the antenna output to ensure that the low-level signals are boosted above the noise produced by subsequent elements. The bandpass filter is followed by a device that squares the corresponding signal voltage (called a square-law detector) producing a fluctuating voltage with a non-zero mean value proportional to the source power. To measure this mean value, the signal is then averaged by passing it through a low-pass filter (integrator), with the mean value, which is always positive, integrating up above the fluctuations (which average to zero). Depending on the details of the system, voltage amplification stages can be included before and/or after the bandpass filter.

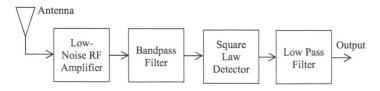

FIGURE 8.6 Basic total power radiometer.

8.3.3 System Temperature

Now consider a radiometer receiving an input in three different ways, as shown in Figure 8.7. In case (a), a resistance at temperature T_S is connected to the receiver input; in case (b), the receiver is connected to an antenna which is contained inside a black body cavity at temperature T_s, and in case (c) the antenna field of view is filled with a scene corresponding to black body temperature T_S. In all three cases, the input signal is from a thermal source with temperature, T_S, and the response must be the same. For both the resistor and the black body source, the power delivered to the receiver within the radiometer bandwidth Δv is $P = kT\Delta v$. For an ideal noiseless radiometer, the output, V_{out}, would be zero for zero input power (equivalent to looking at a source at absolute zero temperature). In reality, there is some noise generated within the radiometer itself which also contributes to the output. It is conventional to regard this noise as if it were generated by a hypothetical thermal source at temperature T_{rec} (called the receiver noise temperature) that injects power into the receiver, as shown in Figure 8.8.

The receiver noise temperature is a measure of the total noise power of the radiometer and is adopted as a figure of merit for it (low T_{rec} means low noise). When viewing a source of brightness temperature T_S, the output voltage (proportional to power) is therefore:

$$V_{out} = Ak_B(T_{rec} + T_S)\Delta v, \qquad (8.20)$$

where A is the overall power gain. Note that the receiver and source noise powers are additive.

V_{out} will always be positive due to noise generated in the receiver. For no source power input, we have

$$T_{rec} = \frac{V_{out}}{Ak_B\Delta v}. \qquad (8.21)$$

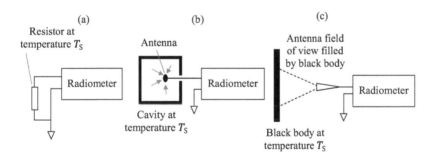

FIGURE 8.7 Concept of system temperature. (a) Resistor at temperature T_S connected to radiometer input. (b) Radiometer coupled to an antenna inside a black body cavity at temperature T_S. (c) Radiometer connected to an antenna with its beam filled by a black body at temperature T_S. The radiometer output is the same for all three cases.

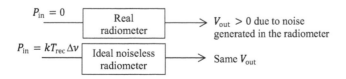

FIGURE 8.8 Equivalence of a real radiometer to an ideal noiseless radiometer with an input at the receiver noise temperature, T_{rec}.

The noise temperature is thus inversely proportional to the radiometer bandwidth. The receiver noise temperature can easily be measured by the so-called Y-factor method. Two known thermal sources, with physical temperatures T_1 and T_2, are alternately placed at the input of the receiver (as in Figure 8.7c), producing two voltage levels, V_1 and V_2, at the output:

$$V_1 = Ak_B(T_{rec} + T_1)\Delta\nu \quad \text{and} \quad V_2 = Ak_B(T_{rec} + T_2)\Delta\nu. \tag{8.22}$$

giving,

$$T_{rec} = \frac{T_1V_2 - T_2V_1}{V_1 - V_2}. \tag{8.23}$$

The overall noise temperature for a measurement often depends on other contributions besides that of the receiver. For instance, the radiometer views the astronomical source through the telescope and, for Earth-based instruments, the atmosphere. Both of these will have some thermal emission giving rise to additional additive contributions, which can be quantified by ascribing to them noise temperatures T_{tel} and T_{atm}, equivalent to their physical temperatures multiplied by their emissivities. The overall noise temperature for the measurement is termed the system temperature, and is the sum of all such contributions, including the source itself:

$$T_{sys} = T_{rec} + T_{tel} + T_{atm} + T_S. \tag{8.24}$$

8.3.4 THE SUPERHETERODYNE RECEIVER

Variants of the basic radiometer described above are used for observations at frequencies up to ~ 100 GHz. However, for frequencies much higher than this, achieving good performance with conventional electronic circuits becomes increasingly difficult and costly. This can be overcome by mixing the signal from the antenna with a constant signal from an oscillator within the receiver (a local oscillator, or LO) to convert the information to a much lower frequency range that can easily be handled by the subsequent electronics. This is the basis of the so-called superheterodyne receiver.

The signal to be detected is combined with the LO signal producing a lower intermediate frequency (IF) signal which is the modulation envelope of the mixed signals. In audio terminology, it would be referred to as the beat frequency as depicted in Figure 8.9. As shown below, the IF signal contains the same information (amplitude and phase) as the source signal.

Figure 8.10 shows the basic features of a superheterodyne radio receiver system. The signal from the antenna may first be boosted by a tuned radio frequency (RF) amplifier, designed to select a particular band of frequencies and amplify it. For high-frequency receivers, this amplifier may be omitted due to the difficulty to build efficient electronic circuits for very high frequencies. The signal, at frequency ω_S, is then combined with a signal from the LO at a precise frequency ω_L which is made to be close – but not equal – to ω_S. The combination takes place in the mixer, which produces an output at the difference frequency due to beating between the two inputs. The output is at a frequency, ω_{IF} ($= \omega_L - \omega_S$ or $\omega_S - \omega_L$), and it retains the amplitude and phase information in the input signal from the antenna. The IF amplifier is designed to work in a band of frequencies, centred on ω_{IF}, and this bandwidth normally determines the band of RF frequencies to which the receiver responds. The term "superheterodyne" was first coined in the early days of AM radio to reflect the fact that even the down-converted IF frequency was still supersonic – i.e., too high in frequency to be audible.

For very high frequencies, the signals from the antenna and the LO may be coupled onto the mixer optically, using lenses, mirrors, or waveguides. At lower frequencies, co-axial cable transmission lines or waveguides may be used. The mixer output is amplified and, depending on the purpose of the receiver, may then be passed to a demodulator, a spectrometer, or a square-law detector. Telecommunication systems would demodulate the encoded information in the RF signal at this

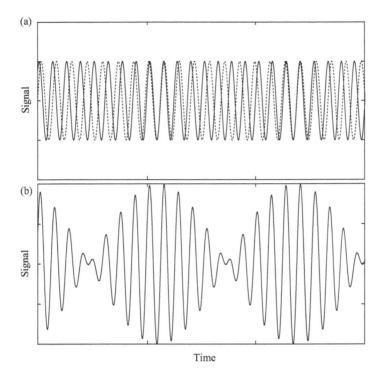

FIGURE 8.9 (a) Two sinusoidal signals with a 10% frequency difference; (b) the superposition of the two, with an envelope at a lower frequency (beat frequency) equal to the difference between the two.

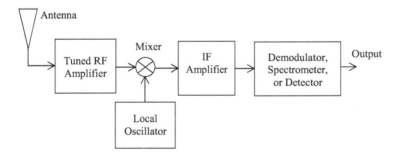

FIGURE 8.10 Essential features of a superheterodyne radio receiver.

stage. Astronomical receivers are used either to measure the spectral content of the RF signal (spectroscopy) or to measure the total power over the full IF band (radiometry).

The heart of this type of receiver is the mixer. This is a component which has a non-linear relationship between voltage and current. Consider, for example, the current–voltage $(I - V)$ relationship for a semiconductor diode, which we use here to illustrate the principles. The voltage–current characteristic of a forward-biased diode is given by

$$I = I_S \left(e^{\beta V} - 1 \right), \tag{8.25}$$

where I_S (the reverse saturation current) and β are constants. The shape of this characteristic is illustrated in Figure 8.11. The diode is forward-biased at some DC operating point (current I_o, voltage V_o).

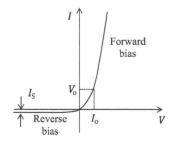

FIGURE 8.11　Current–voltage relationship for a semiconductor diode.

For small current and voltage deviations, i and v, around the operating point, the exponential can be approximated by a low-order polynomial:

$$i(v) = a_0 + a_1 v + a_2 v^2. \tag{8.26}$$

The LO voltage can be represented as

$$v_L(t) = V_L \sin(\omega_L t). \tag{8.27}$$

An incident electromagnetic signal at frequency ω_s, picked up by the antenna, produces a sinusoidal voltage $v_S(t)$ at the mixer input with amplitude V_S and some phase difference, ϕ, with respect to the LO signal. We can write this as

$$v_S(t) = V_S \sin(\omega_s t + \phi). \tag{8.28}$$

The total voltage at the mixer diode is then the sum of all the incident components:

$$v(t) = V_o + v_S(t) + v_L(t). \tag{8.29}$$

The diode current is obtained by substituting into equation (8.26):

$$i(t) = a_0 + a_1 \left(V_o + V_s \sin(\omega_s t + \phi) + V_L \sin(\omega_L t) \right) + a_2 \left(V_o + V_s \sin(\omega_s t + \phi) + V_L \sin(\omega_L t) \right)^2. \tag{8.30}$$

Multiplying this out, using the trigonometric identity $\sin A \sin B = \left[\cos(A - B) - \cos(A + B) \right]/2$, and noting that the mixer is designed to propagate only frequencies much lower than ω_S or ω_L so that terms in ω_S, ω_L, $2\omega_S$, $2\omega_L$ and $\omega_S + \omega_L$, can all be ignored, we get

$$i(t) = V_S V_L \cos\left((\omega_S - \omega_L)t + \phi \right). \tag{8.31}$$

When applied across the input resistance of the IF amplifier, R_{IF}, this generates a corresponding IF voltage given by

$$v_{IF}(t) = V_S V_L R_{IF} \cos\left((\omega_S - \omega_L)t + \phi \right). \tag{8.32}$$

The IF voltage thus has the following important properties:

   i. its amplitude is directly proportional to the amplitude of the astronomical signal, V_S, (with V_L and R_{IF} being constants);

   ii. since $\cos(x) = \cos(-x)$, there are two signal frequencies that produce the same IF frequency, one higher than and one lower than ω_{IF}: $\omega_{S1} = \omega_L - \omega_{IF}$ and $\omega_{S2} = \omega_L + \omega_{IF}$;

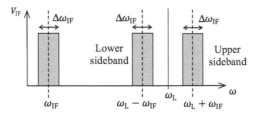

FIGURE 8.12 Sidebands in a heterodyne receiver.

iii. in either case, the frequency of the IF signal depends on the frequency of the astronomical signal, and the phase of the incoming RF signal, ϕ, is also preserved. The preservation of phase as well as amplitude information (coherent detection) is an important property of heterodyne receivers.

The band of radio frequencies to which the receiver responds is set by the bandwidth of the IF amplifier, $\Delta\omega_{IF}$. Thus, two ranges of the astronomical frequency ω_s, each of width $\Delta\omega_{IF}$, and termed sidebands, can be detected simultaneously, as shown in Figure 8.12. The higher-frequency (upper) sideband corresponds to signal frequencies greater than ω_L, and the lower sideband to frequencies less than ω_L.

In the case of continuum observations, for which the source spectrum is smooth across the range covered by the receiver, operation in double-sideband mode is preferred, to get more source signal (from the greater bandwidth) and so enhance the SNR. (A receiver operating in double-sideband mode has a noise temperature a factor of two lower than in single-sideband mode.) However, when it is required to know precisely what frequencies are being detected – for instance to detect unambiguously a particular spectral feature, the receiver must be operated in single-sideband mode. The rejection of one or other of the sidebands is then done using a suitable filter somewhere in the signal train. The IF bandwidth is an important parameter for spectral line receivers – a wide bandwidth ensures that broad spectral features can be resolved and can allow multiple lines to be observed simultaneously or a range of frequencies to be surveyed more quickly.

8.3.5 THE SUPERHETERODYNE TOTAL POWER RECEIVER

Since the IF output voltage is proportional to the signal voltage, V_s, which is, in turn, proportional to the incident signal RF field amplitude, the IF output can be squared to produce an output proportional to electromagnetic power. Finally, as with the basic radiometer, the output signal from the square-law device is passed through a low-pass filter which averages the output over the appropriate integration time to improve the SNR of the measurement.

From equation (8.32), for frequency ω_s the voltage output from the square-law detector is

$$V_{sq} \propto V_s^2 V_L^2 \cos^2\left[(\omega_s - \omega_L)t + \phi\right]. \tag{8.33}$$

Using the trigonometric identity $\cos^2\theta = \left[1 + \cos(2\theta)\right]/2$ we can write this as

$$V_{sq} \propto V_s^2 V_L^2 \left[1 + \cos\left(2(\omega_s - \omega_L)t + 2\phi\right)\right]. \tag{8.34}$$

This is always positive and has a DC term proportional to the signal amplitude squared – i.e., to the signal power. The $\omega_s - \omega_L$ term is removed by the low-pass filter to give

$$V_{out} \propto V_s^2 V_L^2. \tag{8.35}$$

The signal and LO powers are $P_S \propto V_s^2$ and $P_L \propto V_L^2$, so we can write

$$V_{out} = GP_s. \tag{8.36}$$

where G is defined as the conversion gain of the mixer and quantifies how much IF power is produced for a given input signal. The conversion gain is proportional to LO power, so increasing P_L will increase the output signal. However that does not mean that the IF signal can be made arbitrarily large just by increasing LO power. Usually $P_L \gg P_s$, and therefore, the overall current flowing in the mixer is dominated by that due to the LO. This current, I_L, will generate an electron shot noise current proportional to $I_L^{1/2}$, with corresponding noise power $\propto I_L$ so that increasing the LO power results in a proportional increase in this shot noise power. As a result, there is no overall increase in SNR to be gained by increasing the LO power. In practice, other factors (such as mixer efficiency) are also dependent on LO power, and the optimum LO power is found empirically.

8.3.6 Quantum-Limited System Temperature

Even in the case of zero background and a noiseless radio receiver, there is still an ultimate sensitivity limit for coherent detection. This is imposed by Heisenberg's uncertainty principle, which implies that whenever one quantity is measured accurately, a price is paid in the form of a larger uncertainty in our knowledge of something else. A radio receiver measures both the phase and the amplitude of the input signal simultaneously. The uncertainty principle can be expressed as:

$$\Delta E \Delta t \approx \frac{h}{4\pi}. \tag{8.37}$$

Regarding the radiation being detected as a stream of n photons, $E = nh\nu$ and

$$\Delta n \nu \Delta t \approx \frac{1}{4\pi}. \tag{8.38}$$

Putting $\nu = 1/T$, where T is the period, gives:

$$\Delta n \Delta t \approx \frac{T}{4\pi} \Rightarrow \Delta n \left(\frac{2\pi \Delta t}{T} \right) \approx \frac{1}{2}. \tag{8.39}$$

Now $2\pi \Delta t / T$ represents the uncertainty in the phase so that

$$\Delta n \Delta \phi \approx \frac{1}{2}. \tag{8.40}$$

Therefore, measuring the phase accurately introduces an uncertainty of around one photon, or $h\nu$ in energy.

Relating this energy uncertainty to a corresponding system temperature by putting $h\nu = k_B T_{sys}$, we get an expression for the quantum limit to sensitivity of a coherent detection system:

$$T_{sys} = \frac{h\nu}{k_B}. \tag{8.41}$$

This corresponds to a limiting noise temperature of 0.48 times the frequency in GHz. It is a fundamental limit to sensitivity to which incoherent detectors are not subject because they do not measure both phase and amplitude. This means that incoherent systems can have greater sensitivity. However, the ability to measure phase gives heterodyne receivers another potentially important advantage, which is that very high spectral resolution can be achieved.

8.3.7 MINIMUM DETECTABLE TEMPERATURE AND POWER

Consider the measurement of a source in the presence of background temperature T_B, and background power at the detector, P_B, which is the limiting factor for the sensitivity of the system. We can represent the relationship between the background power at the detector and the system temperature as

$$P_B = k_B T_{sys} \Delta v. \tag{8.42}$$

As shown in Chapter 6 (equation 6.29), the noise equivalent power for a thermal background at temperature T is given by

$$\text{NEP}_{ph} = 2 \left[P_B h v \left(\frac{1}{\eta_d} + \varepsilon \eta_o b \right) \right]^{1/2}. \tag{8.43}$$

In the Rayleigh–Jeans regime ($hv \ll k_B T_B$), the second term dominates and the Bose factor b is approximated by

$$b \approx \frac{k_B T_B}{hv}, \tag{8.44}$$

giving

$$\text{NEP}_{ph} = 2 \left[P_B k_B \varepsilon \eta_o T_B \right]^{1/2}. \tag{8.45}$$

We can take $\varepsilon \eta_o T_B = T_{sys}$, giving

$$\text{NEP}_{ph} = 2 \left[P_B k_B T_{sys} \right]^{1/2} = 2 \left[\left(k_B T_{sys} \right)^2 \Delta v \right]^{1/2} = 2 k_B T_{sys} \left(\Delta v \right)^{1/2}. \tag{8.46}$$

This can be compared to the equivalent expression for a photon detector (see Section 6.3):

$$\text{NEP}_{ph} = 2 \left(P_B h v \right)^{1/2} = 2 N^{1/2} h v \left(\Delta v \right)^{1/2}. \tag{8.47}$$

Therefore, for direct detection, the sensitivity is dictated by the background photon rate and photon energy, but for the radio receiver the only property of the background that affects the sensitivity is the temperature.

We can define a noise equivalent temperature, NET, by

$$\text{NEP}_{ph} = k_B \left(\text{NET} \right) \Delta v, \tag{8.48}$$

giving

$$\text{NET} = \frac{2 T_{sys}}{\left(\Delta v \right)^{1/2}}. \tag{8.49}$$

Like the NEP, this corresponds to an integration time of 0.5 second. For an integration time t_{int}, we have a corresponding minimum detectable temperature which is lower by $\left(2 t_{int} \right)^{1/2}$:

$$\Delta T_{min} = \frac{\sqrt{2}T_{sys}}{\left(\Delta \nu t_{int}\right)^{1/2}}.$$

(8.50)

Note that this result is $\sqrt{2}$ higher than the version often quoted, as we are assuming here that half of the total integration time needs to be used for measuring and subtracting the background.

The corresponding minimum detectable power is given by

$$\Delta P_{min} = k_B \Delta T_{min} \Delta \nu = \frac{k_B T_{sys}\left(2\Delta \nu\right)^{1/2}}{t_{int}^{1/2}}.$$

(8.51)

In practice, a typical source antenna temperature (see Section 8.7.1) will be 10^{-3} K, which must be measured with a much larger system noise temperature. Thus, any small gain fluctuation in the receiver will cause large apparent signal fluctuations at the output. For example, a 1% fluctuation in a system with $T_{sys} = 1000$ K is equivalent to 10 K – much larger than the typical source antenna temperature and making the source undetectable. Gain fluctuations can arise in several ways:

  i. power supply voltage fluctuations (regulation to 1 part in 10^4 or better is common);
  ii. receiver environment temperature fluctuations (so temperature stability or control are important);
  iii. flexure in waveguides and mirror supports (rigid structures are used where possible);
  iv. LO power fluctuations (which are minimised by temperature and/or active LO amplitude control).

In the presence of a gain fluctuation ΔA, equation (8.20) for the output voltage is modified to

$$V_{out} = \left(A + \Delta A\right)k_B\left(T_{rec} + T_s\right)\Delta \nu = Ak_B\left(T_{rec} + T_s\right)\Delta \nu + \Delta Ak_B\left(T_{rec} + T_s\right)\Delta \nu.$$

(8.52)

The second term involves the change in gain multiplied by T_{sys}, which is much larger than the source temperature, and so can produce output fluctuations that overwhelm the signal. In the presence of gain fluctuations, the minimum detectable power becomes

$$\Delta T_{min} = \sqrt{2}T_{sys}\left[\frac{1}{\Delta \nu t_{int}} + \left(\frac{\Delta A}{A}\right)^2\right]^{1/2},$$

(8.53)

where ΔA is the fluctuation in the power gain during t_{int}. Gain fluctuations generally have a $1/f$ spectrum, requiring the integration time to be less than the corresponding $1/f$ knee frequency to avoid significant degradation in sensitivity. Even with careful attention to all aspects of the receiver design, it can be extremely difficult to minimise such fluctuations, and without further measures they can dominate the overall sensitivity. In addition, when observing through the Earth's atmosphere, fluctuations in atmospheric emission on timescales faster than the integration time can swamp the signal.

A commonly adopted scheme to cope with these effects involves the receiver input beam being position-switched, on a timescale shorter than receiver or atmospheric fluctuations, between the source and a nearby position on the sky, or sometimes a matched load within the instrument itself (in which case, the technique is known as Dicke switching). A Dicke-switched receiver system is illustrated in Figure 8.13. The use of a phase-sensitive detector (Chapter 6) after the square-law detector results in a positive output only when there is a power difference between the antenna and the load.

For no source in the antenna beam, the output is zero. Gain fluctuations still affect the signal but now apply only to the difference between the two signals rather than to the absolute level. For example, in the case described above, the fluctuation is now only 1% of 10^{-3} K.

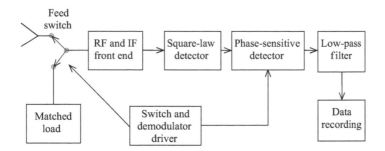

FIGURE 8.13 Block diagram of a Dicke-switched receiver.

Another technique sometimes used, when spectral lines are being observed, to remove atmospheric or receiver gain variations is LO frequency switching: the LO is switched between the spectral line and a nearby frequency with no line emission, relying on the behaviour of the receiver or atmosphere to be the same for both frequencies.

8.4 MIXERS AND AMPLIFIERS

A key component of the superheterodyne receiver described in Section 8.3.4 is the mixer, which relies on a device with a non-linear current–voltage relationship to generate the difference frequency component of the output. There are several basic types of mixer devices.

8.4.1 THE SCHOTTKY DIODE MIXER

Although semiconductor p-n diodes have the required non-linear variation of current with voltage, the junction capacitance limits their speed of response to voltage changes and hence limits the usable bandwidth and operating frequency. A Schottky diode, comprising a low-area metal-semiconductor junction, responds faster than the p-n junction diode since the majority carriers, electrons, spread rapidly on entering the metal with minimal diffusion time, and the contact area can also be made very small. The Schottky diode is therefore commonly used as a mixer in heterodyne receivers for which large IF bandwidths are required to get greater spectral coverage or better photometric sensitivity. In addition, Schottky devices operate well at ambient or moderately cold temperatures, which is further advantageous for sensitivity. Their chief disadvantages for sensitive astronomical receivers are their relatively higher noise, especially compared to superconducting mixers, which are described in the following sections, and their high LO power requirements. Superconducting mixers provide superior performance for frequencies greater than ~ 100 GHz.

8.4.2 SUPERCONDUCTOR-INSULATOR-SUPERCONDUCTOR (SIS) MIXERS

The SIS mixer is widely used for frequencies of ~100 GHz to ~ 1 THz. An SIS junction comprises a small-area (~ 1 μm^2) sandwich with two superconducting films separated by a thin (a few nm) insulating barrier. In a superconductor, electrons are coupled quantum mechanically as Cooper pairs. There is an energy gap, of total width 2Δ, between the Cooper pair energy and the lowest available free electron energy states. When a DC bias voltage, V, is applied across the junction, an electron can tunnel through the insulator if there is an unoccupied state with the same energy on the other side. It must therefore have energy greater than the superconducting energy gap:

$$eV > 2\Delta, \tag{8.54}$$

where e is the electronic charge.

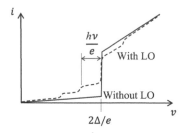

FIGURE 8.14 Current–voltage characteristics of an SIS junction showing non-linear response.

Thus, for voltages below $2\Delta/e$ (which is typically a few mV), the current is zero or very low, and it increases very sharply when V exceeds this value. Above this threshold voltage, tunnelling readily occurs and the two films become normal conductors. This transition represents a highly non-linear response, and so can be used as the basis for mixing. Photon-assisted tunnelling can occur at lower voltages, when the electron gains energy from one or more incoming photons:

$$nh\nu + eV > 2\Delta \quad (n = 1, 2...). \tag{8.55}$$

The general form of the current–voltage characteristics for an SIS junction mixer is shown in Figure 8.14 for the case of no radiation input and with an LO signal applied. Given the highly non-linear nature of the curve, if an RF signal from a telescope is also fed to the device, mixing can occur with the LO signal to produce an IF difference frequency.

The SIS mixer must be operated below the superconducting transition temperature, which for common materials used (e.g. niobium) requires temperatures of a few K. At these temperatures, the thermal noise is also minimal resulting in devices that can approach the fundamental quantum-limited sensitivity for heterodyne detection. Another advantage compared to Schottky diode mixers is that the amount of LO power needed to operate the mixer is much smaller (only a few tens of nW). This is important because it is difficult to make powerful LO sources at high frequencies.

Most high sensitivity receivers operating at frequencies between 100 GHz and around 1 THz use SIS mixers. The cut-off frequency $2\Delta/h$, above which the mixer cannot operate, depends on the superconducting material used – for niobium devices this is ~700 GHz.

8.4.3 Hot Electron Bolometer Mixers

For frequencies above 1 THz, superconducting mixers that do not depend on the superconducting energy gap are used. The Hot Electron Bolometer (HEB) is a fast superconducting bolometric device (similar in principle to the bolometers described in Chapter 9). Typically, an HEB is made in the form of a very small superconducting bridge (e.g. niobium-nitride, NbN) with nanometre-scale dimensions. When operated on the superconducting-normal transition, incoming radiation heats the electron gas to a temperature slightly above that of the crystal lattice. The relaxation time associated with this temperature difference, τ_R, can be ~0.1 ns or lower, enabling the RF signal from the telescope and the LO to produce beat frequencies up to a few GHz.

The RF radiation is coupled into the HEB through a lens or a feed antenna, and the LO is usually injected quasi-optically in the same way. For a high IF bandwidth, a very thin superconducting film is needed. Currently, HEBs are the most sensitive mixers for heterodyne receivers operating in the 1–10 THz region.

8.4.4 High Electron Mobility Transistors

The High Electron Mobility Transistor (HEMT), is a form of a field effect transistor (FET) developed for high-frequency applications. It utilises compound semiconductor layers to create a thin layer with a large concentration of free electrons (a 2-D electron gas) in a highly pure semi-conductor (usually GaAs or InP). Electrons in this layer, being the majority carriers in the pure material, can conduct with scattering of the charge carriers by impurity atoms largely eliminated. This results in very high electron mobility and consequent fast speed of response. HEMTs are suitable for operation at low temperatures and are widely used in low-noise amplifiers. Their high speed of response also makes them suitable as the detectors in radiometers operating at frequencies up to ~100 GHz or to be used as a mixer where high IF bandwidths (60 GHz or more) can be achieved.

8.5   LOCAL OSCILLATORS

In a heterodyne receiver, the LO needs to have a precisely known frequency and be highly stable both in phase and amplitude. To ensure the necessary high stability, the LO source is usually stabilised using a control system known as a phase-locked loop. The LO power level must be high enough to sufficiently drive the mixer to its optimal performance. Tunability over a wide frequency range is also often desirable. In many cases, the primary oscillator is generated at a much lower frequency which is up-converted, via a chain of frequency multiplication stages, to that needed for the LO. Frequency multipliers are based on the non-linear response of an electrical component, often a Schottky diode, in a similar way to the mixing process. When a signal is applied, the non-linear response produces outputs at various harmonics of the input. The second or third harmonics are usually the strongest, so most multipliers are doublers or triplers.

Some LO systems use an accurate frequency standard. An example is the hydrogen maser, which relies on the precise 1,420,405,752-Hz spin-flip transition of atomic hydrogen gas in a suitably tuned cavity. Quartz oscillators, which rely on the production of a piezoelectric signal from the mechanical oscillation of a quartz crystal, are also used. Other LO systems are based on semiconductor oscillators. The Yttrium iron garnet (YIG) oscillator utilises the fact that YIG is a ferrite substance that exhibits microwave-frequency resonance when placed in a magnetic field. The resonant frequency can be adjusted, enabling tuning of the LO over a wide range by modifying the magnetic field strength. It has linear tuning over multi-octave frequencies. YIG oscillators work well up to frequencies of ~20 GHz. The Gunn diode oscillator is based on the transferred electron effect, which can occur in a GaAs crystal producing a regular series of electron avalanches resulting in regularly spaced sharp current pulses. The crystal is placed in a tuned cavity which is maintained in a state of continuous ringing by the current impulses, with the output fed to the mixer through a waveguide. The oscillator can be tuned by a combination of electrical bias adjustment and mechanical tuning (i.e. alteration of the dimensions of the resonant cavity). Gunn diodes can provide frequencies up to ~100 GHz, which can be further multiplied if needed.

Photonic LOs rely on mixing of two infrared (typically 1.5 μm) laser signals in a photodiode, with the LO being the beat frequency between the two lasers. Such systems are very suitable for radio interferometers, which require the same fundamental LO system to be delivered to a number of different telescopes, which can be done using optical fibres to transmit the laser signals before mixing separately at each telescope. The quantum cascade laser (QCL) is a recently developed oscillator for THz frequencies, with the emission arising from inter-sub-band transitions within a stack of semiconductor quantum wells.

For receivers operating at high frequencies (hundreds of GHz or in the THz range), to avoid significant waveguide losses the LO signals are sometimes fed to the mixers quasi-optically through Martin-Puplett diplexers. This technique takes advantage of the fact that a polarising Michelson interferometer (known as a Martin-Puplett interferometer) can be configured to direct two singly polarised signals at its input ports to one output port, with 100% efficiency (LeSurf 1990). The LO power is adjusted to the optimum value for each mixer using a rotatable polarising wire grid in front of the diplexer.

8.6   IF SPECTROMETERS

The IF band mirrors a portion of the RF spectrum, albeit down-converted. For radio observations of spectral features, the IF spectrum must be measured. Various spectrometer types have been developed to analyse this signal to retrieve the spectral information. As noted above, it is desirable that the IF bandwidth be as large as possible to ensure that spectral features can be resolved, and indeed to have the possibility of measuring multiple spectral features at the same time. The role of the spectrometer is thus to split the IF band into many narrow spectral channels to measure the signal in each frequency bin (Figure 8.15). By so doing, a line profile can be measured, allowing valuable information on the physical properties of the emitting region to be determined. The spectrometer is located after the receiver, and hence often referred to as the back-end spectrometer. There are four common types of IF spectrometer: the multichannel (filter bank) spectrometer, the acousto-optic spectrometer, the autocorrelation spectrometer, and the direct Fast Fourier Transform (FFT) spectrometer.

8.6.1   THE MULTICHANNEL (FILTER BANK) SPECTROMETER

The multichannel spectrometer involves simply dividing the IF band (and so the corresponding observing RF band) into N contiguous channels. This requires N contiguously tuned bandpass filters, so that there are no gaps, and N square-law detectors – in effect we create N separate IF sections in the receiver, as shown in Figure 8.16. Typically, $N \geq 1000$. The difficulty with this type of system is in the manufacture of multiple filters, requiring a lot of electronic circuitry and meticulous tuning of each channel to get the complete coverage. This design is therefore not commonly used in modern receivers, although it has played an important role in the development of spectral line astronomy.

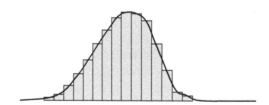

FIGURE 8.15   Splitting the IF into many narrow bins allows a spectral line profile to be measured.

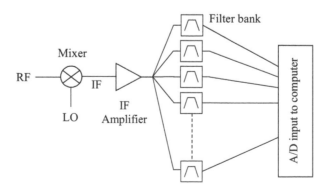

FIGURE 8.16   The multichannel spectrometer.

8.6.2 THE ACOUSTO-OPTIC SPECTROMETER (AOS)

This type of spectrometer is based on the diffraction of light by ultrasonic waves. The IF signals from the receiver mixer are used to drive a piezoelectric transducer to generate an acoustic wave in a crystal. The acoustic wave modulates the refractive index of the crystal and creates a phase grating across its surface; it is termed a Bragg-cell. By illuminating this cell with a collimated laser beam, the angular dispersion of the diffracted light can be measured with a linear detector array (e.g., a CCD), as shown schematically in Figure 8.17. The signals recorded by the CCD array represent an image of the IF spectrum. The spectral resolution depends on the type of crystal used and the focal length of the imaging optics.

Acousto-optic spectrometers are no longer widely used on ground-based telescopes, but are still sometimes appropriate for wide-bandwidth space-borne instruments for which on-board computing power can be very limited. An AOS-based spectrometer was used in the HIFI instrument on board the *Herschel* Space Observatory (de Graauw et al. 2010).

8.6.3 THE DIGITAL AUTOCORRELATION SPECTROMETER

According to the Wiener–Khinchin theorem in Fourier analysis (Rohlfs 1986), the power spectrum of a signal is given by the Fourier transform of its autocorrelation function. The digital autocorrelation spectrometer implements this computationally. The essential features are illustrated in Figure 8.18. The IF signal is first filtered if need be to limit it to bandwidth Δv, before being digitised at a sampling interval $\Delta \tau$ (which must be at least $1/(2\Delta v)$ for Nyquist sampling). Because

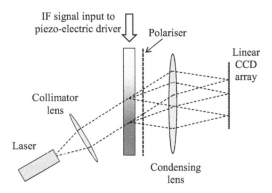

FIGURE 8.17 Acousto-optic spectrometer.

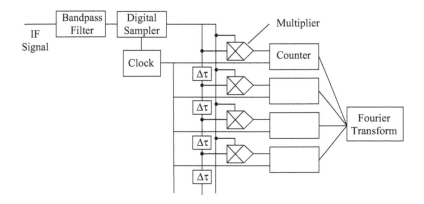

FIGURE 8.18 Architecture of a digital autocorrelation spectrometer.

the IF signal is dominated by noise (from the receiver and/or background), it is not necessary to digitise it with many bits to carry out the correlation. Even one-bit (2-level) sampling, which preserves only the sign of the instantaneous voltage as 0 or 1, involves only a modest SNR penalty by a factor of 0.64, and for three-level (2-bit) sampling, the degradation factor is 0.81 (D'Addario 1989). The ability to use this crude sampling reduces the complexity and power dissipation of the electronics. The digitised signal is then subject to successive delays of $\Delta\tau$ and at each delay is multiplied by the un-delayed signal. The results for each value of delay are accumulated to give the components of the digitised autocorrelation function, which are then Fourier transformed to produce the IF spectrum.

It is instructive to compare this process to the Fourier transform spectrometer (Section 7.6), which uses the same principle except that here a time delay is used and in the FTS a spatial delay is used – in both cases a signal is compared with a delayed version of itself over a range of delays to generate the autocorrelation function. The fundamental arguments about apodisation and spectral limits also apply to both. The advantages of this type of spectrometer are that it is easy to reproduce, unlike the filter bank, has the multiplex advantage, and can be made very flexible in design and operation – e.g. in the optimisation of bandwidth and resolution for a given observation.

8.6.4   The Direct FFT Spectrometer

The autocorrelation spectrometer has the advantage that computational requirements are minimised by using only one or two-bit sampling and generating the autocorrelation function in a simple and convenient way before the FFT is performed. For many years, these features were essential in order to be able to compute the IF spectra over wide bandwidths. More recently, advances in high-speed digital electronics, particularly the availability of fast analogue-to-digital converters (ADCs) and field programmable gate arrays (FPGAs), have made it possible to measure the FFT of the IF signal directly, and such systems are increasingly used on ground-based radio telescopes. An example is the FFT spectrometer for the APEX telescope, described by Klein et al. (2006), which implements 8-bit sampling and a complete real-time FFT signal processing pipeline to decompose a 1-GHz band into 16,384 spectral channels.

8.7   RADIO OBSERVATIONS

8.7.1   Source Antenna Temperature

Consider a radio observation of an astronomical source with an antenna of effective area A_e and normalised beam pattern on the sky $P_n(\theta,\phi)$. As noted in Chapter 3, the throughput of an antenna is

$$A_e\Omega_A = A_e \iint\limits_{4\pi} P_n(\theta,\phi)\,d\theta\,d\phi = \lambda^2. \tag{8.56}$$

The source might not uniformly fill the antenna beam, so we must take into account how the antenna beam and the source brightness vary as a function of position on the sky. The power collected from the source, assuming observation over a narrow frequency band Δv, is then

$$P_S = \frac{A_e\Delta v}{2} \iint\limits_{4\pi} P_n(\theta,\phi)I(\theta,\phi)\,d\theta\,d\phi, \tag{8.57}$$

where we integrate the product of the beam response and the sky intensity, $I(\theta,\phi)$, over 4π steradians to ensure that all the power incident is accounted for. In radio astronomy observations, the signal from the source is characterised by equating this to $kT_A\Delta v$ where T_A is defined as the antenna temperature of the source:

$$T_A = \frac{A_e}{2k_B} \iint\limits_{4\pi} P_n(\theta,\phi) I(\theta,\phi) \, d\theta \, d\phi. \tag{8.58}$$

In the Rayleigh–Jeans regime,

$$I(\theta,\phi) = \frac{2k_B T(\theta,\phi)}{\lambda^2}, \tag{8.59}$$

where $T(\theta,\phi)$ is the sky brightness temperature, giving

$$T_A = \frac{A_e}{\lambda^2} \iint\limits_{4\pi} P_n(\theta,\phi) T(\theta,\phi) \, d\theta \, d\phi = \frac{\displaystyle\iint\limits_{4\pi} P_n(\theta,\phi) T(\theta,\phi) \, d\theta \, d\phi}{\displaystyle\iint\limits_{4\pi} P_n(\theta,\phi) \, d\theta \, d\phi}. \tag{8.60}$$

The antenna temperature is thus the temperature which a source of the same overall brightness would have if it filled the beam of the antenna uniformly.

For an observation of a point source, the integral of the sky intensity over the beam is just the source flux density, S_ν:

$$\iint\limits_{4\pi} P_n(\theta,\phi) I(\theta,\phi) \, d\theta \, d\phi = S_\nu, \tag{8.61}$$

and

$$T_A = \frac{A_e S_\nu}{2k_B}, \tag{8.62}$$

so that

$$S_\nu = \frac{2k T_A}{A_e}. \tag{8.63}$$

The overall gain of the system is the relationship between antenna temperature and source flux density and is

$$\frac{T_A}{S_\nu} = \frac{A_e}{2k_B} \left(\text{usually quoted in K/Jy} \right). \tag{8.64}$$

8.7.2 Sensitivity to a Point Source

The sensitivity of a radio receiver to a point source can be derived from the definition of noise equivalent flux density, NEFD (equation 6.90):

$$\text{NEFD} = \frac{\text{NEP}}{A_{tel} \Delta \nu t_a \eta_o \eta_{Ap}}. \tag{8.65}$$

where A_{tel} is the telescope area, η_{Ap} is the aperture efficiency, and t_a and η_o are the atmospheric and instrument transmission efficiencies. Here $A_{tel}\eta_{Ap} = A_e$, and we do not need to include t_a and η_o as they already contribute to the system temperature. Therefore, we can write

$$\text{NEFD} = \frac{2k_{\mathrm{B}}T_{\mathrm{sys}}(\Delta v)^{1/2}}{A_{\mathrm{e}}\Delta v} = \frac{2k_{\mathrm{B}}T_{\mathrm{sys}}}{A_{\mathrm{e}}(\Delta v)^{1/2}}. \tag{8.66}$$

As usual, this corresponds to a 0.5-second integration time, and the minimum detectable flux density for an integration time t_{int} is thus

$$\Delta S_v = \frac{2\sqrt{2}k_{\mathrm{B}}T_{\mathrm{sys}}}{A_{\mathrm{e}}\Delta v} = \frac{2k_{\mathrm{B}}T_{\mathrm{sys}}}{A_{\mathrm{e}}(\Delta v t_{\mathrm{int}})^{1/2}}. \tag{8.67}$$

As noted above, the basic radio receiver is sensitive to a single polarisation, meaning that half of the source power is lost. However, receiver systems can be designed so that both polarisations are available. For instance, in the case of a feedhorn coupled to a circular waveguide, the waveguide propagates both polarisations. A waveguide component known as an orthomode transducer (OMT) can separate the vertical and horizontal polarised signals so that they can be passed to parallel mixers. Other designs can involve separate feed antennas to pick up the orthogonal polarisations, again feeding parallel mixers. Detecting both polarisations provides improved sensitivity to an unpolarised source (by a factor of $\sqrt{2}$ when the two outputs are co-added) and can also enable polarimetric measurements to be made.

8.7.3 Calibration Methods

A radio receiver can be calibrated using hot and cold black body loads, which fill the beam as described in Section 8.3.3. If we could do this for the complete antenna with receiver, this would provide an absolute calibration, but this is not practical for large antennas. There are only a few astronomical observations in which a source of known brightness fills the beam of the antenna uniformly, so this method has limited usefulness. It is, however, possible to do a piecemeal calibration whereby we build up knowledge of the telescope and receiver efficiencies and then put them together to provide an expectation of the overall performance.

Two techniques are adopted to calibrate radio observations. The first is to use a source of a known brightness temperature (or standard radio sources of known flux density). This source is observed periodically as it moves in elevation to monitor atmospheric absorption and/or remove any systematic gain changes in the antenna (for instance from elevation-dependent gravitationally induced distortion of its structure). This same technique is also used for infrared and optical observations.

The second technique, only used with radio receivers, is to place hot and cold loads in front of the receiver. The corresponding measured receiver outputs can be used to establish the antenna temperature scale. The measured sky temperature adjacent to the source being observed can then be used to determine and correct for atmospheric loss while the temperature scale and beam parameters can be used to determine the source antenna temperature.

8.8 RADIO INTERFEROMETRY

Enhancing angular resolution using multi-antenna interferometry is an important technique in radio astronomy because at radio wavelengths the resolution ($\sim\lambda/D$) of an individual reasonably sized antenna is often inadequate. As discussed in Chapter 4, the essential principles of interferometry are the same for all wavelengths. However, radio receivers are coherent heterodyne systems which measure the wave amplitude and phase rather than recording only its intensity. It is important not to mix concepts here: coherent radio detectors are used to detect incoherent radiation from astronomical sources. Radio interferometers use the coherence property of the detector to determine the phase delay of waves arriving at different antennas, which, as we will see below, provides spatial information on the incoherent emission from a source.

The purpose of interferometric observations is to measure the fine spatial detail of the source brightness distribution, which we would get from a very large antenna, by using a number of smaller antennas. Although the collecting area is much reduced compared to a large filled antenna, it is still possible to recreate the phase delays across the virtual large antenna by deploying the small antennas within a virtual antenna area.

8.8.1 THE TWO-ELEMENT RADIO INTERFEROMETER

It is simplest to begin with a two-element interferometer, as shown in Figure 8.19. The antennas, separated by baseline distance B, both track a particular source on the sky, and the rotation of the Earth causes the zenith angle, θ, and the path difference, $B\sin\theta$, to vary. In the case of a heterodyne system, to retain the phase information in the radio emission from the source, it is essential that a single LO feeds the heterodyne receivers on the two antennas.

Letting $\hat{\mathbf{s}}_0$ be a unit vector towards source, the path difference is $\mathbf{B}\cdot\hat{\mathbf{s}}_0$, and the time difference between the wavefront arrival at the two antennas, known as the geometric delay, is

$$\tau_g = \frac{\mathbf{B}\cdot\hat{\mathbf{s}}_0}{c}. \tag{8.68}$$

The corresponding phase difference, ϕ, between the waves at the two antennas is given by

$$\phi = \frac{2\pi}{\lambda}\mathbf{B}\cdot\hat{\mathbf{s}}_0 = \frac{\omega B\sin\theta}{c} = \frac{2\pi\tau_g c}{\lambda} = \omega\tau_g. \tag{8.69}$$

For observation of monochromatic on-axis point source at frequency ω, and assuming identical receivers, the voltages from the two receivers can be written as

$$v_1(t) = Ve^{i\omega t} \quad \text{and} \quad v_2(t) = Ve^{i(\omega t+\phi)} \tag{8.70}$$

where V is the source-induced voltage amplitude. The voltage from Antenna 2, which intercepts the wave first, leads the voltage from Antenna 1.

The signals from the two antennas can be combined electronically in two ways. In an adding interferometer (Figure 8.20a), the voltages are added and then squared to produce a signal proportional to the total power, which is then time-averaged (integrated). In a multiplying or correlation interferometer (Figure 8.20b), the signals are multiplied together and then time-averaged.

For the adding interferometer, the mathematical treatment of the process of adding, squaring, and time averaging the two signals is identical to the treatment of the Michelson spectral interferometer in Section 7.6.1 for one value of the optical path difference. In both cases, the signal is

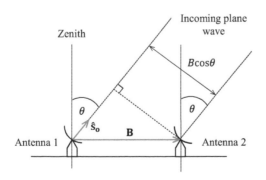

FIGURE 8.19 A two-beam radio interferometer system with baseline $\mathbf{B}$ observing a source at zenith angle θ.

(a)   (b)

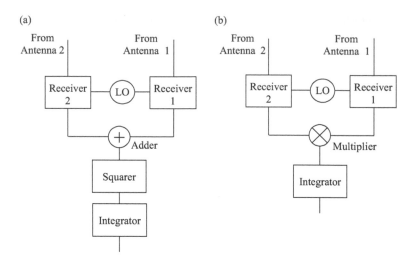

FIGURE 8.20   Basic adding interferometer (a) and correlation interferometer (b).

added to a delayed version of itself. The output as a function of the phase difference between the two beams is (from equation 7.80)

$$V_{\text{out}}(\phi) = V^2(1 + \cos\phi). \tag{8.71}$$

Therefore, in the adding interferometer, the fringes (second term) are seen in addition to the source power. Unfortunately, with a radio receiver, the noise fluctuations of the receiver itself are also seen as an additive noise power after the squarer and are indistinguishable from the source power. The receiver noise power is normally much greater than the source power, and receiver gain fluctuations can therefore mask the fringe information. Techniques based on Dicke switching (described in Section 8.3.7) can be used whereby an additional phase delay of $\lambda/2$ is added into one arm of the interferometer to difference between the in-phase power and out-of-phase power at a frequency above the low-frequency gain changes. This was effective in early interferometer systems but at the expense of added complexity and a $\sqrt{2}$ loss in sensitivity.

The correlation interferometer, based on multiplying the signals and integrating to produce the time average, avoids these shortcomings. In this case, the multiplier output is

$$v_{\text{m}}(t) = V^2 \sin(\omega t)\sin\left(\omega\left(t + \tau_g\right)\right). \tag{8.72}$$

Using the trigonometric identity $\sin A \sin B = \left[\cos(A - B) - \cos(A + B)\right]/2$ gives

$$v_{\text{m}}(t) = \frac{V^2}{2}\cos\left(\omega\tau_g\right) - \cos\left(2\omega\left(t + \tau_g\right)\right). \tag{8.73}$$

The term in 2ω is rejected by the integrator (low pass filter) giving

$$v_{\text{o}}\left(\tau_g\right) = \frac{V^2}{2}\cos\left(\omega\tau_g\right) = \frac{V^2}{2}\cos\phi. \tag{8.74}$$

The receiver voltage amplitude, V is proportional to the incoming electric field amplitude, so that $V^2 \propto S_v$, the source flux density:

$$v_{\text{o}}\left(\tau_g\right) = KS_v \cos\left(\omega\tau_g\right) = KS_v \cos\phi, \tag{8.75}$$

where K is a constant, that depends on size of the antennas, overall atmospheric transmission, etc.

Expressing the phase difference in terms of the zenith angle,

$$v_0(\theta) = KS_v \cos(\omega B \sin\theta) = KS_v \cos\left(\omega \frac{\mathbf{B} \cdot \hat{\mathbf{s}}_0}{c}\right). \tag{8.76}$$

The correlator output for pointing offset $\Delta\theta$ is

$$v_0(\theta + \Delta\theta) = KS_v P_n(\Delta\theta)\cos(\phi + \Delta\phi) = KS_v P_n(\Delta\theta)\cos\left[\phi + \frac{\omega B \cos\theta}{c}\Delta\theta\right], \tag{8.77}$$

where $P_n(\Delta\theta)$ is the normalised beam response at position offset $\Delta\theta$.

The average value of the output is zero. Its amplitude is proportional to the source flux density, and its phase encodes positional information. Importantly, only signals that are correlated influence the integrator output. Receiver noise, atmospheric fluctuations, extended astrophysical backgrounds, and gain fluctuations do not.

The fringe period, $\Delta\phi = 2\pi$, corresponds to a change in source direction of

$$\Delta\theta = \frac{\lambda}{B\cos\theta}. \tag{8.78}$$

This means that the angular resolution is the wavelength divided by the projected baseline, as opposed to the wavelength divided by the antenna size for a single antenna.

8.8.2 The Effect of Finite Bandwidth

If the receiver responds to a range of frequencies $\Delta\omega$ centred on ω_0, equation (8.75) must be integrated over that band:

$$v_0(\tau_g) = KS_v \int_{\omega_0 - \Delta\omega/2}^{\omega_0 + \Delta\omega/2} \cos(\omega\tau_g)d\omega = (KS_v\Delta\omega)\mathrm{sinc}\left(\frac{\Delta\omega\tau_g}{2}\right)\cos(\omega_0\tau_g). \tag{8.79}$$

The fringe amplitude is proportional to the bandwidth, so a larger bandwidth provides a bigger signal. However, the fringe pattern is now also multiplied by a sinc function envelope that depends on the bandwidth and the geometrical delay. To avoid the fringes being washed out (due to loss of coherence over the passband), the sinc function must be kept close to unity, which means that τ_g must effectively be made very small. However, the rotation of the Earth causes the source to move on the sky, changing the zenith angle θ and the time delay. This is compensated by delay tracking – introducing a time delay into the signal from Antenna 2, which is varied as the source moves to compensate exactly for the geometrical time delay.

8.8.3 Fringe-Stopping

The movement of the source on the sky also changes the phase of the fringes. From equation (8.69), the rate of change of phase with θ is

$$\frac{d\phi}{d\theta} = \frac{\omega B\cos\theta}{c} = \frac{2\pi B\cos\theta}{\lambda}. \tag{8.80}$$

Since $B\cos\theta \gg \lambda$, the fringe phase is extremely sensitive to the source position, and without correction very fast-fringe oscillations are seen at the output, with an angular frequency of

$$\omega_{\text{fringe}} = \frac{d\phi}{dt} = \frac{2\pi B \cos\theta}{\lambda} \frac{d\theta}{dt}, \tag{8.81}$$

where $d\theta/dt = 7.3 \times 10^{-5}$ rads s^{-1} is the angular rotation rate of the Earth (the sidereal rate). For example, for a source at the zenith $(\cos\theta = 1)$ observed at 100 GHz with a baseline of 100 m, $\omega_{\text{fringe}} = 15.3$ Hz. This would be very inconvenient as the integrator time constant would need to be short compared to the fast-fringe period to avoid the fringes being averaged out.

A technique known as phase-tracking or fringe-stopping cancels out the on-axis phase shift and removes the fast fringes. It can be implemented by the same delay-tracking process described above, which continually adjusts the phase delay between the two receiver signals as the source moves. Alternatively, fringe-stopping can be done by continually adjusting the phase difference between the two LO signals. This second method allows the two processes of delay tracking (to maximise coherence over the passband and prevent fringe attenuation) and fringe-stopping (to suppress the fast fringes which contain no useful information) to be optimised separately.

With phase-tracking, the output when the source is on-axis is just $v_o(0) = KS_v$, and for positional offset $\Delta\theta$, it is

$$v_o(\Delta\theta) = KS_v P_n(\Delta\theta) \cos\left[\frac{\omega B \cos\theta}{c} \Delta\theta\right]. \tag{8.82}$$

It is useful to distinguish between the different sky positions, known as centres, which are defined in interferometry (Formalont and Perley 1999):

  i. The beam-tracking (pointing) centre: this is the source position on the sky at which the antennas are pointed, defined by the direction of maximum antenna gain.
 ii. The delay-tracking centre: this is the position on the sky for which the geometrical delay is compensated to maximise coherence over the observation passband.
iii. The phase-tracking centre: this is the sky location for which the fringes are stopped, and is the position defined by the vector $\hat{\mathbf{s}}_0$.

Often, the three centres are at the same position. The phase-tracking and delay-tracking centres have to be identical for an RF interferometer (no frequency down-conversion; control only by the geometrical delay compensation), but they can be adjusted separately in a heterodyne system, for which the phase-tracking can be carried out separately by adjusting the LO phase.

8.8.4 Beam Synthesis

The interferometer output (equation 8.82) is plotted in Figure 8.21a for $D = 2$ m, $B = 10$ m, and a monochromatic observing frequency of 10 GHz. The cosinusoidal fringe pattern is multiplied by the envelope representing the primary beam – the beam profile of either of the antennas.

The fringe pattern can be made more position-specific by introducing more antennas to provide additional baselines and adding the correlator outputs, which are all in phase for zero offset but, because of the different fringe frequencies, tend to cancel out for most off-axis angles. Figure 8.21b shows the resulting beam for three antennas in a line with central positions 0, 3.3, and 10 m. Adding further antennas sharpens the beam and suppresses the sidelobes and gives an increasingly well-defined beam: the fringe pattern for five antennas with central positions 0, 3.33, 5.33, 7.5, and 10 m is shown in Figure 8.21c. Effectively, this synthesises a narrow beam (the beam expected from a large telescope of diameter 10 m) in the direction $\Delta\theta = 0$.

However, in the orthogonal direction, perpendicular to the baseline, there is no improvement in angular resolution, which remains at the value determined by the physical antenna size. Arranging

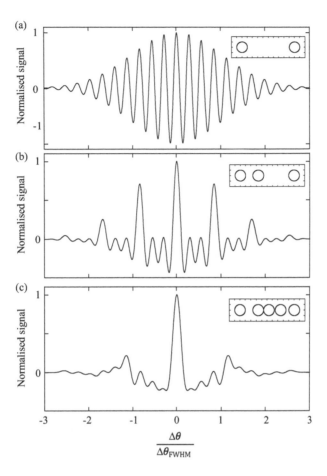

FIGURE 8.21 (a) Two-element interferometer fringe pattern for a 10-GHz observation with antenna diameters of 2 m and baseline 10 m; (b) pattern for three antennas in a line with central positions 0, 3.3, and 10 m; (c) pattern for five antennas with central positions 0, 3.33, 5.33, 7.5, and 10 m.

a number of antennas in a two-dimensional array provides baselines in many directions allowing a well-defined beam to be obtained. With N antennas, there are $N(N-1)/2$ baselines.

8.8.5 INTERFEROMETER OBSERVATION OF AN EXTENDED SOURCE

Consider the two-element interferometer viewing an extended source, as depicted in Figure 8.22, with the two antennas tracking a region of sky. A particular position is designated as the phase-tracking centre, defined by unit vector $\hat{\mathbf{s}}_0$, which will be the centre position of the synthesised image. The projected baseline is in the plane indicated by the dashed line, known as the uv plane. An arbitrary source position is defined by position vector $\mathbf{s} = \hat{\mathbf{s}}_0 + \boldsymbol{\sigma}$.

The source intensity distribution is specified by $I_\nu(\boldsymbol{\sigma})$. Let $d\Omega$ be a small solid angle of the source at that position, and $P_n(\boldsymbol{\sigma})$ be the beam profile (assumed to be the same for both antennas). The correlator output is then given by

$$
\begin{aligned}
\nu_0 &= \int_{\text{Beam}} P_n(\boldsymbol{\sigma}) I_\nu(\boldsymbol{\sigma}) \cos\left(\omega \frac{\mathbf{B} \cdot \hat{\mathbf{s}}}{c}\right) d\Omega \\
&= \int_{\text{Beam}} P_n(\boldsymbol{\sigma}) I_\nu(\boldsymbol{\sigma}) \cos\left(\omega \frac{\mathbf{B} \cdot \hat{\mathbf{s}}_0}{c} + \omega \frac{\mathbf{B} \cdot \boldsymbol{\sigma}}{c}\right) d\Omega.
\end{aligned}
\tag{8.83}
$$

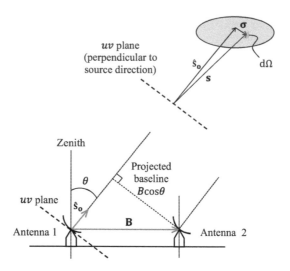

FIGURE 8.22   Geometry for interferometer observation of an extended source.

If fringe-stopping is implemented, the $\mathbf{B} \cdot \hat{\mathbf{s}}_0$ term is continuously nulled, and

$$v_0 = \int\limits_{\text{Beam}} P_n(\boldsymbol{\sigma}) I_v(\boldsymbol{\sigma}) \cos\left(\omega \frac{\mathbf{B} \cdot \boldsymbol{\sigma}}{c}\right) d\Omega. \tag{8.84}$$

This is actually the same for positive or negative $\mathbf{B} \cdot \boldsymbol{\sigma}$ as cos is an even function. Therefore, we cannot distinguish between regions of the source on either side of the source centre. To break this degeneracy, an additional observation is made with a 90° phase shift introduced in one of the arms so that the fringe pattern becomes sinusoidal instead of cosinusoidal:

$$v_0 = \int\limits_{\text{Beam}} P_n(\boldsymbol{\sigma}) I_v(\boldsymbol{\sigma}) \sin\left(\omega \frac{\mathbf{B} \cdot \boldsymbol{\sigma}}{c}\right) d\Omega. \tag{8.85}$$

A complex correlator simultaneously computes the sine and cosine outputs and combines them to produce a quantity known as the complex visibility (see also Section 4.8):

$$V = \int\limits_{\text{Beam}} P_n(\boldsymbol{\sigma}) I_v(\boldsymbol{\sigma}) \left(\cos\left(\omega \frac{\mathbf{B} \cdot \boldsymbol{\sigma}}{c}\right) - i \sin\left(\omega \frac{\mathbf{B} \cdot \boldsymbol{\sigma}}{c}\right)\right) d\Omega = \int\limits_{\text{Beam}} P_n(\boldsymbol{\sigma}) I_v(\boldsymbol{\sigma}) e^{-i\omega \frac{\mathbf{B} \cdot \boldsymbol{\sigma}}{c}} d\Omega. \tag{8.86}$$

The complex visibility is thus the two-dimensional Fourier transform of the sky intensity weighted by the beam profile.

8.8.6 The *uv* Plane and Aperture Synthesis

Consider the projected interferometer baseline, as viewed from the source. Adopting a reference frame perpendicular to $\hat{\mathbf{s}}_0$, let baseline vector $\mathbf{B}$ have coordinates (u, v), expressed in wavelengths. Let (x, y) be the coordinates of the point on the source defined by $\boldsymbol{\sigma}$. We then have

$$\frac{\mathbf{B} \cdot \boldsymbol{\sigma}}{\lambda} = ux + vy, \tag{8.87}$$

and

$$V = \int_{\text{Beam}} P_n(x,y) I_\nu(x,y) e^{-i2\pi(ux+vy)} \, dx \, dy. \tag{8.88}$$

This is the two-dimensional Fourier transform of $P_n(x,y) I_\nu(x,y)$, the source intensity distribution weighted by the beam profile. The uv plane (sometimes called the aperture plane) represents the plane of possible projected baselines, with one uv point corresponding to one particular baseline. To recover the source image with high fidelity, the uv plane must be well sampled. It is clearly impossible to sample it completely (as a filled aperture does), but with good coverage of the uv plane over a range of baseline lengths up to some maximum dictated by the largest antenna separations, a high-quality image can be obtained. The use of a set of antennas arranged to do this is the basis of aperture synthesis imaging. Sampling of the uv plane is assisted by the rotation of the Earth, which results in a range of values (a uv track) being sampled by each pair of fixed antennas.

Inevitably the sampling of the uv plane will be incomplete, resulting in a synthesised beam pattern with significant sidelobe structure compared to what would be achieved with a filled aperture of equivalent size. This is commonly termed the Dirty Beam. The corresponding raw image of the sky (the Fourier transform of the measured visibility function) is known as the Dirty Image and is the convolution of the Dirty Beam with the True Image of the sky. The CLEAN algorithm (Högbom 1974) is a widely used process that exploits knowledge of the Dirty Beam to deconvolve the data and generate the best achievable representation of the True Image. Assuming that the image is a superposition of point sources, it iteratively identifies the brightest point in the image and subtracts from the data a model image based on the response of the Dirty Beam to a point source at that position. This process is continued until the limit imposed by noise in the image is reached. The image is then reconstructed by adding together the various point sources convolved with a Clean Beam (the Dirty Beam without the sidelobes). CLEAN works most effectively when the image is indeed a collection of point or compact sources but also improves images of more complex structure. Some examples of uv plane sampling and the application of the CLEAN algorithm are given in Jackson (2008) and Perley (2019).

8.8.7 Interferometer Sensitivity

For a radiometer coupled to a single dish, the minimum detectable flux density from a point source is given by equation (8.67). In this system, the signal is voltage multiplied by itself so that the multiplier output is proportional to V_s^2. The noise at the multiplier output is also proportional to the square of the input noise voltage. In the case of a two-element interferometer with identical antennas, the two signal voltages are multiplied in the correlator, giving an output again proportional to V_s^2. However, now the noise voltages are uncorrelated and they combine in quadrature so that the output noise level is reduced by a factor of $\sqrt{2}$. The overall SNR is thus better by a factor of $\sqrt{2}$ than that of the single-antenna system.

An N-element interferometer is equivalent to $N(N-1)/2$ independent two-element interferometers, so the overall sensitivity is further improved by $\sqrt{N(N-1)/2}$, i.e., it is $\sqrt{N(N-1)}$ times better than for the single dish. Equation (8.67) thus becomes

$$\Delta S_\nu = \frac{2k_B T_{\text{sys}}}{A_e \left(N(N-1) \Delta \nu t_{\text{int}} \right)^{1/2}}. \tag{8.89}$$

For large N,

$$\Delta S_\nu \approx \frac{2k_B T_{\text{sys}}}{A_{\text{tot}} \left(\Delta \nu t_{\text{int}} \right)^{1/2}}, \tag{8.90}$$

where $A_{tot} = NA_e$ is the total collecting area. The sensitivity of a multi-element interferometer therefore approaches that of a single dish with the same total collecting area.

8.9 CASE STUDIES

8.9.1 THE 4-MM RECEIVER ON THE GREEN BANK TELESCOPE

The US National Radio Astronomy Observatory (NRAO) Green Bank Observatory, located in Green Bank, West Virginia, operates a number of radio telescopes including the 100-m diameter Robert C. Byrd Green Bank Telescope (GBT), shown in Figure 8.23. The telescope has an unblocked aperture with an 8-m diameter secondary mirror. A suite of receivers, mounted at the prime focus and the Gregorian focus, operate at frequencies between 0.1 and 130 GHz (wavelengths of 3 m to 2.3 mm). The main dish of the telescope comprises over 2000 independently adjustable panels. In operation, the surface is actively adjusted to compensate for thermally induced and elevation-dependent gravitational deformations in order to maintain an accurate parabolic surface to within 250 μm rms. Its huge collecting area, high surface accuracy, and sensitive receivers make the GBT one of the world's most powerful radio astronomical observatories.

One of its instruments is a 4-mm (68–92 GHz) single-sideband receiver, commissioned in 2011 and designed primarily for sensitive spectral line observations of molecules that trace dense gas in star-forming clouds in our own and nearby galaxies (GBT 4-mm Receiver Project Book 2017). For maximum observing efficiency, the receiver has two beams on the sky, separated by 286″ in azimuth so that the source in one beam and a nearby region of background sky can be observed simultaneously. The telescope secondary mirror can be used to switch the source alternately between the two beams. The two beams are provided by a pair of corrugated feedhorns, which illuminate the

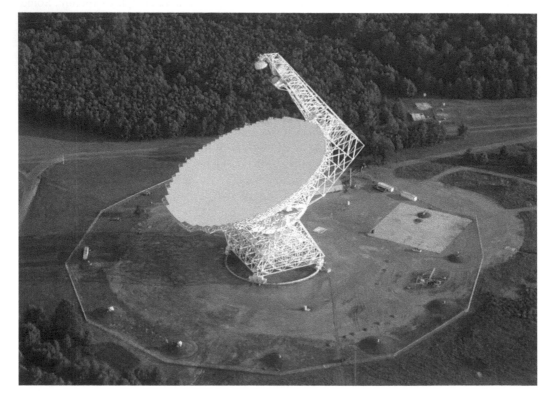

FIGURE 8.23 The Robert C. Byrd Green Bank Telescope (GBT). (Credit: NRAO/AUI.)

telescope primary with an edge taper of approximately 14 dB, giving a diffraction-limited FWHM beamwidth (equation 8.14) varying between 10.7″ at 68 GHz and 7.9″ at 92 GHz.

The signal in each beam is detected in both polarisations – behind each feedhorn is an OMT which separates two orthogonal polarisations. Each OMT output is first amplified by a low-noise HEMT amplifier and then fed to a Schottky diode mixer. The LO signal is provided by a 16–17 GHz LO system (common to a number of other GBT receivers), which is multiplied up by a factor of 4. For optimum performance of the receiver front end, the feedhorns, HEMT amplifiers, and mixers are located in a cryostat and cooled to around 15 K by a closed cycle mechanical cooler. The cryostat vacuum windows, located just in front of the feedhorns, are made from anti-reflection-coated quartz.

A rotatable six-position calibration wheel (Figure 8.24) is located in front of the cryostat to allow the beams to view the sky or hot and cold calibration loads. One of the wheel positions contains a ¼ wave plate which converts linear to circular polarisation (a feature employed when the receiver is used for very long baseline interferometry (VLBI) observations).

The overall receiver band is divided into four sub-bands: 67–74, 73–80, 79–86, and 85–93 GHz, one of which can be observed at any given time. The IF bandwidth is 4 GHz for 73–93 GHz and up to 6 GHz for 67–74 GHz. The IF signals are fed to the GBT's digital FFT spectrometer system for processing.

Typical system temperatures for the GBT 4-mm receiver are under 100 K over much of the band, as shown in Figure 8.25, which also illustrates how the telescope main beam and aperture efficiencies vary across the band (Frayer et al. 2015a). Figure 8.26 shows a portion of a spectral survey of the Kleinmann–Low nebula, an actively star-forming region in the Orion molecular cloud (Frayer et al. 2015b).

8.9.2 The GREAT Spectrometer On Board the SOFIA Airborne Observatory

The Stratospheric Observatory for Infrared Astronomy, SOFIA, is an airborne observatory operated by NASA and the German space agency, DLR. A modified Boeing 747 aircraft carries a 2.7-m Cassegrain telescope to an altitude of approximately 14 km, where atmospheric absorption (mainly due to water vapour) in the infrared – submillimetre range is much less than that even at the best ground-based sites. The telescope views the sky through an open door on the side of the aircraft fuselage, as shown in Figure 8.27.

SOFIA carries a suite of instruments designed to take advantage of the high atmospheric transparency to carry out imaging and spectroscopic observations from optical wavelengths up to 240 μm. One of these is GREAT, the German REceiver for Astronomy at Terahertz frequencies

Position	Beam 1	Beam 2
0	Sky	Sky
1	Cold	Warm
2	¼ Wave Plate	Sky
3	Sky	Sky
4	Warm	Cold
5	Sky	¼ Wave Plate

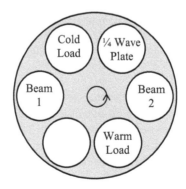

FIGURE 8.24 Positions of the 4-mm receiver calibration wheel. The beams are opposite each other on the wheel. For instance, in the observing position, both beams are on the sky, in the Cold-1 position, Beam-1 views the cold load and Beam-2 sees the ambient load, and in the Cold-2 position, Beam-2 views the cold load and Beam-1 views the ambient load. In positions 2 or 5, the 1/4-wave plate can be placed in one of the beams. (Adapted from GBT Observers' Guide for the 4-mm receiver, 2014.)

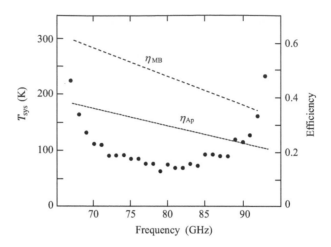

FIGURE 8.25 System temperature (black dots; left-hand scale) and telescope aperture efficiency (dotted lines; right-hand scale) for the GBT 4-mm receiver (Frayer et al. 2015a).

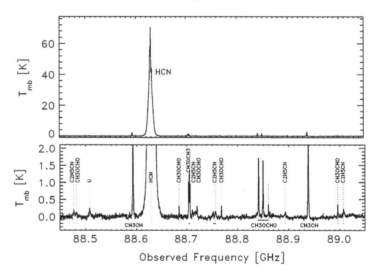

FIGURE 8.26 Portion of a 67–94-GHz spectral survey of Orion-KL made with the GBT 4-mm receiver, from Frayer et al. (2015b). The brightest feature is due to the 1-0 rotational transition of the HCN molecule. The two panels show the same data with the temperature scale on the bottom panel expanded to show up the weaker spectral features due to various organic molecules. A line of unknown origin is labelled U. (From Frayer et al. (2015b). © AAS. Reproduced with permission.)

(Heyminck et al. 2012; Risacher et al. 2018), also shown in Figure 8.26. GREAT has heterodyne spectrometers which can be used for very high-resolution ($v/\Delta v$ up to 10^8) spectral line observations at frequencies between 0.49 and 4.75 THz, which are not accessible from the ground. Sensitive heterodyne observations at such high frequencies rely on recent technical advances in mixer and LO technology, many of which have been associated with the development of the HIFI heterodyne instrument for the *Herschel* Space Observatory (de Graauw et al. 2010).

GREAT has three separate cryostats, one housing four receivers (4GREAT) covering bands between 0.49 and 2.7 THz, with their beams co-aligned on the sky. The other two cryostats house recently installed receivers (known as upGREAT) each with multiple feedhorn-coupled mixers for increased observing speed when mapping extended sources. The low-frequency array (LFA) accepts orthogonal polarisations with two 7-pixel arrays (central circular feedhorn aperture surrounded by

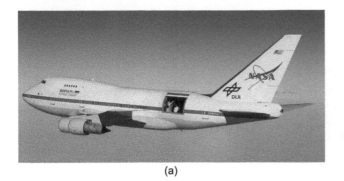

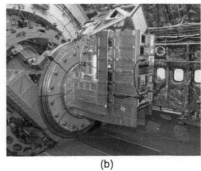

(a) (b)

FIGURE 8.27 (a) The SOFIA airborne observatory. (Credit: NASA/Jim Ross.)(b) GREAT instrument mounted on the SOFIA telescope. (Credit: NASA/Tom Tschida.)

six neighbours) co-aligned on the sky. It operates between 1.83 and 2.07 THz, accommodating important astrophysical transitions from ionised carbon and atomic oxygen. The high-frequency array (HFA) is a single-polarisation seven-pixel array operating at 4.75 THz (63 μm), specifically targeted at measuring line emission from atomic oxygen.

The GREAT mixers are HEBs made from thin NbTiN superconducting films and operating at a temperature close to 4 K. The HEBs are mounted in waveguides and coupled to the telescope through corrugated feedhorn antennas. The mixers are followed by cold low-noise cryogenic amplifiers to boost the signals before they are conveyed to the warm processing electronics. The LO sources for 4GREAT and the LFA are multiplier chains based on Schottky diodes, driven by 10–20 GHz frequency synthesisers, and for the HFA, a quantum cascade laser is used.

The backend for each mixer is a high-resolution FFT spectrometer with 4 GHz bandwidth and up to 64,000 frequency channels, providing a frequency resolution of 44 kHz.

GREAT has angular resolution limited by diffraction at the telescope primary. The main beam efficiency is in the range 0.5–0.6 depending on frequency and is limited mainly by the central blockage. The GREAT bands and the main chemical species probed are summarised in Table 8.1 (from Heyminck et al. 2012 and Risacher et al. 2018).

Calibration of GREAT observations is based on regular observations of hot and cold loads. As with many major observatories, there is an on-line calculation tool to allow observers to predict sensitivities and integration times for the various SOFIA instruments, including GREAT: https://dcs.arc.nasa.gov/proposalDevelopment/SITE/index.jsp.

TABLE 8.1

GREAT Channels

Band	Frequency Range (THz)	DSB T_{rec} (K)	Beam FWHM (″)	Main Species Probed
4GREAT	0.49–0.64	<150	50	NH$_3$, C, CO, CH
	0.89–1.09	>600	25	CO, CS
	0.99–1.09	300		
	1.24–1.40	1100	19	N, CO, OD, SH, H2D$^+$, HCN
	1.43–1.53			
	2.49–2.59	3300	12	OH
upGREAT LFA	1.84–2.07	1000	15	O, C$^+$, CO, OH
upGREAT HFA	4.75	1250	6	O

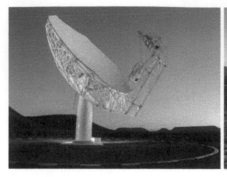 

FIGURE 8.28   A MeerKat antenna (receptor unit) and an aerial photograph of a number of MeerKAT antennas during the construction of the array. (Credit: SKA/SA; Angus Flowers.)

8.9.3   THE MEERKAT RADIO INTERFEROMETER

MeerKAT (Booth & Jonas 2012; Booth et al. 2012a) is an aperture synthesis interferometer facility located in South Africa, covering frequencies between 1 and 14.5 GHz. MeerKAT has 64 interlinked antennas (termed "receptors"). Each receptor comprises a 13.5-m offset Gregorian antenna with receivers and associated amplifiers and digitisation system. The offset Gregorian design has minimal blockage and low spillover, and is suited to good rejection of radio-frequency interference (important as communication systems operate at similar frequencies). Baseline lengths range between 30 m and 8 km, with 48 of the dishes concentrated in a central region of 1 km diameter. The antennas are made up of 40 aluminium panels aligned to give an overall surface accuracy of 0.6 mm rms (equivalent to $\lambda/40$ at the highest frequency of 14.5 GHz). Figure 8.28 shows a receptor antenna and an aerial view of the array during its construction.

Each MeerKAT receiver has a single feedhorn coupled to an OMT, for sensitivity to both polarisations, with the OMT outputs followed by low-noise amplifiers. For maximum sensitivity, the amplifiers are cooled to ~20 K by closed cycle coolers. Separate receivers cover three bands: UHF (0.58–1.02 GHz), L (0.9–1.67 GHz); and X (8–14.5 GHz). The receivers on the different receptors are all synchronised to a single clock, and the digitised signals are sent over fibre-optic links to a single correlator station for processing. An advanced feature of the MeerKAT receivers is the use of very high-speed electronics to eliminate the need for an IF stage by direct sampling and digitisation of the RF signal immediately after initial amplification.

The MeerKAT receptors are optimised for maximum sensitivity, which means minimising the product of antenna effective area and system temperature $\left(A_e T_{sys}\right)$. The specification is 200 m^2 K^{-1} per receptor, and this has already been exceeded in L-band (with 300–400 m^2 K^{-1} achieved).

MeerKAT is one of several scientific and technical precursors to the Square Kilometre Array (SKA; https://skatelescope.org/), an ambitious and powerful radio interferometer currently in development, and MeerKAT itself will eventually be incorporated into the SKA.

9 Far-Infrared to Millimetre Wavelength Instrumentation

9.1 INTRODUCTION

The far-infrared-millimetre wave part of the spectrum covers wavelengths from about 30 μm to about 3 mm. In the longer-wavelength (lower frequency) end of this range, radio receivers, as described in Chapter 8, are often used. However, as frequency increases, radio detection becomes more and more challenging for a number of reasons. With the wavelength becoming smaller in relation to component size, the behaviour and performance of circuits and components are influenced by phase changes within the system and become more sensitive to the detailed design. Stray or undesired capacitances, impossible to avoid completely, also have the effect of shorting signals to ground. Finally, the quantum limit on noise temperature means that even the theoretically best sensitivity inevitably degrades as frequency increases. For these reasons, as wavelength decreases, it becomes more attractive to abandon coherent detection – the simultaneous measurement of amplitude and phase – and to use an incoherent or "direct" detection technique in which the detector responds to the incident electromagnetic power, with no phase information retained. It also becomes more natural to consider such detectors as responding to photons rather than a travelling electromagnetic wave. The cross-over region between the suitability of coherent and incoherent detection is typically in the submillimetre region but depends strongly on the exact application and requirements (especially sensitivity and spectral resolution), and there are circumstances in which either technique can be used.

Both ground-based and space-borne observatories are used in the far-infrared to millimetre range. Current major ground-based telescopes operating at submillimetre and millimetre wavelengths include the 12-m diameter Atacama Pathfinder Experiment (APEX) at altitude 5105 m in Llano de Chajnantor, Chile, the 15-m James Clerk Maxwell Telescope (JCMT) at altitude 4092 m on Maunakea in Hawai'i, the 30-m Institut de Radioastronomie Millimétrique (IRAM) telescope at altitude 2850 m on Pico Veleta in Spain, and the 50-m Large Millimeter telescope (LMT) at 4850 m altitude on Sierra Negra in Mexico. For a ground-based observatory, there is an unavoidable thermal background from the atmosphere and the telescope, which can be no colder than the ambient temperature. Space-borne telescopes must be much smaller (the largest yet launched is the *Herschel* Space Observatory which had a 3.5-m primary aperture) but have the advantage of access to parts of the spectrum for which the atmosphere is opaque or strongly attenuating. Passive and/or active cooling can also be used to minimise the telescope thermal background. The *Herschel* telescope was cooled to approximately 80 K by passive radiation, and *Spitzer* and other far-infrared observatories with smaller (less than 1 m in diameter) telescopes have had apertures cooled to liquid helium temperatures (a few K).

In this chapter, we describe the basic features of various kinds of direct detectors used in this part of the spectrum – bolometers, kinetic inductance detectors, and photoconductors – and consider the pros and cons of different ways of coupling the telescope beam to the detectors.

9.2 DIRECT DETECTION INSTRUMENTS

The essential features of a direct detection instrument are shown in Figure 9.1. The telescope and optical system collect the radiation to be detected and bring it to the detector(s).

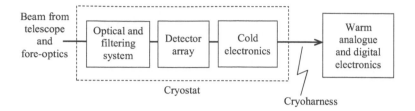

FIGURE 9.1 Essential features of a direct detection instrument.

For both ground and space-borne applications, the instrument optics must be cooled under vacuum, usually to helium temperature (1.7–4 K), and the detectors may require sub-kelvin cooling. A filtering system is needed to select the wavelength range incident on the detector and to protect the colder parts of the instrument from excessive radiant power, which could result in temperatures becoming too high or cryogens boiling off too quickly. If very narrow-band detection (spectroscopy) is required, a spectrometer such as a grating or a Fabry-Perot may be incorporated, as described in Chapter 7. If polarimetric measurements are to be made, optical components such as polarisers or waveplates may be used to control or modulate the polarisation.

Infrared and submillimetre detectors are operated at ultra-low temperatures inside cryogenic systems. The detector signal levels are usually low and vulnerable to interference or microphonic disturbances or to losses in long cables to the warm electronics. For instance, if the detector has a high resistance, then the signal can be attenuated by the RC filtering effect introduced by the cable capacitance. For these reasons, the front-end readout electronics are normally located as physically close as possible to the detectors to amplify the detector signals and/or to transform the signals to a lower source impedance. It is important that the readout electronics do not introduce significant additional noise and that they do not disturb the cryogenic environment of the detectors (for example by dissipating too much electrical power).

A cryoharness connects the detectors and cold readout electronics to the warm analogue and digital electronics, which provide appropriate bias and control functions to the detectors and further process and record the detector signals. For arrays with many detectors, a cold multiplexer near the focal plane may be used to reduce the total number of wires in the cryoharness. This is an important consideration because the thermal load from wiring connections on the ultra-low-temperature cryogenic stage on which the detectors are located can pose significant design challenges.

9.3 BOLOMETRIC DETECTORS

Bolometric detectors are based on sensing the heating effect of the incoming radiation and are widely used at wavelengths above around $30\,\mu m$. A bolometer converts electromagnetic power to a change in temperature. The principle was first demonstrated at the end of the 18th century by William Herschel, who used a prism to produce a spectrum of the Sun in order to investigate the heating effects of different colours of visible light. As illustrated in the artist's impression in Figure 9.2, he placed the bulb of a mercury thermometer in different colours of the spectrum and observed the greatest rise in temperature when the bulb was placed in the red light. However, he observed an even higher temperature increase when the thermometer was positioned beyond the red end. In so doing, he discovered infrared radiation.

He also employed an important general technique in sensitive measurement, which is still widely used in many detector systems today – the use of reference detectors (additional thermometers not exposed to the solar spectrum) to monitor fluctuations in the ambient temperature so that they could be subtracted if necessary.

The essential features of a modern bolometric detector are shown schematically in Figure 9.3. The electromagnetic power is captured in an absorber. Typically, this is a thin resistive metal film

FIGURE 9.2   Discovery of infrared radiation by William Herschel. (Credit: Universal History Archive/UIG/ Shutterstock.)

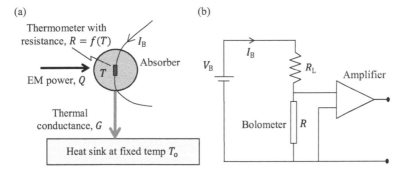

FIGURE 9.3   Essential features of a bolometric detector (a) and typical readout circuit (b).

coated on some suitable supporting electrically insulating and thermally conducting material. The incoming radiant power is converted to kinetic energy of free electrons in the absorber, and collisions between the electrons and atoms in the lattice result in the energy being converted to lattice vibrations or heat. This "thermalisation" of the incident power occurs on a very short time-scale (typically < 0.1 ms; Turner et al. 2001). Heat is allowed to flow from the absorber to a heat sink at a fixed temperature T_o by a thermal conductance, G (the higher G, the more rapidly the heat leaks away). A thermometer is attached to the absorber, to measure its temperature. The thermometer resistance is strongly temperature-dependent. A bias current, I_B, is passed through the thermometer, and the corresponding voltage, V, is measured. The bias current dissipates electrical power

$P = I_{\mathrm{B}}^2 R$, which heats the bolometer to a temperature, T_1, slightly higher than T_0. If no radiant power is incident, the absorber will be at this temperature T_1. If a steady radiant power, Q, is incident, the absorber is heated up to a new equilibrium temperature $T > T_1$. The value of T depends on the amount of power absorbed and the thermal conductance between the absorber and the heat sink.

The principles of a typical bolometer bias and pre-amplifier circuit are also shown in Figure 9.3. A constant voltage source in series with a fixed load resistor R_{L} provides the bias current. In practice, the bias point can be set by either a DC voltage source, as shown here, or an AC source with frequency much higher than the 3-dB frequency of the bolometer responsivity. AC biasing has the advantage that it eliminates the $1/f$ noise from all electronic components following the bolometer by increasing the signal frequency.

A practical bolometer design is illustrated schematically in Figure 9.4. In the feedhorn-coupled design, the radiation from the telescope is coupled to the bolometer by a feedhorn antenna. The bolometer itself is mounted in an integrating cavity to ensure maximum absorption. The radiant power dissipated in the absorber raises the temperature of the whole absorber/thermometer combination. The alternative absorber-coupled design involves dispensing with feedhorns and having an array of square bolometers, with each one acting as an absorbing pixel. A quarter-wavelength backshort (a reflecting plane $\lambda/4$ behind the detector array) maximises the absorption efficiency of the pixels. A wave that passes through the absorber without absorption is reflected with a 180° phase change and returns to the absorber in phase, providing another opportunity for absorption.

Each configuration has its advantages and disadvantages, which are discussed in more detail in Section 9.8.

The electrical wires often provide a controlled heat conduction path (and so determine the thermal conductance G) to the heat sink (often a cryogenic bath such as a vessel of liquid helium). The thermal conductance of the electrical wires is set by choice of the metal and the dimensions (length and cross-sectional area). At low temperatures, the thermal conductivity of metal, k, is directly proportional to temperature. The thermal conductance of a metal wire of cross-sectional area A and length L is given by

$$G = \frac{kA}{L}. \tag{9.1}$$

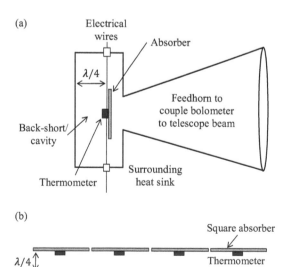

FIGURE 9.4   Schematic design of feedhorn-coupled (a) and absorber-coupled (b) bolometers.

There are two commonly used types of thermometer: semiconductor thermometers are small crystals of doped germanium or silicon whose resistance is very sensitive to temperature, and transition edge superconducting (TES) thermometers are thin metal films which are on the sharp transition between the normal and superconducting states, resulting in a very strong dependence of resistance on temperature. Both types are described in more detail in the following sections. A figure of merit for the thermometer is the temperature coefficient of resistance, α, defined as

$$\alpha = \frac{\mathrm{d}\log(R)}{\mathrm{d}\log(T)} = \frac{T}{R}\frac{\mathrm{d}R}{\mathrm{d}T}. \tag{9.2}$$

A high value of α means a very high sensitivity of the resistance to temperature.

The equilibrium condition of the bolometer is determined by the balance between the total power dissipated and the power flow to the heat sink due to the temperature difference between it and the absorber. The total dissipated power, W, is the sum of the electrical power, P, from the bias current, and the radiant power absorbed, Q. The power flowing to the heat sink is equal to the thermal conductance multiplied by the temperature difference. The heat balance equation is therefore

$$W = P + Q = G(T - T_\mathrm{o}) = G\Delta T. \tag{9.3}$$

9.3.1 BOLOMETER RESPONSIVITY

To find the responsivity, we need to calculate the change in the output, ΔV, corresponding to a small change in the absorbed radiant power, ΔQ. Let Q undergo a step change from Q to $Q + \Delta Q$. This leads to corresponding changes in T, R, and P. These changes will not occur instantaneously, as we shall see below, so let the final value of ΔT be denoted ΔT_f, etc. Applying the heat balance equation to the situation after the step change in Q gives:

$$(P + \Delta P_\mathrm{f}) + (Q + \Delta Q) = G(T + \Delta T_\mathrm{f} - T_\mathrm{o}). \tag{9.4}$$

Therefore,

$$\Delta P_\mathrm{f} + \Delta Q = G\Delta T_\mathrm{f}. \tag{9.5}$$

For the change in temperature, ΔT_f, the change in resistance is

$$\Delta R_\mathrm{f} = \alpha R\Delta T_\mathrm{f}. \tag{9.6}$$

Therefore,

$$\Delta P_\mathrm{f} = I_\mathrm{B}^2\Delta R_\mathrm{f} = I_\mathrm{B}^2\alpha R\Delta T_\mathrm{f} = \alpha P\Delta T_\mathrm{f}, \tag{9.7}$$

giving

$$\alpha P\Delta T_\mathrm{f} + \Delta Q = G\Delta T_\mathrm{f}. \tag{9.8}$$

The amount by which the temperature rises is therefore

$$\Delta T_\mathrm{f} = \frac{\Delta Q}{G - \alpha P}. \tag{9.9}$$

Now that we know ΔT_f, we can find ΔR_f, and hence find the change in the output voltage, ΔV_f:

$$\Delta V_f = I_B \Delta R_f = I_B \alpha R \Delta T_f. \tag{9.10}$$

Therefore,

$$\Delta V_f = \frac{I_B \alpha R \Delta Q}{G - \alpha P}. \tag{9.11}$$

The responsivity is

$$S = \frac{\Delta V_f}{\Delta Q} = \frac{\alpha V}{G - \alpha P}. \tag{9.12}$$

A high responsivity requires large α and/or small G, producing a large temperature change for a given change in absorbed power, resulting in a large change in bolometer resistance.

9.3.2 BOLOMETER TIME CONSTANT

The change in bolometer temperature, in response to a change in radiant power, does not occur instantaneously because the bolometer has some heat capacity and a finite amount of thermal energy (power × time) must flow for the temperature to change by ΔT_f.

Letting C be the heat capacity of the bolometer (i.e. the energy required for a unit increase in temperature; SI units J K^{-1}), and recalling that G is the thermal conductance to the heat sink (SI units W K^{-1}). The amount of heat that must flow (either into the bolometer or away to the heat sink) is $C\Delta T_f$. The rate at which heat leaks away to the heat sink is proportional to G. In terms of its thermal response, the bolometer is a first-order system, analogous to an RC circuit. The heat capacity is the equivalent of the electrical capacitance (both symbolised by C), and the inverse of the thermal conductance is the equivalent of the resistance. The time constant is RC for the RC circuit and for the bolometer

$$\tau \approx \frac{C}{G}. \tag{9.13}$$

Therefore, if G is small (making it difficult for heat to flow), or if C is large (meaning that a lot of heat needs to flow), then the device tends to be slow to respond. The transient response to a step change in absorbed power, and the roll-off of the responsivity with increasing modulation frequency are as described in Chapter 5 (the equivalents of equations 5.5 and 5.10):

$$\Delta V(t) = \Delta V_f e^{-t/\tau} \tag{9.14}$$

and

$$S(\omega) = \frac{S(0)}{\left[1 + (\omega\tau)^2\right]^{1/2}}. \tag{9.15}$$

Because high responsivity requires G to be small (equation 9.12), a compromise is necessary between high responsivity and good speed of response (which requires high G). A bolometer must be optimised in this respect for the particular application. Time constant requirements for practical bolometer detectors are typically in the range of ~ 1–100 ms.

9.3.3 BOLOMETER NOISE AND NEP

Three main noise mechanisms occur in bolometers: (i) photon noise from the random arrival rate of photons; (ii) Johnson noise from the thermometer resistance; and (iii) phonon noise from the quantised heat transport along the link between the bolometer and its heat sink.

The photon noise NEP contribution (for the case of two detectors for simultaneous measurement of source + background power, P_B, and background alone) is given by equation (6.29):

$$\text{NEP}_{\text{ph}} = 2\left[P_B h\nu \left(\frac{1}{\eta_d} + \varepsilon\eta_o b \right) \right]^{1/2}. \tag{9.16}$$

The Johnson noise component is equal to the Johnson noise voltage spectral density (equation 6.64) divided by the responsivity:

$$\text{NEP}_J = \frac{(4k_B TR)^{1/2}}{S}. \tag{9.17}$$

The phonon noise component arises from the flow of energy between the bolometer and the heat sink, which is quantised in the form of phonons – packets of lattice vibration energy – that flow like particles to the heat sink. The flow rate of phonons has a statistical fluctuation about its mean value that results in fluctuations in the transport of power to the heat sink, and so there are corresponding detector temperature fluctuations. The contribution of this phonon noise to the overall NEP is given by (Mather 1982)

$$\text{NEP}_{\text{phonon}} = \left(4k_B T_o^2 G \right)^{1/2}. \tag{9.18}$$

The combination of the Johnson and phonon noise terms constitutes the inherent bolometer NEP. The three NEP contributions are uncorrelated and so combine in quadrature to give the overall NEP:

$$\text{NEP}_{\text{tot}} = 2\left(P_B h\nu \left(\frac{1}{\eta_d} + \varepsilon\eta_o b \right) + \frac{4k_B TR}{S^2} + 4k_B T_o^2 G \right)^{1/2}. \tag{9.19}$$

The phonon and photon noise contributions are both white (independent of modulation frequency) but because the responsivity decreases with frequency (equation 9.15), the Johnson noise contribution increases with modulation frequency.

If the bolometer NEP is close to the photon noise NEP, then there is little to be gained by improving it any further. In a well-optimised device, operated in a suitably low signal frequency range, the responsivity is made high enough that the phonon noise component dominates over the Johnson noise, so neglecting the latter provides a reasonable approximation to the inherent bolometer NEP. To achieve a low NEP, we therefore need to make T_o and G small. However, as we discussed in the previous section, reducing G too much could result in the bolometer becoming too slow, so there is a compromise between sensitivity and speed of response.

9.3.4 DEPENDENCE OF ACHIEVABLE BOLOMETER NEP ON OPERATING TEMPERATURE

If a particular experiment requires some maximum value of the time constant, τ_{max}, then we can estimate the dependence of the best achievable NEP on the operating temperature of the bolometer. Achieving the required time constant means making the thermal conductance greater than some minimum value, G_{min}, so that

$$\tau_{max} = \frac{C}{G_{min}}. \tag{9.20}$$

For a given τ_{max}, $G_{min} \propto C$. The heat capacity depends on the materials from which the bolometer is made – generally a combination of dielectric and metallic substances. Dielectric heat capacity reduces with temperature according to $C \propto T^3$ (Debye law), whereas the heat capacity of the metallic components reduces according to $C \propto T$ (being dominated by the electron gas). If we assume that the overall heat capacity is dominated by dielectric components, then the dependence of the phonon noise NEP on $T_o G^{1/2}$ translates to

$$\text{NEP} \propto T^{5/2}. \tag{9.21}$$

This rule of thumb shows that the achievable NEP thus improves very significantly for low operating temperatures, and that is the basic reason why bolometers are cooled using cryogenic systems. Cooling with liquid ^{4}He achieves a temperature of 4.2 K at atmospheric pressure, or as low as ~1.5 K if the pressure over the liquid is reduced by a vacuum pump or if the helium vessel is in the vacuum of space. Using ^{3}He as the cryogen allows lower temperatures to be achieved: 0.25–0.3 K. Still, lower temperatures (0.1 K or less) can be attained using adiabatic demagnetisation refrigerators (ADRs) or ^{3}He-^{4}He dilution refrigerators. The development of compact and reliable ultra-cold refrigeration systems means that the bolometric detectors in modern ground-based and space-borne instruments are usually operated at sub-kelvin temperatures.

9.3.5 SEMICONDUCTOR BOLOMETERS

Although semiconductor bolometers have largely been superseded by superconducting detectors, (see the following sections) we describe them here as they were widely used until around 2010, and formed the basis of many successful and scientifically influential instruments.

Semiconductor bolometers use doped germanium or silicon crystals as the thermometric element. Most germanium bolometers use neutron transmutation doped (NTD) Ge thermometers. NTD germanium is p-type with Ga atoms acting as the acceptors. It is made from a highly pure single crystal in which the doping is effected not by adding Ga impurity atoms to the material but by irradiating the crystal with neutrons causing some Ge atoms to transmute to Ga in situ, with the final doping concentration determined by the total neutron dose. The main advantages of this doping technique are that there are no crystal lattice imperfections associated with the introduction of additional atoms, and that the doping level is uniform throughout the material (because neutrons are so penetrating). After irradiation, the crystal is annealed to cure radiation damage, and the radioactivity associated with the transmutation process (half-life ~ 11 days) must be allowed to decline. NTD germanium bolometers provide a very close approximation to ideal bolometers in two key respects: the resistance depends primarily on temperature, with negligible dependence on the applied electric field, and the material also exhibits very low levels of excess noise over and above the theoretical Johnson noise level, even down to very low frequencies. Because the detector noise levels are low, NTD Ge bolometers do not need to have a high responsivity to approach the photon noise limit, and the extremely low $1/f$ noise allows operation at very low signal frequencies.

Conventionally doped silicon can also be used as the thermometer element, but produces material that does have non-ideal characteristics including electric-field effects and higher noise. This requires larger responsivity to ensure close to photon noise-limited operation.

Typically, the resistance of a semiconductor bolometer increases exponentially with decreasing temperature, as

$$R(T) = R^* e^{(T_g / T)^{1/2}}, \tag{9.22}$$

where R^* is a constant that depends on the geometry of the individual thermometer and T_g characterises the resistivity-temperature dependence of the material.

The temperature coefficient of resistance is easily shown to be

$$\alpha(T) = -\frac{T_g^{1/2}}{2T^{3/2}}. \tag{9.23}$$

A typical value of α for semiconductor bolometers is 5–10.

A typical $R - T$ curve for a doped germanium bolometer is shown in Figure 9.5. Note that the resistance increases as the temperature is decreased. The form of the voltage–current characteristic, or load curve, is also shown in Figure 9.5 for a typical bolometer with both zero and a finite value of incident background power. At low bias current (i.e. low electrical power), the bolometer has a resistance that decreases with the absorbed radiant power. As the electrical bias current is increased, the resistance decreases due to the heating effect of the bias power. The voltage reaches a peak and then decreases as the bias current is further increased. The two curves converge for high bias current as the electrical power dissipation dominates over the absorbed power.

Semiconductor bolometers are usually biased at near-constant current by having a fixed voltage source in series with a load resistance, R_L, much greater than that of the bolometer. The dotted straight line in Figure 9.5b represents a typical load line giving the allowed values of bolometer voltage and current, as dictated by the value of the bias voltage, V_B, and the load resistance:

$$V = V_B - I_B R_L. \tag{9.24}$$

The operating point is the intersection of the load curve and this load line. Maximum responsivity and minimum NEP are achieved by choosing an operating point just before the voltage peak.

The detailed theory of operation and optimisation of semiconductor bolometers is described in detail by Mather (1982), Grannan et al. (1997), and Sudiwala et al. (2002). Sensitive semiconductor bolometer technology is exemplified by the NTD Ge spider-web bolometer (Turner et al. 2001). The absorber is a silicon nitride "spider-web" structure, which is coated with a thin metal film with surface impedance matched to that of free space for maximum absorption. The thermometer is a small NTD germanium crystal attached to the absorber. The spacing of the spider-web pattern is much smaller than the wavelength being detected, so it acts as a plane surface to the incident radiation. Besides reducing the overall heat capacity, the web structure also minimises the geometrical cross section for ionising radiation, which is important for operation in space. NTD bolometers operating at temperatures down to around 0.1 K approach quite closely the performance of an ideal detector. Spider-web bolometers have achieved NEP $= 1.5 \times 10^{-17}$ W Hz$^{-1/2}$ with time constant < 100 ms at 300 mK, and 1.5×10^{-18} W Hz$^{-1/2}$ with $\tau = 65$ ms at 100 mK.

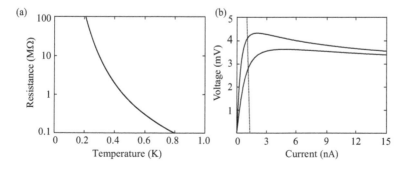

FIGURE 9.5 (a) Typical resistance-temperature dependence for a semiconductor bolometer. (b) Typical voltage–current characteristics for a semiconductor bolometer with both zero (top curve) and finite (bottom curve) radiant power loading.

9.3.6 SEMICONDUCTOR BOLOMETER READOUT ELECTRONICS

A typical bolometer operating resistance is a few MΩ. The combination of Johnson and phonon noise from such a device at a typical operating temperature of 0.3 K is around 20 nV Hz$^{-1/2}$, and the readout amplifier must have significantly lower noise than this if sensitivity is not to be degraded. Silicon junction field effect transistors (JFETs) are the best option. They achieve suitably low noise levels when operating the detector in AC-biased mode at modulation frequencies of ~100 Hz. The optimum operating temperature for the JFETs is about 100 K (at lower temperatures, the carriers in silicon start to freeze out, and at higher temperatures, the noise is too high). The need to have transistors operating at 100 K in close proximity to detectors at sub-kelvin temperatures is a major inconvenience. A typical JFET readout circuit is shown in Figure 9.6 and uses a matched pair of JFETs and a balanced arrangement with the load resistance spilt in two equal resistors. Another inconvenience of JFET readout is that, given the low noise levels, multiplexing the detectors before transmission of the signals to the warm electronics is not possible because multiplexers tend to be noisy and so need to be fed with signals for which the absolute noise levels are significantly higher. This means a high number (typically ~6) of connections from the warm electronics to the cold focal plane per detector, which introduces additional complications. Solutions to these problems can be implemented provided the overall number of detectors is not too large, and JFET-based readouts are regarded as worth the effort given the excellent performance achieved, giving overall noise dominated by the detector.

9.3.7 TES BOLOMETERS

An alternative to the semiconductor thermometer is the superconducting transition edge sensor, TES, described in detail by Irwin and Hilton (2005). TES devices are now generally preferred over semiconductor bolometers as they can achieve better sensitivity (lower NEP) and are easier to multiplex when implemented as arrays. A superconducting material exhibits a sharp transition between the normal (resistive) and superconducting (zero resistance) states at a well-defined critical temperature, T_C. Therefore, there is a very strong dependence of the resistance on temperature in the transition region between the two states (Figure 9.7). The TES is a thin superconducting film, biased to operate in this normal-superconducting transition region. It is usually implemented as a bi-layer comprising thin normal metal and superconducting metal layers. The superconducting transition temperature is suppressed by the proximity effect, whereby the superconducting gap is reduced near the interface with the normal metal and can be controlled by adjusting the thicknesses of the normal and superconducting films.

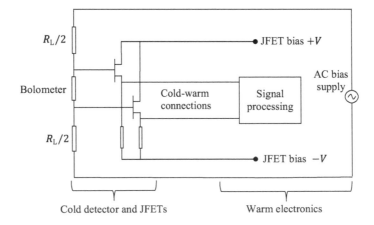

FIGURE 9.6   Typical semiconductor bolometer readout electronics.

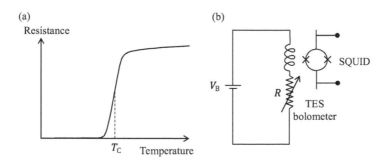

FIGURE 9.7 (a) Variation of resistance with temperature for a superconducting material as it goes through the transition between the normal to superconducting states; (b) essential features of a TES biasing and read-out circuit incorporating inductive coupling to a SQUID.

Because TES devices have low impedances (typically a few mΩ), it is more appropriate to use a constant voltage bias rather than the constant current bias used for semiconductor bolometers. Operation on the sharp transition means that there is a fixed value of total power, equal to the power needed to warm the TES from the heat sink temperature to the transition temperature. The sum of the bias power and radiant background power is thus always equal to this value. Changes in the absorbed radiant power due to the signal are compensated by opposite changes in the bias power (i.e. in the current for a fixed voltage). This change in current is read out using a sensitive current amplifier known as a superconducting quantum interference device (SQUID), which is inductively coupled to the TES, as shown in Figure 9.7.

On the normal-superconducting transition, a small change in temperature corresponds to a large change in resistance, giving rise to very high values of α (typically 50–1000, which is much higher than for semiconductor bolometers ($\alpha < \sim 10$). This leads to very high values of responsivity, boosting both the phonon noise and the signal such that the Johnson noise contribution to the NEP becomes negligible. An additional advantage is that the voltage biasing leads to a strong electro-thermal feedback effect, which results in a fast speed of response. In addition, TES detectors are easier to fabricate than semiconductor bolometer arrays because they can be made using standard lithographic techniques.

A typical measured TES resistance vs temperature characteristic is illustrated in Figure 9.8 for a titanium TES detector developed for the ground-based BICEP2 CMB polarisation instrument.

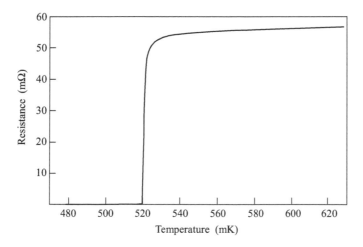

FIGURE 9.8 Resistance vs temperature for a titanium TES bolometer with a transition temperature of 520 mK, developed for the BICEP2 instrument. (From Ade et al. (2014); © AAS, reproduced with permission.)

9.3.8 TES BOLOMETER READOUT ELECTRONICS

Most TES-based systems involve multi-detector arrays and require cold multiplexing to reduce the number of wiring connections to the cold focal plane. SQUID multiplexers have been developed to read out the signal from TES bolometer arrays, based on both time division multiplexing (TDM) and frequency division multiplexing (FDM) schemes (Irwin 2002). In TDM, a number of bolometers are connected in sequence to the same readout line by turning SQUIDs on and off in turn – in effect the bolometer signals are encoded by an on-off modulating function, and the individual TES signals can subsequently be decoded by multiplying the multiplexer output by the same modulation function. In the case of FDM, sinusoidal bias voltages are applied at different frequencies to the individual TES bolometers. This enables a number of bolometer signals to be combined in a single readout line. The largest multiplexing factor (number of detectors read out per line) implemented with TDM multiplexing to date is 64 (Henderson et al. 2016), while 528 has been demonstrated using an FDM multiplexer (Henderson et al. 2018).

Both TDM and FDM multiplexers involve complex readout electronics and software to control the multiplexer and demodulate the signals, but for the multipixel arrays required in modern instruments, this is a much more attractive option than having a very large number of wiring connections to the ultra-cold bolometer stage.

Figure 9.9 shows the architecture of a TDM SQUID multiplexer for an $M \times N$ TES array, developed at the US National Institute of Standards and Technology, NIST, (Irwin 2002). N TES bolometers in a column share the same DC bias line. Each TES is coupled to a first-stage SQUID switch (shaded grey), which is closed unless the address current, I, for its row is on. Boxcar currents I_1, I_2 etc. are applied sequentially. When a current is on, its row of M first-stage SQUIDS is turned on. An address resistor $(R_A \sim 1\ \Omega)$ shunts each first-stage SQUID, and the current through R_A is coupled inductively to a second-stage SQUID shared by N SQUIDs in a column. Each second-stage SQUID

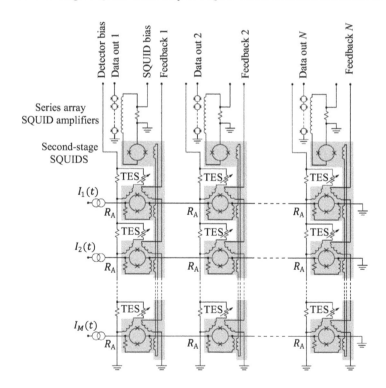

FIGURE 9.9 SQUID time-domain multiplexer. (From Irwin (2002) reproduced with permission from Elsevier.)

feeds a series array SQUID amplifier, the output of which goes to the warm electronics. A feedback signal needs to be applied to the first-stage SQUIDs to linearise them. Only one SQUID per column is on at any time, so one feedback line is common to all N SQUIDs.

9.4 KINETIC INDUCTANCE DETECTORS

A recently developed superconducting detector for FIR-millimetre wavelengths relies on a different physical principle to the bolometric effect used in the TES. Figure 9.10 illustrates the energy levels and density of states of a superconductor. The superconducting ground state can only be occupied by paired electrons of opposite spin – Cooper pairs. There is an energy gap of typically 1 meV around the Fermi level, E_F, between the superconducting ground state and the first allowed free electron energy level. At zero temperature, all of the free electrons have formed Cooper pairs which are in the superconducting ground state with binding energy Δ per electron. At a finite temperature, some Cooper pairs are broken due to thermal excitation, resulting in some single free electrons above the energy gap, which are termed quasiparticles. The number of thermally generated quasiparticles decreases exponentially with temperature and so can be suppressed by cooling the superconductor well below its transition temperature (typically to $0.1T_c$).

Quasiparticles can also be created by the absorption of electromagnetic energy, either due to high-frequency currents in the case of an antenna-coupled detector or due to direct photon absorption in the film (producing many quasiparticles per photon in the case of an optical, UV, or X-ray detector). This change in the populations of Cooper pairs and quasiparticles affects the electrical properties of a superconducting film. A current-carrying superconductor exhibits the phenomenon of kinetic inductance because the flowing Cooper pairs have inertia which the changing electric field must overcome. Pair breaking changes the densities of Cooper pairs and of quasiparticles, leading to changes in both kinetic inductance and resistance. The kinetic inductance is inversely proportional to the number density of Cooper pairs (Meservey and Tedrow 1969) and is therefore changed by the breaking of pairs to form quasiparticles.

This change is most easily detected by making the superconductor part of a resonator (equivalent to an LC resonant circuit), with inductance L and capacitance C, which has a resonant frequency given by

$$\omega = \left(\frac{1}{LC}\right)^{1/2}.$$
(9.25)

An increase in the absorbed EM power increases the kinetic inductance, leading to a decrease in resonant frequency. A corresponding increase in the surface resistance (due to a change in the quasiparticle concentration) also leads to a broadening of the resonance (decrease in Q-factor), as shown in Figure 9.11a. In a KID, the resonator is capacitively coupled to a microwave feed-line, illustrated

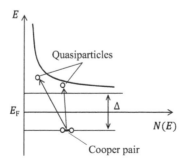

FIGURE 9.10 Energy (E) – density of states, $N(E)$, diagram for a superconductor. When a Cooper pair is broken (e.g., by a photon of energy $\geq 2\Delta$), two free electrons (quasiparticles) are created.

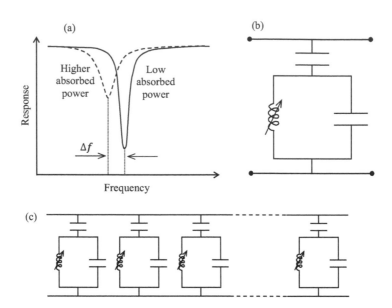

FIGURE 9.11 (a) Modification of the response of a KID resonator by a change in the absorbed power. Breaking of Cooper pairs leads to an increase in kinetic inductance leading to a decrease in the resonant frequency. This is accompanied by an increase in the surface resistance of the film, due to the increased quasiparticle population, leading to a lower Q-factor (a broader and less sharp response). (b) Equivalent circuit of a single KID pixel. (c) Many such pixels can be connected in parallel and read out using one coaxial cable through frequency division multiplexing, with each KID having a different resonant frequency.

in Figure 9.11b, and a signal propagating along the feed-line near the resonant frequency will be affected in amplitude and phase by changes in the kinetic inductance.

Besides being straightforward to manufacture with simple lithographic processes, KID arrays have the advantage that they are well-suited to reading out the signals using FDM. Each KID in a line of pixels (Figure 9.11c) is designed to have a different resonant frequency, and they are all coupled to a single readout line which is excited by a comb of frequencies matched to those of the pixels. The change in kinetic inductance of each pixel only affects the amplitude and phase of its own readout frequency. Up to ~ 1000 pixels can be read out using a single feed-line so that a large-format array can be accommodated with a relatively small number of wires connecting the warm electronics to the cold focal plane.

9.5   CHOICE OF BOLOMETRIC AND KID DETECTORS FOR THE FAR INFRARED AND SUBMILLIMETRE

Until the 2010s, virtually all direct detections systems in the far infrared and submillimetre used semiconductor bolometers. Ground-based examples included cameras on the JCMT, the Caltech Submillimetre Observatory (CSO, the IRAM 30-m, and APEX telescopes. In space the *Planck*-HFI CMB photometer and the *Herschel*-PACS camera and SPIRE camera and spectrometer all used semiconductor bolometers which achieved their performance requirements. However, the drive towards large-format focal planes, with thousands rather than hundreds of pixels, for which semiconductor bolometer technology is not well-suited, and for lower NEPs, meant that new technologies were called for. Superconducting detector systems, initially TES-based and now also KID-based, have provided the ability to implement much larger focal planes and also to achieve the improved sensitivity needed for the next generation of cold-aperture space missions.

TES bolometer arrays have become a well-developed and mature technology for ground-based, balloon-borne and aircraft-borne instruments. Examples include the SCUBA-2 camera (Holland et al. 2013; described in a case study in Section 9.9.3), CMB polarisation experiments BICEP and SPIDER (Ade et al. 2015), and ACTPol (Grace et al. 2014), and the HAWC+ camera and polarimeter on the SOFIA airborne observatory (Harper et al. 2018). These instruments take advantage of the suitability of TES bolometers for large-format arrays, with NEP requirements of a few $\times$ 10^{-17} W Hz$^{-1/2}$. Future space instruments with cooled telescopes will require much lower NEPs to take advantage of the extremely low-background photon noise. TES bolometers have been made with NEPs of a few $\times$ 10^{-19} W Hz$^{-1/2}$ (van der Kuur et al. 2015) and are in principle capable of achieving a few $\times$ 10^{-20} W Hz$^{-1/2}$.

KID-based instruments are currently at a somewhat earlier stage of development, but have also been developed and operated in ground-based instruments, including the MUSTANG2 camera on the Green Bank 100-m telescope (Dicker et al. 2014), and the NIKA2 camera on the IRAM 30-m (Catalano et al. 2018). Instruments are in development for the LMT and APEX telescopes, and for the balloon-borne BLAST-TNG experiment. The feasibility of a KID-based system for future low-background space applications requiring NEP levels of a few $\times$ 10^{-19} W Hz$^{-1/2}$ has also been demonstrated (Baselmans et al. 2017).

9.6 PHOTOCONDUCTIVE DETECTORS

For wavelengths shortward of around 200 µm, photoconductive detectors based on doped semiconductors can be used. The simplest photoconductive detector is just a piece of the material with two ohmic (conducting) contacts. The band structure of intrinsic and extrinsic semiconductors was described in Chapter 3. The intrinsic band gaps of silicon and germanium are 1.1 and 0.67 eV respectively, corresponding to cut-off wavelengths in the near infrared: 1.1 µm for Si and 1.9 µm for Ge. Introducing impurity atoms into the crystal lattice creates additional states for the electrons which can be much shallower (smaller band gap) and therefore allow response to longer wavelengths. The promotion of charge carriers into the conduction band increases the conductivity of the material, so the current flowing in response to an applied voltage is proportional to the photon detection rate.

9.6.1 RESPONSIVITY

Consider an extrinsic photoconductor of thickness (distance between electrical contacts) l. The probability that a single incident photon will liberate an electron or a hole is characterised by η_d, the detector RQE. A liberated charge will drift under the influence of an applied electric field, before it recombines with an opposite charge. Let the mean free path (average distance travelled before recombination) of the liberated charge be l_m. In Figure 9.12, in which $l_m < l$, photon 1 does not contribute to the signal because its liberated electron recombines before reaching the contact, but the electron released by photon 2 makes it to the contact and contributes to the current. On average, photons absorbed in the darker-shaded region produce an output and those absorbed in the

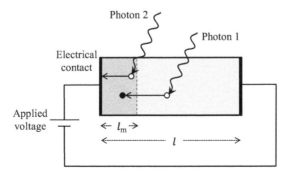

FIGURE 9.12 Photon absorption and contribution to the output signal for a photoconductive detector with photoconductive gain less than unity.

lighter-shaded region do not. The fraction of the incident photons that contribute to the photocurrent is thus

$$g = \frac{l_m}{l},$$

(9.26)

known as the photoconductive gain. It depends on the properties of the material and also on the magnitude of the applied electric field. The photoconductive gain can be less than unity, as illustrated here, or it could be greater than unity if $l_m > l$, in which case when the charge leaves on one side another charge must enter on the other to preserve charge neutrality. This charge may also travel across the detector making another contribution to the signal.

For a stream of n_v photons per unit time per unit radiation bandwidth, and a bandwidth Δv, we get $\eta_d n_v \Delta v$ charge carriers created per second and a photocurrent of

$$i_P = \eta_d g n_v \Delta v e.$$

(9.27)

By definition, the detector responsivity is the change in the output current divided by the change in the input radiant power:

$$S = \frac{di_P}{dQ}.$$

(9.28)

The incident power is

$$Q = n_v \frac{hc}{\lambda} \Delta v,$$

(9.29)

so that

$$S = \eta_d g \frac{e\lambda}{hc} \Delta v.$$

(9.30)

Maximising the product $\eta_d g$ maximises the responsivity. Because the photoconductive gain, g, depends on the applied electric field, the responsivity increases with bias voltage. A small contact separation also leads to high g, tending to increase S. However, the reduced thickness of the detector could counteract that by reducing the absorption efficiency.

Equation (9.30) also shows that the responsivity of a photoconductor increases linearly with wavelength. This is because to first order it responds equivalently to any photon with wavelength shorter than the cut-off, but photons with wavelength just below the cut-off carry lower energy, and so the responsivity for such photons is the highest. For $\lambda > \lambda_c$, the responsivity falls to zero. An idealised plot of responsivity vs. wavelength has the general shape indicated in Figure 9.13. Maximum responsivity is achieved just below the cut-off wavelength.

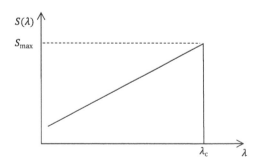

FIGURE 9.13   Responsivity vs. wavelength for an ideal photoconductor.

9.6.2 Dark Current

In the absence of any signal photons, the current should ideally be zero. In practice, this will not be true: the detector will have a "dark current". This constitutes an undesirable background for the measurement, and it is therefore important to make it as small as possible. The dark current arises from the thermal generation of electron-hole pairs in the material.

The equilibrium free electron and hole concentrations, n_o and p_o (carriers per unit volume) are given by (Sze 2002)

$$n_o = K_1 T^{3/2} \exp\left(-\frac{E_C - E_F}{k_B T}\right); \qquad p_o = K_2 T^{3/2} \exp\left(-\frac{E_F - E_V}{k_B T}\right), \tag{9.31}$$

where E_F is the Fermi level, E_C and E_V are the energies of the conduction and valence band edges, and K_1 and K_2 are constants that depend on the material. In an intrinsic semiconductor, the Fermi level is near the centre of the gap. In a doped material, it is controlled by the doping level. In n-type material, it moves up towards the conduction band by an amount that depends on the doping, and, likewise, in p-type material it moves down towards the valence band. Higher doping thus leads to higher carrier concentration. The dark current is proportional to the thermally generated carrier concentration, and the exponential Boltzmann factor (which represents a much stronger temperature dependence than the $T^{3/2}$ term) dictates that the thermal carrier concentration decreases exponentially with decreasing temperature. As with the radiant background, dark current must be subtracted from the measurement in order to recover the true signal, and – more problematically – fluctuations in the dark current due to the random generation and recombination of electrons and holes cause generation-recombination noise. Dark current can therefore be minimised by making sure that the detector temperature is low enough that the thermal generation of carriers across the band gap becomes unlikely. For large band gaps (i.e. short-wavelength detectors), only modest cooling is required. For instance, indium antimonide (InSb) infrared detectors can be operated at 77 K (the boiling point of liquid nitrogen). Small band gap detectors (operating at longer wavelengths) need much lower temperatures. For example, doped germanium detectors such as germanium:gallium (Ge:Ga) operate in the 2–4-K range.

Equation (9.31) would imply that the carrier concentration, and thus also the dark current, decrease inexorably as temperature is reduced. However, in practice, dark current tends to flatten off at some residual level due to the phenomenon of hopping conduction, in which charges can move by tunnelling from one localised impurity state to another allowing a current to flow even though there are no free electrons in the conduction band. Once hopping conduction has become the dominant contribution to the dark current, further reducing the temperature will be of no benefit. Figure 9.14 shows dark current vs. temperature curve for a typical Ge:Ga photoconductive detector. The exponential decline levels off with a residual dark current of around 100 electrons s^{-1} at temperatures below about 2 K.

9.6.3 Noise

Photon noise due to random photon arrival will manifest itself as a noise current in the photoconductor. For $n_v \Delta v$ incident photons s^{-1}, $\eta_d n_v \Delta v$ electrons (or holes) s^{-1} are generated. Assuming Poisson statistics (random photon arrival), the rms fluctuation in the number of charge carriers created in a time t_{int} is then $\left(\eta_d n_v \Delta v \, t_{int}\right)^{1/2}$. For a 1-Hz post-detection bandwidth, the integration time is $t_{int} = 1/2$s. The corresponding noise current spectral density is therefore

$$i_{n-ph} = ge\left(\frac{\eta_d n_v \Delta v}{2}\right)^{1/2}. \tag{9.32}$$

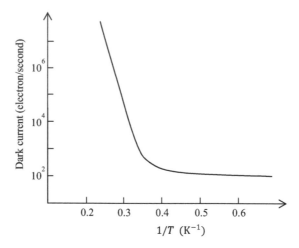

FIGURE 9.14 Dark current vs. inverse temperature for a typical gallium-doped germanium photoconductive detector. (From Church et al. 1992.)

This noise contribution is increased by a factor of $\sqrt{2}$ due to the random recombination of the charge carriers, giving a photon noise current spectral density of

$$i_{n-ph} = ge\left(\eta_d n_v \Delta v\right)^{1/2}. \tag{9.33}$$

The detector also has a certain value of resistance, R_d, which gives rise to Johnson noise. The Johnson noise current spectral density is just the noise voltage spectral density divided by R:

$$i_{n-J} = \frac{e_{n-J}}{R_d} = \frac{\left(4kTR_d\right)^{1/2}}{R_d} = \left(\frac{4kT}{R_d}\right)^{1/2}. \tag{9.34}$$

Generation-recombination noise associated with the combination of the detector dark current, I_d, results in a noise current spectral density given by (Section 6.7.3)

$$i_{n-GR} = \left(4egI_d\right)^{1/2}. \tag{9.35}$$

9.6.4 NEP

Taking all of the noise contributions into account, the overall NEP of the photoconductor is

$$NEP = \frac{i_{n-tot}}{S}, \tag{9.36}$$

where i_{n-tot} is the quadrature sum of the separate noise contributions:

$$i_{n-tot} = \left(i_{n-tot}^2 + i_{n-GR}^2 + i_{n-J}^2\right)^{1/2}. \tag{9.37}$$

The G-R noise increases in proportion to the square root of the total current and so depends on the square root of the applied bias voltage. It can therefore be made to dominate over the Johnson noise component – so we can neglect the Johnson noise term in evaluating the detector sensitivity.

Since the responsivity increases linearly with bias voltage, the NEP will decrease (i.e. the SNR will increase) with bias. However, in practice, there is always an optimum bias voltage that is determined by the maximum bias that can be applied before the detector starts to become excessively noisy and eventually breaks down due to impact ionisation of atoms in the lattice liberating a flood of charge carriers. This optimum point is found empirically and depends on temperature, the photon background, and the purity of the semiconductor.

It is useful to consider two limiting situations for the sensitivity of a photoconductor: low and high photon flux.

In the case of low photon flux, we take the photon noise to be negligible, and the signal current is also very small. The total current is thus dominated by the dark current, and the total noise is dominated by the G-R component:

$$i_{n-tot} \approx i_{n-GR} = \left(4egI_d\right)^{1/2}. \tag{9.38}$$

Therefore,

$$NEP = \frac{i_{n-tot}}{S} \approx \left(\frac{2hc}{\eta_d\lambda}\right)\left(\frac{I_d}{eg}\right)^{1/2}. \tag{9.39}$$

Note that (i) NEP decreases (gets better) with increasing wavelength (because the responsivity increases with wavelength); (ii) NEP is proportional to the square root of the dark current (so dark current must be minimised); (iii) making η and/or g large improves (lowers) NEP because both affect the detector responsivity. For very low dark currents, a NEP $\sim 10^{-18}$ W Hz$^{-1/2}$ can be achieved.

If there is a high level of background radiation incident on the detector, it will produce a photocurrent much larger than the dark current, and its associated fluctuations will dominate the overall noise:

$$i_{n-tot} \approx i_{n-ph} = ge\left(\eta_d n_v \Delta v\right)^{1/2}. \tag{9.40}$$

Dividing by the responsivity gives

$$NEP = \frac{i_{n-ph}}{S} \approx \frac{ge\left(\eta_d n_v \Delta v\right)^{1/2} hc}{\eta_d g e \lambda}. \tag{9.41}$$

Therefore,

$$NEP = \frac{\left(n_v \Delta v\right)^{1/2} hc}{\eta_d^{1/2} \lambda} = \left(\frac{P_B h v}{\eta_d}\right)^{1/2}, \tag{9.42}$$

where $P_B = n_v hv\Delta v$ is the background power incident on the detector. The dependence on $\eta_d^{-1/2}$ is to be expected – lowering η_d reduces the signal in proportion to η_d but reduces the photon noise in proportion to $\eta_d^{1/2}$ – so overall sensitivity only gets worse in proportion to $\eta_d^{1/2}$. As usual, it is also necessary in practice to subtract off the large background, so a further degradation factor of 2 will apply in practice: $2^{1/2}$ due to the need to observe the source for only half the time, and another $2^{1/2}$ due to the need to subtract the background measurement from the (source + background) measurement.

9.6.5 Photoconductor Readout Electronics

The simplest readout circuit for a photoconductive detector is a voltage divider, as for the bolometric detector readout shown in Figure 9.3. The load resistance should be both cold and much larger than the detector resistance to prevent it from adding significantly to the noise. This arrangement is suitable for low-resistance detectors, but for low-background applications, the detector resistance can be extremely large – 10^{10} Ω or higher. Such a high output impedance and small signal level make the signal vulnerable to RC attenuation and the effects of microphonic noise or interference. A readout circuit that can transform the detector signal from the high-impedance output of the detector itself to a much lower impedance is the transimpedance amplifier (TIA). A typical TIA circuit is shown in Figure 9.14 together with its small-signal equivalent. The detector output signal is fed to the inverting input of a high-gain operational amplifier, with the non-inverting input at earth. The output of the op-amp is coupled back to the detector through a feedback resistor, R_f. The high gain of the op-amp results in the inverting input being at a voltage only just above ground (a virtual earth). This means that the bias voltage across the detector is always fixed at V_{bias} regardless of the detector current. In addition, due to the high input impedance of the FET or op-amp, the detector current can only flow through the feedback resistor. The output voltage, v_{out}, is therefore just the detector current times the feedback resistance.

$$v_{out} = -i_d R_f. \tag{9.43}$$

The feedback resistor is usually located very close to the detector, and, in the circuit shown, a pair of matched FETs, also located as close as possible to the detector, acts as the input stage. The dual FET converts the high-impedance detector output to a much lower impedance signal that can propagate along the potentially long cryoharness without degradation. For extremely high resistance detectors, with low dark current and low photocurrent, the TIA is not an ideal readout because the noise of the feedback resistor can dominate. For such low-background applications, an integrating readout is preferred. One form of integrating readout is the CTIA amplifier which involves replacing the feedback resistor of the TIA with a capacitor, C_f, as in the inset in Figure 9.15. As before, the current in the feedback impedance must match the detector current. The capacitor charges up at a rate proportional to the detector current and is discharged at the end of each integration.

An alternative configuration is a simple integrating amplifier, the essential features of which are shown in Figure 9.15. The detector current builds up charge on a capacitor during the integration

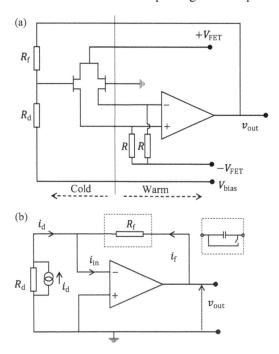

FIGURE 9.15 (a) Example of a practical TIA circuit. (b) TIA small-signal equivalent circuit; the inset shows the alternative CTIA configuration in which the feedback resistor is replaced by a capacitor.

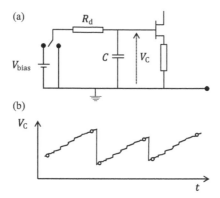

FIGURE 9.16 Essential features of an integrating detector readout (a); a sequence of integration ramps separated by reset operations (b).

time. The capacitor voltage is usually read out through a high-input impedance FET. The integrating capacitance, C, may be formed by that of the detector itself plus the input capacitance of the FET. The time constant of the detector resistance and integrating capacitance, $R_d C$, must be large compared to the integration time so that the increase in charge on the capacitor is in the initial linear regime of the RC charge-up curve. The capacitor voltage and detector voltage are related by (equation 5.4):

$$V_C(t) = V_{in}\left(1 - e^{-t/R_d C}\right) = I_d R_d \left(1 - e^{-t/R_d C}\right). \tag{9.44}$$

Therefore, for $t \ll R_d C$,

$$V_C(t) \approx I_d R_d \frac{t}{R_d C} = \frac{I_d}{C}t. \tag{9.45}$$

The capacitor voltage is thus proportional to the integral of the detector current, which can be estimated by calculating the slope of the linear integration ramp. When the integration is finished, the capacitor is discharged allowing a new integration to start (Figure 9.15). The exact value of V_C at the end of each reset is not the same each time, being subject to kTC noise. The signal can be measured by taking the difference in voltage between the start and end points of the integration (correlated double sampling) or by continuously monitoring the output voltage ramp. The longer the integration time, the greater the accuracy with which the slope can be measured in the presence of noise on the ramp. The kTC noise associated with the discharge of the capacitor is eliminated, as described in Section 6.7.6. Even though the reset process may leave an indeterminate amount of charge on the capacitor, as long as this is not too large it will have a negligible effect on the detector bias, and the slope of the ramp, which constitutes the signal, will be unaffected.

The maximum length of an integration is limited by the need to avoid saturation or unacceptable non-linearity of the output, leading to a maximum acceptable value of V_C and a corresponding maximum accumulated charge – referred to as the full well capacity and quoted as a number of electrons that can be built up before a reset is deemed necessary.

9.7 CHOICE OF PHOTOCONDUCTORS FOR THE FAR INFRARED

Detectors for FIR wavelengths are not as well-developed as those for shorter wavelengths (to be described in Chapter 10). P-type doped germanium photoconductors are normally used, in the form of large volumes of material (typically ~ 1 mm cubed) to achieve high absorption efficiency. For the 30–60 µm region, Ge:Be can be used, and for 50–115 µm, Ge:Ga is the detector of choice. The response of Ge:Ga can be extended beyond 200 µm through the technique of applying a strong

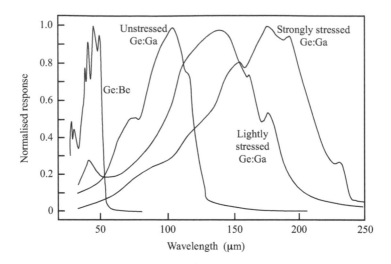

FIGURE 9.17 Spectral response curves for four photoconductive detectors used in the ISO Long Wavelength Spectrometer instrument: a Ge:Be detector with a cut-off at ~50 μm, an unstressed Ge:Ga detector with a cut-off at ~115 μm), a lightly stressed Ge:Ga detector with a cut-off at ~170 μm, and a strongly stressed Ge:Ga detector with a cut-off at ~210 μm. (From Church et al. 1992.)

compressional mechanical stress on the crystal. This changes the ionisation potential of the acceptor impurities from the unstressed value of 10.8 meV ($\lambda_c = 115$ μm) to ~ 5.9 meV ($\lambda_c = 210$ μm) for a pressure of 70 kg mm^{-2}, close to the mechanical strength of the crystal.

Figure 9.17 shows typical spectral response curves for Ge:Be, unstressed Ge:Ga, and stressed Ge:Ga detectors used in the Long Wavelength Spectrometer (LWS) on board ESA's Infrared Space Observatory (ISO) satellite. Ge:Be and unstressed Ge:Ga detectors operate best at a temperature of ~3 K. The smaller energy gap of stressed Ge:Ga detectors means that they need to be cooled to a lower temperature (<2 K) to suppress the dark current. The FIR spectrometer in the *Herschel*-PACS instrument (Poglitsch et al. 2010) used 16 × 25 element arrays of slightly stressed Ge:Ga detectors for 55–110 μm and highly stressed detectors for 110–210 μm. Each array was constructed from 25 linear modules each with 16 detectors all stressed by the same stressing mechanism.

Doped Ge detectors can achieve very low instantaneous NEP, but they tend to exhibit a number of unwelcome characteristics that can adversely affect their usability and sensitivity in very low background applications. These include memory effects, whereby the responsivity and noise can depend on the previous history of illumination of the detector, and sensitivity to ionising radiation both in the form of charge released by individual photon or particle events, and long-term changes in responsivity with accumulated radiation dose. After a period of irradiation, the original detector properties can be restored through bias boosting or through a short period of high FIR illumination. Despite these unwelcome complications, Ge:Ga photoconductors have the advantage that, when carefully set up and operated, they can achieve very low NEP – a few × 10^{-18} W Hz$^{-1/2}$, at operating temperatures of 1.6 K or higher – much more conveniently achieved than the < 0.3 K temperatures needed for bolometers and KIDs. For wavelengths above the cut-off of stressed Ge:Ga, no high-quality photoconductive detectors are available, and bolometric or KID detectors are used.

9.8 COUPLING FIR AND SUBMILLIMETRE DETECTOR ARRAYS TO THE TELESCOPE

The diameter of the Airy diffraction disk at wavelength λ is given by $2.44F\lambda$ for an optical system with focal ratio F (equation 4.32). A typical value for F is ~5, giving a diffraction disk on the order of 10 mm in size at submillimetre wavelengths. To couple most efficiently to a point source, a pixel must not be much smaller than this. This is one of the main reasons why we do not have mega-pixel cameras for submillimetre wavelengths – the detector arrays would be unmanageably big. This is compounded by the need to operate the detectors at very low temperatures. The cryogenic systems needed to achieve such temperatures have limited cooling power, and it is difficult to cool large or heavy focal planes or to accommodate a large number of wires connecting the focal plane to the warm electronics. Only recently have arrays with thousands of pixels been developed.

The two methods of coupling bolometric detectors to the focal plane of a telescope, antenna coupling and absorber coupling are both widely used in FIR and submillimetre instruments. The optimum choice for a given application depends on the detailed design of the instrument and on practical constraints such as available mass and power.

The large diffraction spot size led to the traditional approach of mounting the bolometer in an integrating cavity and coupling it to the incident beam by means of an antenna such as a feedhorn. The simplest feedhorn is a smooth-walled conical horn with a suitably sized circular entrance aperture, which views the telescope and channels the collected radiation to a short section of circular waveguide connecting its much smaller exit aperture to the detector. The waveguide is often single-moded – i.e., it propagates only the fundamental waveguide mode. The aperture efficiency (i.e. the fraction of the power from a point source that is coupled to the detector) of a single-mode feedhorn is shown in Figure 9.18 as a function of horn aperture diameter. The aperture diameter is expressed as an angle on the sky in terms of λ/D (corresponding to a physical diameter of $F\lambda$). Figure 9.18 also shows the equivalent for a square absorbing pixel coupling to the Airy disk (repeated from Figure 4.26).

These two curves are very different. For the absorber-coupled pixel, the intercepted power continues to increase monotonically as the pixel size is increased, but for the circular feedhorn, it rises to a maximum of about 0.75 at a diameter of $2\lambda/D$ and then falls again. The reason for this behaviour is that the feedhorn does not view the telescope uniformly, but with its own tapered (roughly Gaussian) illumination profile, and so it sees the centre part of the telescope with higher efficiency than the outer parts. The width of the feedhorn beam profile is inversely proportional to its size – a small aperture means that its field of view comfortably encompasses the whole telescope. However, a small aperture intercepts only a small fraction of the Airy disk, so the overall efficiency is low. As

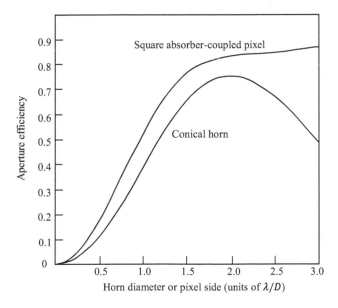

FIGURE 9.18 Aperture efficiencies versus square pixel side for a square absorber-coupled pixel and horn aperture diameter for a smooth-walled conical horn. The pixel sides and aperture diameter are expressed in terms of λ/D.

the aperture increases, aperture efficiency increases as the feedhorn collects more and more of the power in the Airy disk, but this is counteracted by a tendency for the efficiency to decrease as the beam of the feedhorn narrows leading to less efficient collection of radiation from the outer parts of the telescope (due to a reduction in the taper efficiency, as illustrated in Figure 8.4). This second effect starts to dominate at an aperture diameter of around $2\lambda/D$, where maximum efficiency is reached. Further increasing the feedhorn size reduces efficiency as less and less of the telescope area is effectively used.

9.8.1 Antenna-Coupled Arrays

Feedhorn arrays are often designed to maximise the aperture efficiency per pixel by adopting a feedhorn aperture of $2\lambda/D$ (physical diameter $2F\lambda$). The feedhorns are usually close-packed in a hexagonal arrangement in the focal plane to fit as many as possible into the area available. The telescope illumination is tapered as shown in Figure 9.19. The outer parts of the telescope are not fully used, and the beam is therefore slightly broader than the achievable diffraction limit of the telescope, but sidelobes due to the cut-off at the edge of the primary are suppressed.

The beam spacing on the sky is equivalent to the detector centre-to-centre spacing of $2\lambda/D$, but the FWHM of each beam is only $\sim \lambda/D$. Therefore, beams on the sky do not fully sample the image (this would require a horn diameter of $0.5F\lambda$ or less). Several separate telescope pointings or some sort of telescope scanning scheme are therefore needed to create a fully sampled image. In principle, for $2F\lambda$ horns 16 separate array pointings are required, as illustrated in Figure 9.20.

9.8.2 Absorber-Coupled Arrays

The other approach to coupling the detectors to the telescope is to dispense with any kind of antenna and simply have bare absorbing pixels, typically of side $0.5F\lambda$, in the focal plane. Such pixels have a wide field of view ($\sim \pi$ sr). Signal is maximised and background is minimised when the detector field of view is restricted to the telescope alone. To achieve this, it is therefore necessary to include a cold-stop in the optical system to define a solid angle matched to that of the telescope, otherwise the detectors could be flooded with unwanted background radiation. The telescope illumination is flatter than the tapered profile of a feedhorn, as shown in Figure 9.16, resulting in more efficient use

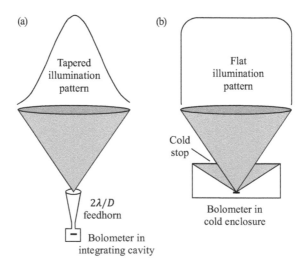

FIGURE 9.19 (a) Illumination of the telescope with a tapered, near-Gaussian, profile by a $2\lambda/D$ feedhorn-coupled bolometer. (b) Illumination of the telescope by a bare absorber-coupled pixel, which has a wide field of view, with a cold-stop defining its uniform illumination of the telescope.

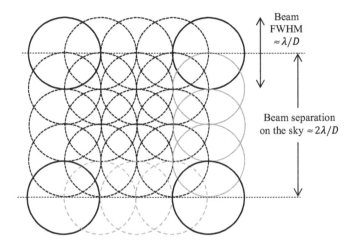

FIGURE 9.20 Telescope pointing pattern needed to achieve a fully sampled map with square-packed $2F\lambda$ feedhorns. The solid circles represent the FWHM beams on the sky of four adjacent feedhorns. The dashed circles show the additional 15 positions of the top left feedhorn. The solid grey circles show three of the positions of the top right feedhorn, and the dashed grey circles show three of the positions of the bottom left feedhorn. The 16 positions achieve a fully sampled map of the rectangular region shown. In the case of hexagonal close packing, the pointing pattern is slightly different, but 16 steps are still needed.

of the collecting area of the telescope. Close packing square $0.5F\lambda$ pixels achieves instantaneous sampling of the image. Although the aperture efficiency of an individual pixel is low (only about 0.16 – see Figure 9.17), signal not picked up by one pixel will be collected by adjacent pixels, so that, in principle, all of the power falling on the focal plane is intercepted. This makes a $0.5F\lambda$ absorber-coupled array more efficient in mapping than a $2F\lambda$ feedhorn array. Detailed calculations, taking into account aperture efficiency, throughput, detector numbers, and observing strategy (Griffin, Bock, & Gear 2002), show that, in the photon-noise-limited case, the advantage in terms of observing speed (the inverse of the time needed to achieve a given sensitivity) is a factor of 3–3.5. On the other hand, for observation of a point source, it is always advantageous to put as much of the signal as efficiently as possible onto a single pixel, which is what the $2F\lambda$ feedhorn does. For this case, in the photon noise limit, the feedhorn configuration is faster by a factor of ~3.

In practice, the choice of absorber-coupled or antenna-coupled architecture is often a complex trade-off. Practical considerations such as the wavelengths of observation, background power levels, achievable per-pixel sensitivity, stray light suppression requirements, multiplexing capabilities, power dissipation, available data rate, instrument cryogenic design, etc. may be important in deciding among the different options. The main advantages of feedhorn arrays are: (i) maximum efficiency for detection of a point source with known position; (ii) minimum number of detectors needed for a given field of view (a factor of 16 fewer than for $0.5F\lambda$ pixels); (iii) higher (and so more easily achieved) required detector NEP to achieve the photon noise limit, due to the higher background per pixel; (iv) good stray light rejection – the bolometer field of view is restricted to the telescope; (v) low sidelobes, as a result of the tapered illumination of the telescope; (vi) good rejection of electromagnetic interference – the horn plus integrating cavity act as a Faraday enclosure. The main disadvantages are that the feedhorns can add greatly to the amount of mass to be cooled to very low temperature, the observing modes are complicated (multiple pointings or scanning) and, more importantly, the overall efficiency for mapping is less than the ideal: the price paid for the well-defined feedhorn profile is that some of the signal power incident on the horn array is actually reflected back out.

The main advantages of absorber-coupled arrays are (i) maximum efficiency for mapping, arising from the fact that no power falling on the focal plane is rejected; (ii) instantaneous full sampling of the image, making for simpler observing modes, and (iii) a generally less massive and more compact focal plane since no antennas are needed. The main disadvantages are the larger pixel numbers and the increased susceptibility to stray light.

9.9   CASE STUDIES

9.9.1   *HERSCHEL*-SPIRE: A SPACE-BORNE FIR-SUBMILLIMETRE CAMERA AND SPECTROMETER

The *Herschel* Space Observatory (Figure 9.21) carried a large-aperture (3.5 m diameter) Cassegrain telescope and three cryogenically cooled instruments (HIFI, PACS, and SPIRE) designed to carry out imaging photometry and spectroscopy in the far infrared and submillimetre. To minimise the thermal background on the detectors, the telescope was passively cooled (by radiation into cold space) to around 85 K and was designed to have very low emissivity (typically less than 1%) over its operational wavelength range (approximately 50–700 μm). The *Herschel* instruments' cold focal plane units (FPUs) were located in a cryostat containing 2200 L of liquid helium at a temperature of 1.7 K. *Herschel* was launched in May 2009 and operated until April 2013, when its supply of liquid helium was exhausted.

The *Herschel*-SPIRE FPU (Griffin et al. 2010) contained a three-band submillimetre camera and an imaging Fourier transform spectrometer (FTS), illustrated in Figure 9.21. The focal plane unit was connected to the instrument warm electronics by a cryoharness, approximately 5 m in length.

FIGURE 9.21   The *Herschel* Space Observatory. (© ESA-AOES Medialab.)

(a)

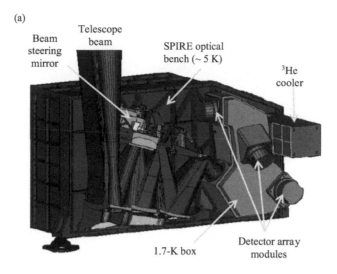

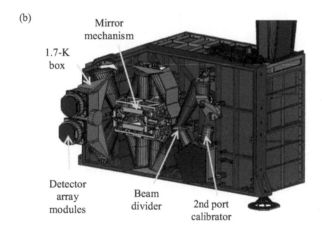

FIGURE 9.22 The *Herschel*-SPIRE FPU camera side (a) and spectrometer side (b). (Credit: *Herschel*-SPIRE consortium.)

The FPU was supported from the 10-K cryostat optical bench by thermally insulating mounts, and had three temperature stages: the *Herschel* cryostat provided temperatures of 4.5 and 1.7 K via high thermal conductance straps to the instrument, and an internal ^{3}He refrigerator cooled the detector arrays to approximately 0.3 K. The camera and spectrometer (Figure 9.22) occupied two separate compartments, with some common fore-optics on the camera side shared by both. The beam steering mirror allowed the telescope pointing to be adjusted in two dimensions without moving the telescope itself. After the spectrometer beam was diverted to the other instrument compartment, the camera-side beam was imaged onto three bolometer arrays, contained inside a 1.7-K enclosure with the detectors themselves cooled by a thermal strap connected to the ^{3}He cooler. The three arrays were used for broadband photometry $(\lambda/\Delta\lambda \sim 3)$ in spectral bands centred on approximately 250, 350 and 500 μm. Quasi-optical filters defined the bands and minimised the thermal load on the low-temperature stages by reflecting short-wavelength radiation. The bands were defined by a combination of filters in front of the detectors and the cut-off wavelengths of the feedhorn output waveguides.

The overall transmission profiles for the three camera bands are shown in Figure 9.23. The aperture efficiency functions for the detector feedhorns are also shown. The choice of $2F\lambda$ feedhorn diameter ensured that the efficiency was close to maximum across each band.

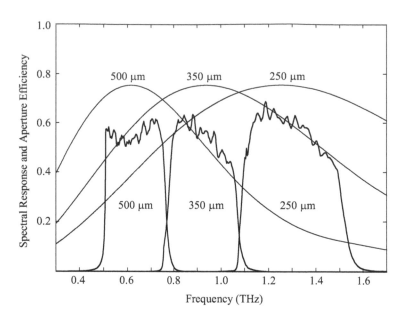

FIGURE 9.23 Transmission profiles for the three SPIRE camera bands (solid lines). The dashed lines show the aperture efficiencies of the $2F\lambda$ detector feedhorns, which are matched to the bands. (From Griffin et al. (2013), reproduced by permission of Oxford University Press.)

The photon noise levels for the SPIRE bolometers were dominated by the thermal emission of the telescope, with absorbed background power levels of 1–2 pW and photon noise limited NEPs of 3–6×10^{-17} W Hz$^{-1/2}$. Compatible detector NEPs were achieved using NTD spider-web bolometers operating at 300 mK. The bolometers were coupled to the telescope by hexagonally close-packed $2F\lambda$-diameter single-mode conical feedhorns, giving diffraction-limited beams of FWHM 18″, 25″ and 36″ for the 250, 350, and 500-μm bands, respectively. The three arrays contained 43 (500 μm), 88 (350 μm), and 139 (250 μm) detectors. The same field of view of 4×8 arcminutes was observed simultaneously by the three arrays through the use of two fixed dichroic beam-splitters inside the 1.7-K enclosure. Dual JFET preamplifiers (as in Figure 9.6) were used to read out the bolometer signals. The JFET unit was attached to the 10-K optical bench next to the 4.5-K enclosure, with the JFETs themselves heated internally to their optimum operating temperature of ~ 120 K.

The bolometers were AC-biased with a frequency of 130 Hz, reducing $1/f$ noise from the JFET readout, and giving a $1/f$ knee for the system as low as a few mHz. The main observing mode for the SPIRE camera was scan mapping, whereby the telescope scanned across the sky at 0.5 arcminutes s^{-1}, and the low $1/f$ noise knee allowed large maps to be made with good fidelity to faint extended structures. To give the beam overlap needed for full spatial sampling over a strip defined by one scan line, and to provide a uniform distribution of integration time over the area covered by the scan, the scan angle was tilted with respect to the array axes.

The FTS side of the instrument is also shown in Figure 9.20 and used two broadband intensity beam-splitters in a Mach-Zehnder configuration. A single back-to-back scanning roof-top mirror served both interferometer arms. It had a frictionless mechanism and a Moiré fringe position sensing system. One input port of the interferometer viewed a 2-arcminute diameter field of view on the sky and the other was fed by an on-board reference source. Two 0.3-K feedhorn-coupled bolometer arrays, similar to those used for the camera, were placed at the output ports and were sensitive in overlapping bands providing complete wavelength coverage between 194 and 672 μm. The FTS spectral resolution was set by the total optical path difference and could be adjusted between 0.04 and 2 cm^{-1} (corresponding to $\lambda/\Delta\lambda = 1000 - 20$ at 250 μm).

9.9.2 *Spitzer*-MIPS: A Space-borne FIR Camera

The *Spitzer* Space Telescope was launched by NASA in August 2003. It had a 0.85-m telescope cooled to ~5 K by an on-board tank of liquid helium. Its three instruments (IRAC, IRIS, and MIPS) performed imaging and spectroscopy between 3.6 and 160 μm. The multiband imaging photometer, MIPS, had three imaging arrays operating at 24, 70, and 160 μm, with the 70-μm array also used for low-resolution spectroscopy between 55 and 95 μm. The instrument layout is shown in Figure 9.24. The telescope beam was directed by the pick-off mirrors to the field mirror, which created pupil images at the scan mirror. The latter had various facets and positions used to direct beams onto the different arrays for imaging or spectroscopy. All of the MIPS detector arrays were filled arrays with pixel sizes of $0.5F\lambda$ or less. The 24-μm band used a 128×128 pixel Si:As BIB detector array (see Section 10.2.4) with integrating readouts, operating at a temperature of 5.2 K, with a 5×5 arcminute field of view. The individual detector DQE was 60% with a dark current of 3 electrons s^{-1} and a read noise of only 50 electrons. Each detector was read out continuously and non-destructively with a 0.5-s sampling interval, with integration times up to 30 s.

For the 70-μm channel, the detector array was a 32×32 array of unstressed Ge:Ga photoconductive detectors with a 2.5×5 arcminute field of view. The longest wavelength 160-μm band had a 2×20 pixel array of stressed Ge:Ga photoconductors with a 0.5×5 arcminute field of view. In both cases CTIA readout amplifiers were used. Filters defined passbands with characteristic widths of 4.7, 19, and 35 μm for the 24, 70, and 160 μm bands, respectively. The FWHM beam widths, dictated by the telescope size, were (6, 18, 40)″ at (24, 70, and 160) μm.

The performance of the MIPS 24-μm channel was excellent, but the large detector sizes needed for the longer-wavelength channels ($0.75 \times 0.5 \times 2$ mm for the 70-μm array and $0.8 \times 0.8 \times 1$ mm for the 160-μm array) made them more susceptible to ionising particle hits in space which caused some sensitivity degradation. In addition, the inherent non-linear behaviour of Ge:Ga detectors under low background introduced complications in the data analysis and calibration. Despite these problems, the overall performance of the MIPS instrument was impressive, providing major advances over previous far-infrared missions.

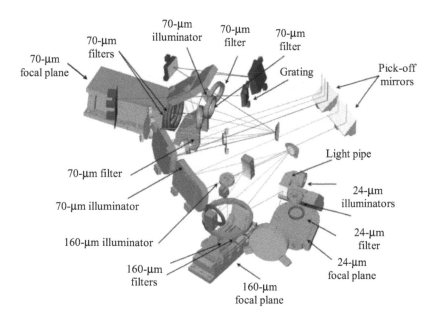

FIGURE 9.24 Layout of the Spitzer-MIPS instrument focal plane. (From the *MIPS Instrument Handbook*; credit NASA, MIPS Instrument and MIPS Instrument Support Teams.)

9.9.3   SCUBA-2: A GROUND-BASED SUBMILLIMETRE CAMERA

SCUBA-2 is a TES bolometer camera operating at submillimetre wavelengths on the 15-m diameter James Clerk Maxwell Telescope (JCMT), on Maunakea, Hawai'i (Figure 9.25). At an altitude of 4.2 km, the JCMT and other telescopes on Maunakea are able to operate above most of the atmosphere's water vapour, providing usable transmission in the submillimetre windows at 350, 450 and 850 μm. SCUBA-2 was designed to replace the previous generation JCMT camera, SCUBA, which observed at 450 and 850 μm with 37 and 91 feedhorn-coupled bolometers, respectively, operating at 0.1 K, and with a field of view of around 2 arcminutes diameter (~3 square arcminutes). SCUBA-2 also observes simultaneously at 450 and 850 μm with 0.1-K detectors, but with a much larger pixel count and field of view: it has two arrays of 5120 absorber-coupled bolometers, at each of 450 and 850 μm, and a field of view of 45 square arcminutes.

Free from the tight mass, volume and power constraints that apply to space instruments, SCUBA-2 is a much larger instrument. The SCUBA-2 optical layout is shown in Figure 9.26. The 45 square arcminute field of view is the largest that could be relayed through the telescope elevation bearing to the instrument cryostat, which due to its large size (2.3 × 2.1 × 1.7 m with a weight of over 3 tonnes) has to be located at the Nasmyth focus of the telescope. A second optical relay brings the focal plane to a position just inside the cryostat window. Cold optics inside the cryostat then re-image the focal plane onto the detector arrays. To minimise thermal power loading on the bolometer arrays, the last three of the re-imaging mirrors are cooled to temperatures less than 10 K.

FIGURE 9.25   The James Clerk Maxwell Telescope (JCMT) on Maunakea, Hawai'i. (Credit UKATC, Royal Observatory Edinburgh.)

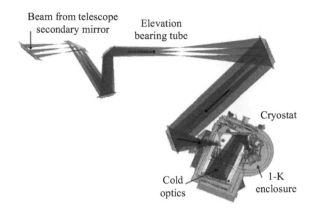

FIGURE 9.26 Optical layout of the SCUBA-2 instrument. (Reproduced from Holland et al. (2013) with permission of Oxford University Press.)

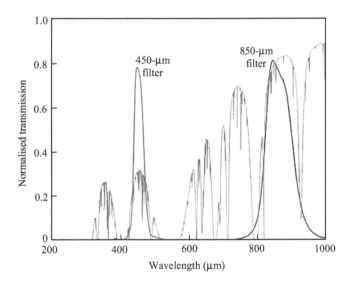

FIGURE 9.27 SCUBA-2 450- and 850-µm spectral passbands overlaid on an atmospheric transmission plot for Maunakea with 1 mm precipitable water vapour. (Reproduced from Holland et al. (2013) with permission of Oxford University Press.)

The SCUBA-2 spectral bands, defined by a series of filters along the optical chain within the cryostat, are matched to the atmospheric windows, as shown in Figure 9.27. The cryostat needs liquid cryogens only for pre-cooling. In operation, two pulse tube coolers maintain the radiation shields at ~50 K and the cold optics at ~4 K.

The SCUBA-2 detector array architecture is shown in Figure 9.28. Each array is configured as four quadrants butted up against each other, each with 32×40 square pixels. The detector pixels are $0.5F\lambda$ in size at 850 µm, providing instantaneous full sampling of the image, and $1.0\ F\lambda$ at 450 µm, requiring some telescope motion to sample the image fully. Each detector chip is indium bump-bonded to a SQUID multiplexer chip. This arrangement allowed the same basic detector and multiplexer chip design to be used for both bands. The signals are further amplified by a series array of SQUIDS at the 1-K temperature level. Figure 9.27 also shows the design of a single pixel. The upper surface of the silicon detector wafer is implanted with phosphorous to provide a resistive absorbing coating. The Mo-Cu TES bi-layer is formed on the back of the wafer, and also acts as a

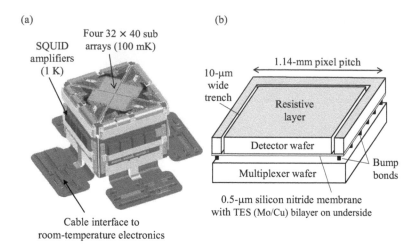

FIGURE 9.28 (a) SCUBA-2 detector array architecture. (Reproduced from Holland et al. (2013) with permission of Oxford University Press.) (b) Schematic diagram of a single SCUBA-2 pixel. (From Audley et al. (2004) with permission from Elsevier.)

highly reflecting backshort. The wafer itself has a thickness of $3\lambda/4$ (450 μm) or $\lambda/4$ (850 μm) to maximise the overall absorption. A thin silicon nitride membrane supports the pixels and provides the weak thermal conductance to the 0.1 K bath. Each bolometer is isolated by a 10-μm wide trench etched around it in the detector wafer, which is connected electrically to the SQUID multiplexer wafer via indium bump bonds. The superconducting transition temperatures for the 450- and 850-μm bolometers are 130 and 190 mK, respectively, selected to optimise performance over the range of background powers corresponding to different atmospheric conditions.

10 Infrared to UV Instrumentation

10.1 INTRODUCTION

In this chapter, we consider the detectors and techniques commonly used in the infrared, optical, and UV regions of the spectrum, covering wavelengths from around $30\,\mu m$ down to around $0.2\,\mu m$. It is appropriate to consider the radiation as a stream of discrete photons, which can interact with electrons in a detector, either promoting them to higher energy levels or liberating electrons by the photoelectric effect. Compared with the far-infrared region, detectors at these shorter wavelengths are considerably more advanced for several reasons: (i) the high transparency of the Earth's atmosphere in the visible region and the well-established field of visible-wavelength astronomy fostered the development of astronomical detectors with greater sensitivity and spectral bandwidth than the human eye; (ii) it is not necessary to operate detectors and instruments at such low temperatures, thereby greatly simplifying or eliminating cooling systems; and (iii) there are many more industrial, consumer, and military applications than there are for far-infrared detection so that the level of development and investment over many years has resulted in a much more mature technology. In particular, various kinds of large-format (mega-pixel) detector arrays are common, which can provide huge imaging and spectroscopic power when used with space-borne or large ground-based telescopes. In the infrared, most detectors are based on arrays of photodiodes, while in the visible and ultraviolet, large-format arrays of charge-coupled devices (CCDs) are often used. Photoemissive detectors, which rely on photoelectric currents flowing in vacuo, have been superseded by electronic detectors for most applications, but are still not completely obsolete.

As described in Chapter 3, the Earth's atmosphere is opaque in some parts of this spectral region, requiring observations from space, and transparent in others. Space-borne observations allow the best overall optical performance and the greatest stability and quality of the data, at the expense of smaller telescopes than can be used on the ground, and, of course, much higher cost. Even in parts of the spectrum in which atmospheric transparency is high, such as the visible region, ground-based observations still need to cope with absorption, emission, scattering, and turbulent distortion caused by the Earth's atmosphere. In the optical and near infrared (NIR), adaptive optics systems have been developed to counteract the effects of atmospheric turbulence, achieving much improved angular resolution and sensitivity than would otherwise be possible.

10.2 INFRARED DETECTORS

10.2.1 PHOTODIODES

For infrared wavelengths up to around $14\,\mu m$, the most commonly used detectors use arrays of photodiodes formed from semiconductor p-n junctions. A p-n junction is a single crystal of semiconductor with a sharp transition between p- and n-type materials. The structure of a p-n junction can be understood by first imagining an n-type and a p-type slab being joined together to form an abrupt transition. Figure 10.1a shows the situation before contact is made. The p-type material has a population of free holes and an equivalent population of fixed negatively ionised acceptor atoms, which have captured electrons to create the holes. Conversely, the n-type material has many free electrons and an equivalent population of fixed positively ionised donor atoms which have lost electrons.

When the two pieces are brought together, the situation depicted in the middle panel of Figure 10.1 is rapidly established. The sharp concentration gradients in hole and electron density at the junction lead to diffusion – the spreading out of particles through random thermal motions, tending to even out a concentration gradient. Holes diffuse from the p-side to the n-side where they are

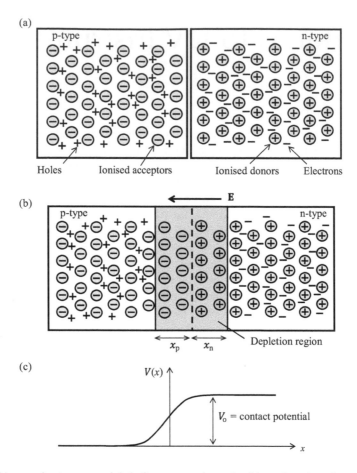

FIGURE 10.1   (a) p- and n-type materials before contact is made; (b) p-n junction after contact is made; (c) potential vs position across the junction.

neutralised by electrons; likewise, electrons diffuse from the n-side to the p-side where they are neutralised by holes. Therefore, the n-side becomes positive and the p-side becomes negative. This creates an electric field **E** that opposes further diffusion, and a dynamic equilibrium is established. A depletion region is created in which the free carriers have been neutralised, leaving the impurity atoms to constitute electrically charged zones on either side. The dynamic equilibrium is between opposite currents due to diffusion and to the electric field caused by the charged regions. Electron–hole pairs are continually being created and recombining. If an electron on the p-side finds itself in the transition region, it is quickly swept across to the n-side by the electric field. Likewise, a hole appearing in or wandering into the transition region on the n-side is quickly swept across to the p-side by the field.

The original p- and n-type slabs were electrically neutral, and no charge has been added to the system. Charge neutrality therefore requires that the total charges on the two sides must cancel out. For a junction of area A,

$$N_A x_p A e = N_D x_n A e. \tag{10.1}$$

where N_A and N_D are the acceptor and donor concentrations (number of impurity atoms per unit volume) on the p and n sides, respectively, and x_p and x_n are the p- and n-side widths of the depletion layer. This means that the thicknesses of the depletion layers on the two sides are inversely proportional to their original carrier concentrations:

$$\frac{x_{\mathrm{p}}}{x_{\mathrm{n}}} = \frac{N_{\mathrm{D}}}{N_{\mathrm{A}}}. \tag{10.2}$$

If one of the sides is made heavily doped, the depletion region will be much more extensive on the opposite side. Such a junction with a heavily doped p-side is designated as p^+n.

Because of the electric field between the oppositely charged zones on either side, the p-n junction has an in-built potential difference between the two sides, called the contact potential, as shown in Figure 10.1c.

Consider a p-n junction to which an external bias voltage is applied, as shown in Figure 10.2a. By convention, the external voltage across the junction is measured from the p-side to the n-side, and positive current flows from the p-side to the n-side. The depletion region has much higher resistance than the rest of the material, so nearly all of the voltage appears across it.

Forward bias (positive V_{B}) opposes the contact potential V_{o}, decreasing the potential difference across the junction, and reverse bias (negative V_{B}) enhances the contact potential, increasing the overall potential difference between the two sides.

Under forward bias, the diode is highly conducting because the potential barrier that opposes diffusion of charges is much reduced, so the diffusion current is strongly increased. The drift current in the opposite direction due to the electric field is relatively insensitive to the height of the potential barrier. Any electron entering into the transition region on the p-side (or hole entering on the n-side) will be swept over to the other side regardless of the barrier height. Therefore, the drift current is limited by the density of minority carriers available, not by the height of the barrier. With reverse bias, the potential barrier is increased, and diffusion is suppressed. This leaves only the small negative drift current, which is independent of the applied bias.

The current–voltage characteristic of a diode is illustrated in Figure 10.2b. The p-n junction is referred to as a diode as it conducts very well for one voltage polarity and hardly at all for the other. With reverse bias, the small current, known as the reverse saturation current, is independent of the bias voltage up to some limit beyond which the material suffers electrical breakdown resulting in a large increase in current.

A p-n junction can be used to detect photons. (This is also the principle behind the photovoltaic solar cell.) If a photon with energy greater than band gap is absorbed in the depletion region, an electron–hole pair is generated. The electron and hole are swept across the junction by the electric field, as shown in Figure 10.3, generating a photocurrent, flowing from n to p, proportional to the incident photon rate. No bias voltage is needed, and there is no recombination noise as the electron and hole can escape from the detector on a timescale shorter than the recombination time. Photons absorbed in the neutral material outside the depletion region can also produce a photocurrent as long as the characteristic diffusion length of the liberated charges is greater than the thickness of the neutral region because the electron or hole may wander through random thermal diffusion into the depletion region. If the absorption is too far away from the depletion region, then it becomes more

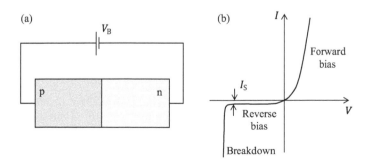

FIGURE 10.2   p-n junction with applied bias voltage (a); current–voltage characteristic (b).

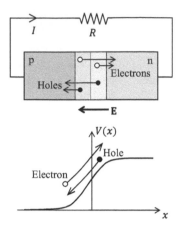

FIGURE 10.3 Operation of a p-n junction as a photon detector.

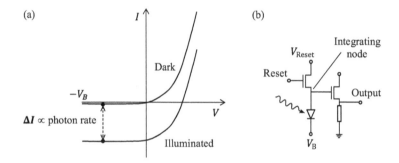

FIGURE 10.4 Reverse-bias operation of a photodiode (a); essential features of an integrating readout circuit for a photodiode (b).

likely that recombination will occur and the charges will be lost. For best quantum efficiency and speed of response, it is desirable that the absorption take place inside the depletion region.

For best speed of response and linearity, photodiode arrays are normally operated in reverse bias mode (negative V_B), as shown in Figure 10.4a, and with an integrating readout. The integrating node is initially negatively charged by the bias voltage and is progressively discharged during the integration by the detector current. The corresponding voltage is read out (non-destructively sampling up the ramp if desired) by a FET source follower. As described in Chapter 9, the integrating capacitance needs to be discharged (reset) at the end of each integration. The basic detector and readout circuit are shown in Figure 10.4b. MOSFETs, as symbolised here, are normally used as the FETs for the readout and for the circuitry used to control the array and multiplex the outputs. An advantage of MOSFETS is that they only need to be switched on while read and switching operations are in progress.

10.2.2 Infrared Photodiode Materials

Photon energies in the NIR vary from around 1.24 eV (1 μm) to 0.25 eV (5 μm). The band gaps of silicon (~1.1 eV) and germanium (~ 0.7 eV) are generally too large for photo-excitation, and doped Si and Ge tend to have energy levels too close to the band edges, and so are more appropriate for longer wavelengths. Compound semiconductors such as indium antimonide, InSb, and mercury cadmium telluride, HgCdTe (sometimes known as Mer-Cad-Tel or MCT) have narrower gaps than Si or Ge and are the best materials for NIR arrays. InSb has an energy gap of 0.23 eV at 77 K (cut-off

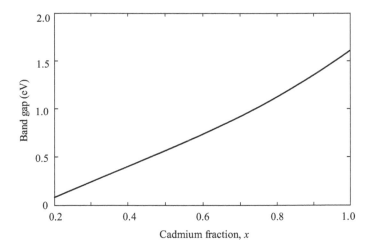

FIGURE 10.5 Band gap energy vs. cadmium fraction, x, for $Hg_{1-x}Cd_xTe$ at a temperature of 77 K. (Based on the formula given in Hansen et al. (1982).

at 5.3 μm). HgCdTe can exist in a range of forms depending on the relative proportions of mercury and cadmium. The composition is expressed in the form $Hg_{1-x}Cd_xTe$, with the total number of mercury and cadmium atoms equalling the number of tellurium atoms. An important and very useful feature of the material is that the energy gap varies with the cadmium fraction, x, as shown in Figure 10.5. The higher the cadmium fraction, the larger the band gap and so the shorter the cut-off wavelength. The material can therefore be "tuned" for the best performance over the wavelength range of interest, between around 1.25 and 14 μm. Another attractive property of HgCdTe is that the lattice constant (the interatomic spacing) varies hardly at all with x. This means that devices can be made combining different compositions without encountering lattice mismatch problems such as unwanted energy states at the interfaces that could trap charges. HgCdTe is normally grown using the technique of molecular beam epitaxy (MBE) in which layers of material are grown on a substrate in a precisely controlled way.

10.2.3 INFRARED PHOTODIODE ARRAYS

Infrared photodiode arrays come in various architectures, with materials and structures optimised for different applications. Figure 10.6 illustrates some of the essential features of a typical infrared HgCdTe array (the same principles apply to InSb arrays).

Photons are incident from the top side and pass through the transparent CdTe substrate with an anti-reflection coating to minimise first-surface reflectance. They are absorbed in a layer of n-type HgCdTe, typically 10–20 μm thick, and generate electron–hole pairs which diffuse within the neutral material. The individual pixels are formed by p-n junctions created by p-type regions in a wide-band HgCdTe layer. The electrons are collected by a common contact, and the holes diffuse into the depletion region of the junction and are swept across by the junction electric field. The passivation layer prevents charge recombination at surface energy states. The p-n junction current charges up the integrating capacitance at a MOSFET readout transistor (one for each pixel) in the silicon read-out chip, which is connected to the photodiode chip by indium bump bonds. In this process, small blobs of indium are placed on the contact pads of one chip. The two chips are very closely aligned (within ~ 1 μm) and brought into contact under high pressure, causing the indium bumps to cold-weld to the pads making the electrical contacts permanent. The multiplexing chip reads out the output voltages of the pixels along rows and columns using additional MOS electronics.

Cooling of infrared arrays is essential to reduce the photodiode dark current, but not to such low temperatures as are needed for far-infrared detectors because the band gaps are greater. The

(a)

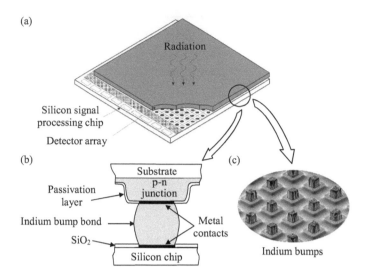

FIGURE 10.6 (a) Structure of a hybrid array with the photodiode array bump-bonded to the silicon signal readout and multiplexer chip; (b) cross section of a single pixel; (c) photo of detector array with indium bumps attached. (From Rogalski (2020) reproduced with permission of Elsevier.)

required operating temperature depends on the band gap (and thus on the cut-off wavelength) and on how low the dark current needs to be. For very low background applications such as high-resolution spectrographs or space-borne instruments with cold telescopes, dark current must be suppressed to very low levels. As usual, due to the exponential dependence of charge carrier density on temperature, dark current is extremely sensitive to the operating temperature.

Figure 10.7 shows dark current vs. temperature for typical InSb and HgCdTe materials with different cut-off wavelengths (approximate values as the cut-off wavelength changes slightly with temperature). If the dark current requirement were < 0.1 electron s^{-1}, the required operating temperatures would be < 10 K for InSb, < 70 K for HgCdTe ($\lambda_c \sim 5\,\mu m$) and < 145 K for HgCdTe ($\lambda_c \sim 1.7\,\mu m$).

InSb or HgCdTe large-format arrays are used on virtually every major ground-based or space-borne infrared telescope. The two major commercial suppliers are Teledyne and Raytheon. The performance of a Teledyne $2k \times 2k$ H2RG™ array with a 2.5-μm cut-off wavelength is summarised in Table 10.1. With the negligible dark current contribution, the combination of 18 electrons read

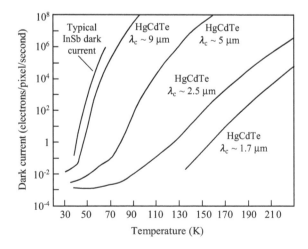

FIGURE 10.7 Typical dark current vs temperature for 18-μm photodiode pixels using different materials. (From Sprafke & Beletic (2008), reproduced with permission of the Optical Society of America.)

TABLE 10.1

Performance Specifications for the Teledyne H2RG™
Array with Cut-Off Wavelength of 2.65 μm

Array size	2048 × 2048
Pixel pitch	18 μm
Frame rate	Up to 74 Hz
Quantum Efficiency	>70%
Dark current	<0.05 electrons s^{-1}
Readout noise	<18 electrons with CDS
Well capacity	>80,000 electrons
Crosstalk	<2%

Source: From Teledyne H2RG™ data sheet, Teledyne Imaging Sensors, 2017.

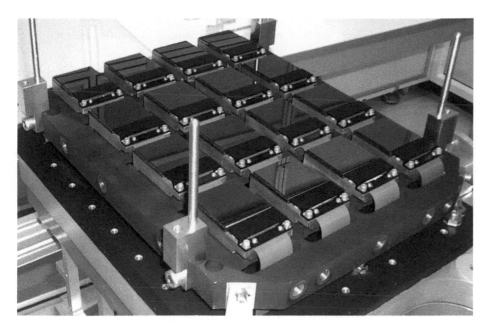

FIGURE 10.8 Photograph of VISTA infrared camera focal plane comprising 16 2k × 2k arrays. (Credit: ESO/VISTA.)

noise and quantum efficiency of 0.7 means that around 25 signal photons can be detected in a 1 second integration, corresponding to a detector NEP of ~ 5×10^{-18} W Hz$^{-1/2}$ per pixel at 1 μm. Similar arrays are available with cut-offs at 1.75 and 5.3 μm. For survey instruments, multiple arrays can be tiled together in a large focal plane. For example, the 0.84–2.5 μm camera on the VISTA infrared survey telescope (Figure 10.8) has 16 2k × 2k Raytheon HgCdTe arrays in the focal plane with a total of 67 million 0.34″ pixels.

10.2.4 BLOCKED IMPURITY BAND (BIB) PHOTOCONDUCTIVE DETECTORS

HgCdTe-based detectors do not work longward of around 14 μm as the energy gap is too large. The best technology for the region between 14 and around 28 μm uses doped silicon photoconductors. For silicon-based detectors, high infrared absorption efficiency requires either high doping levels or large detector thickness. A thick absorbing region is undesirable as it makes charge collection more

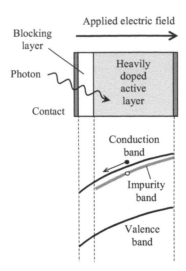

FIGURE 10.9 Structure and energy bands of an n-type blocked impurity band (BIB) detector.

difficult and, for space applications, it also makes the detector more susceptible to ionising radiation. On the other hand, heavy doping leads to the formation of an impurity band rather than localised impurity states, and hopping conduction by charges in the impurity band results in a high dark current. Blocked impurity band (BIB) detectors (also known as impurity band conduction, IBC, detectors) have high doping levels, allowing for small detector volume, but also suppress the dark current by means of a thin blocking layer of pure material. The basic structure and energy bands of an n-type BIB are shown in Figure 10.9. The active region is the heavily doped layer in which an impurity band exists just below the conduction band. Infrared photons can promote electrons from the impurity band to the conduction band. The blocking layer of pure silicon prevents dark current conduction occurring in the impurity band but does not prevent the photo-current flowing in the conduction band. The detector must still be cooled to reduce dark current due to thermal excitation of electrons from the impurity band to the conduction band. The substrate, contacts, and blocking layer can be made transparent in the IR, so the illumination of the detector can be either from the front side or from the rear side. Since silicon BIB arrays perform well down to $5\,\mu m$, they constitute a second high-performance option for the $5–14\,\mu m$ region, together with HgCdTe photodiode arrays.

The state of the art in Si BIB detectors is represented by the 1024×1024 (1k × 1k) pixel arsenic-doped silicon (Si:As) arrays (Love et al. 2005; Rieke et al. 2015) developed for the MIRI instrument which will fly on the James Webb Space Telescope (JWST), which is due to be launched in 2021. MIRI (Wright et al. 2015) is a camera and spectrometer operating in the $5–28\,\mu m$ range and has three such arrays. The pixels are $25 \times 25\,\mu m$ in area and have a well capacity of about 2.5×10^5 electrons and peak responsive quantum efficiency >0.6, with anti-reflection coating. The detector chips are indium bump-bonded to silicon readout and multiplexer chips, which provide a read noise per pixel of ~14 electrons. The optimum operating temperature for the arrays is <6.7 K, with a dark current ~ 0.2 electrons s^{-1}.

10.3 OPTICAL AND UV DETECTORS

10.3.1 CHARGE-COUPLED DEVICES (CCDs)

For many imaging applications, detector arrays based on silicon metal-oxide-semiconductor (MOS) technology have now become the technology of choice. In astronomy, the most commonly used detector type for optical, UV, and soft X-ray wavelengths is the CCD. The basic unit of a CCD is a MOS capacitor, the p-type variant of which is illustrated in Figure 10.10a. A slab of p-type silicon has a thin (~ $0.1\,\mu m$) surface coating of silicon dioxide (an insulator), on top of which a

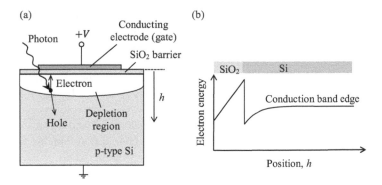

FIGURE 10.10 MOS capacitor as the basic unit of a CCD (a); energy of the conduction band edge in the oxide and the silicon (b).

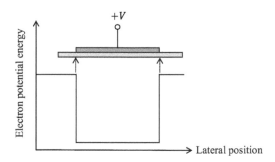

FIGURE 10.11 Depiction of MOS capacitor as a potential well in which electrons are trapped.

conducting layer forms a gate electrode. A positive voltage applied to the gate repels holes, the majority carriers in the material, leaving a depletion region devoid of free carriers underneath the gate. If an electron/hole pair is created in or near the depletion region, then the hole is neutralised by an incoming electron from ground and the electron moves towards the positive electrode until it gets stuck at the Si-SiO$_2$ barrier. Electron–hole pairs can arise either due to thermal generation (which can be reduced to a low level by cooling) or due to the absorption of incident photons of sufficient energy.

The energy band structure of the MOS capacitor is shown in Figure 10.10b. The conduction and valence band edges in the p-type material are decreased in energy near the oxide interface by the influence of the positive gate potential (positive voltage means lower electron energy). The conduction band in the oxide is unpopulated and at higher energy than in the semiconductor and also decreases in energy with proximity to the conducting electrode. The minimum potential energy, to which any electrons in the semiconductor conduction band will be attracted, is at the oxide-semiconductor interface.

Figure 10.11 shows how electrons, accumulated in proportion to the total number of photons absorbed and collected in the potential energy minimum at the interface, can be regarded as trapped in a potential well. The CCD array is a regular two-dimensional grid of such capacitors, configured so that charge can be moved from one to another, and eventually to a readout point, where the charge is measured. For simplicity, most arrays have just one readout amplifier, so each packet of charge has to be transferred to it sequentially.

10.3.1.1 Charge Transfer and Readout

Charge transfer is best illustrated by considering the three-phase CCD. In this device, each pixel has three gate electrodes. During charge collection, the outer two are maintained at 0 V or a negative potential to ensure that all the collected charge is located under the central gate. Electrons released by incoming photons in any part of the pixel will be drawn into the central potential well. Every third gate along a row of such pixels is connected to the same voltage supply line (or phase). By controlling the voltages applied to the gates, the charge packets, which are effectively a record of the amount of light incident, can be moved along the array.

The charge transfer process between two pixels is shown in the simplified illustration in Figure 10.12a. The pixels have gates labelled 1a,b,c and 2a,b,c as shown. In step 1, there are charge packets stored under the two middle gates of the pixel. In step 2, the voltages on the c gates are increased to V_g causing the charge packets to be shared between the b and c gates. Then, in step 3, the voltages on the b gates are reduced to zero leaving all the charge stored under the c gates. Thus, the charges have been shifted one gate to the right without getting mixed up.

Charge transfer along an entire row can be effected by connecting each of three control lines (phases) to every third gate as shown in Figure 10.12b. The voltage waveforms for the three phases are carefully synchronised as shown in Figure 10.12c to maximise the charge transfer efficiency and keep the charge packets separate.

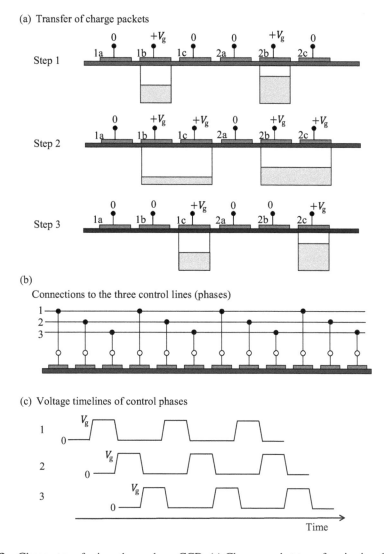

FIGURE 10.12   Charge transfer in a three-phase CCD. (a) Charge packet transfer via signals applied to the CCD gate electrodes. (b) Connections to the three control lines. (c) Voltage timelines corresponding to the three steps in (a).

The whole array can be read out by transferring charges in this way and introducing each charge packet to the input of a single readout amplifier, as shown in Figure 10.13. The charges are moved in the x direction by one pixel into the output register, a line of pixels at the bottom edge devoted to the readout process. The contents of the output register are then shifted sequentially in the y direction to the readout amplifier, which has input capacitance C. A charge ΔQ generates a corresponding voltage input of $\Delta V = \Delta Q/C$ at the amplifier for digitisation and recording before the capacitor is discharged in readiness for the next pixel. Typically, $C \sim 0.05$ pF, so that each electron at the gate contributes 5–10 μV to the measured voltage. The voltage is converted to digital form and digitised and recorded in analogue-to-digital units (ADUs), with 1 ADU typically corresponding to a few electrons.

The readout process is repeated until the entire array has been read out. Knowing the sequence in which the pixels have been read out, the two-dimensional charge distribution (analogous to the astronomical image) can be reconstructed.

CCDs are made in very large arrays, with thousands of pixels on a side. The charge transfer process, therefore, has to be very efficient. Any charge left behind in a pixel after a transfer operation will add to the charge being transferred from the next pixel in line. The result is smearing of the image. For a pixel-to-pixel charge transfer efficiency of η (the fraction of the charge packet that gets transferred each time), a fraction $1 - \eta$ of the charge gets left behind at each transfer. For an $N \times N$ pixel array, the overall fraction of charge lost for the last pixel to be read out (top left-hand corner in Figure 10.13) is

$$\eta_{\text{tot}} = \eta^{2N}. \tag{10.3}$$

For $N = 2048$ and $\eta = 99.99\%$ (four nines), this is only 66%. For $\eta = 99.999\%$ (five nines), it is 96%. Clearly, the charge transfer process has to be very efficient – at least 99.999%.

If the exposure time is too long for the brightness of the source, then some pixels become saturated (their electron storage capacity is exceeded), and electrons can spill out into neighbouring pixels. This results in streaking of the images of bright point sources, known as blooming. It can be eliminated by making sure that the exposure time is sufficiently short, but that may be undesirable – for instance, when a field of faint sources contains a bright star that is not of interest, in which case over-exposure of the star's image is allowable but blooming contamination of other parts of the image is not. Blooming can be eliminated in the CCD array design by incorporating anti-blooming gates in each pixel, which drain off excess charge when the well capacity is reached. This is at the cost of some reduction in fill factor due to the area occupied by the extra gates.

The charge transfer efficiency of the simple "surface channel" CCD depicted in Figure 10.12 is not good enough because, inevitably, there are crystal lattice imperfections at the boundary between the silicon and silicon dioxide, materials that have different crystal structures. These lattice defects have associated energy states, or traps, where otherwise free electrons can get stuck. This is why it is not a good idea to try and move charge along the ragged interface between the two materials,

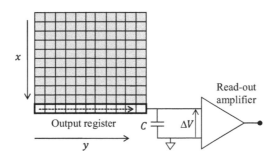

FIGURE 10.13 CCD charge readout.

especially for low light-level applications. To avoid that, practical CCDs are constructed in such a way that the charge is transported in a "buried channel" at a level below the Si-SiO$_2$ interface, within the bulk material.

10.3.1.2 Buried Channel CCDs

In a surface channel CCD, the maximum in electric potential is right at the interface so electrons collect there where their potential energy is minimised. In a buried channel CCD, an additional thin layer of n-type silicon is located under the gate, and forms a p-n junction with the p-type substrate, as shown in Figure 10.14a. The p-n junction is reverse biased by a positive voltage applied to the n-type layer, generating an internal electric field pointing from n to p in a depletion layer around the junction (Figure 10.1). This internal electric field and the electric field produced by the gate voltage combine to produce a maximum in the electric potential (and so a minimum in the electron energy) which is within the n-type material, just outside the n-side depletion region instead of at the SiO$_2$ interface as previously. The conduction band edge as a function of position below the gate is shown in Figure 10.14b. Electrons liberated by photons in the p-type substrate are transported across the junction to the neutral n-region but are then held there because the bias voltage on the n-side is more positive than V_g. The charge storage thus occurs inside the bulk silicon material, and the electron packets can be moved freely along the array without charge trapping. Buried channel CCDs have much better charge transfer efficiency than the surface channel CCD but at the cost of lower charge storage capacity.

10.3.1.3 CCD Quantum Efficiency and Spectral Response

Silicon has a refractive index varying between ~4 and 6 over the visible region, and to avoid substantial reflection losses, scientific CCDs must be anti-reflection coated. They can be illuminated either from the front side or from the backside. In general, backside illumination is better because the surface is uncomplicated by the electrodes, making it easier to apply the anti-reflection coating, and it avoids potential losses due to imperfect transparency of the electrodes.

The optical absorption of silicon varies greatly between UV and infrared wavelengths. The absorption length (corresponding to 63% absorption) is plotted against wavelength between 0.25 and 1.2 μm in Figure 10.15. At UV wavelengths (less than around 0.4 μm), Si is highly absorbing, with photons penetrating less than 10 nm. However, between 0.4 and 1 μm, the absorption length changes by four orders of magnitude, and the material becomes increasingly transparent as the wavelength increases further. This makes it difficult to optimise a CCD over the whole of the visible or NIR band.

Good long-wavelength response requires a thick substrate for good absorption, but with the disadvantage that the electric field is small far away from the gate and electrons produced there may recombine before they are collected. For shorter wavelengths, a thin substrate is best to ensure that

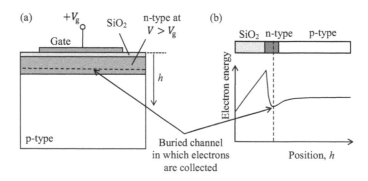

FIGURE 10.14 Structure of a buried channel CCD (a); electron energy distribution with its minimum within the bulk n-type material (b).

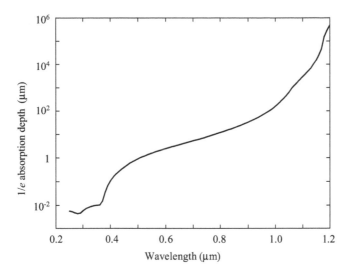

FIGURE 10.15   Absorption depth (1/*e*) of silicon vs. wavelength at near UV to infrared wavelengths. (Based on data tabulated in Green & Keevers 1995.)

absorption occurs close to the charge collection level, minimising recombination losses. Another advantage of a thin substrate is lower susceptibility to ionising radiation (particles or γ-rays), which is important for operation in space. Although thinned CCDs are more difficult (and so more expensive) to produce, their superior performance makes them the technology of choice for most astronomical applications. However, they do pose a problem with poor long-wavelength response. One way to circumvent this is to make a thicker (~50 μm) CCD for better red absorption, but using very high-resistivity silicon as the substrate. The depletion layer then extends much more deeply into the material, so there is a large depletion layer volume with high electric field throughout. This means that liberated electrons are quickly transported to the collection level.

The overall responsive quantum efficiency of a CCD array depends on reflection and absorption losses and recombination losses and inevitably varies with wavelength. No single CCD device will be well optimised across the entire visible range. Examples of spectral performance of thinned back-illuminated astronomical CCDs are shown in Figures 10.16 and 10.17. Figure 10.16 shows quantum efficiencies at −100°C for some of the Teledyne-e2v CCD231-84 4k × 4k family of

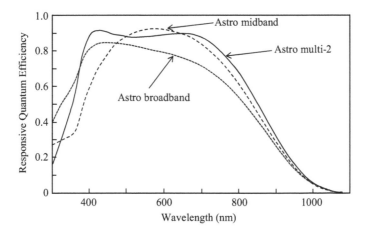

FIGURE 10.16   Typical responsive quantum efficiency vs. wavelength for Teledyne-e2v CCD231-84 anti-reflection coated deep-depletion CCDs. (Adapted from Teledyne-e2v CCD231-84 data sheet (2018), with permission from Teledyne-e2v.)

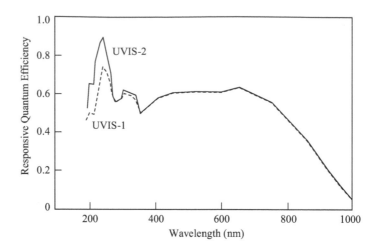

FIGURE 10.17 Responsive quantum efficiency vs wavelength for the UV-optimised e2v CCDs used in the short wavelength channel of the WFC3 HST instrument. (From the WFC3 Instrument Handbook (Dressel 2019); credit NASA/STScI.)

deep-depletion antireflection-coated CCDs. CCD231-84 arrays are used in several major ground-based optical instruments, such as the Multi-Unit Spectroscopic Explorer (MUSE) instrument on the ESO Very Large Telescope and the High-Resolution Spectrograph (HRS) on the South African Large Telescope (SALT). Three quantum efficiency vs. wavelength curves are shown, with optimisations (CCD structure and antireflection coating) for different purposes. Figure 10.17 shows measured quantum efficiency vs wavelength for two UV-optimised CCDs (also manufactured by e2v) in the WFC3 instrument, which was installed in the Hubble Space Telescope in 2009. These 4k × 2k devices are designed for best performance in the 0.2–0.6 µm range.

10.3.1.4   CCD Noise

To derive a general expression for the sensitivity of an observation with a CCD array, we first consider an idealised observation by an array of noiseless pixels, under the following assumptions:

  i. each detector array pixel has responsive quantum efficiency η_d;
  ii. the signal from the source, n_{vS} photons per unit time per unit bandwidth, is spread over a number of pixels, n_{pix} (e.g. by diffraction, source extension or both) and these pixels are summed to derive n_{vS};
  iii. the observation is made in the presence of some background level per pixel, n_{vB} photons/second/Hz/pixel.

If a single pixel captured all of the signal, the SNR would be

$$\text{SNR} = n_{vS}\left(\frac{\eta_d t_{int}\Delta v}{2n_{vB}+n_{vS}}\right)^{1/2} = n_{vS}\left(\frac{\eta_d t_{int}\Delta v}{\left(n_{vS}+n_{vB}\right)+n_{vB}}\right)^{1/2}, \tag{10.4}$$

where Δv is the accepted passband, and t_{int} is the integration time. This is the same as equation (6.12) with the additional $\eta_d^{1/2}$ term taking the detector RQE into account. In the second version of the above expression, the two n_{vB} - contributions from the (source+background) and the background only measurement are separated for clarity.

For the measurement in which the n_{vS} is derived by summing over n_{pix} pixels, we must also account for the increased noise due to the contributions to the background from all those pixels:

$$\text{SNR} = n_{vS} \left(\frac{\eta_d t_{int} \Delta v}{\left(n_{vS} + n_{pix} n_{vB} \right) + n_{pix} n_{vB}} \right)^{1/2}. \tag{10.5}$$

The first n_{pix} term in the denominator of equation (10.5) corresponds to the uncertainty in the number of background photons registered together with the signal, and the second one represents the uncertainty in number for the background alone.

The need to coadd pixels degrades the SNR because the background noise is increased as the square root of the number of pixels. However, with a large-format array, it is possible to reduce the uncertainty in the background-alone measurement by averaging over a larger number of pixels than are used for the signal estimation.

The background can be measured from the same image as the signal, from a suitable set of pixels in the vicinity of the source (e.g. in an annulus around it). For example, if n_{bck} pixels are used instead of n_{pix}, the background uncertainty is reduced by a factor of $\left(n_{pix}/n_{bck} \right)^{1/2}$. The second term in the denominator of equation (10.5) then becomes $\left(n_{pix}/n_{bck} \right) n_{pix} n_{vB}$, giving

$$\text{SNR} = n_{vS} \left(\frac{\eta_d t_{int} \Delta v}{\left(n_{vS} + n_{pix} n_{vB} \right) + \left(n_{pix}/n_{bck} \right) n_{pix} n_{vB}} \right)^{1/2}. \tag{10.6}$$

Therefore,

$$\text{SNR} = n_{vS} \left(\frac{\eta_d t_{int} \Delta v}{n_{vS} + n_{pix} n_{vB} \left[1 + \dfrac{n_{pix}}{n_{bck}} \right]} \right)^{1/2}. \tag{10.7}$$

Maximising n_{bck} thus improves the SNR by reducing the uncertainty in the background. If $n_{bck} \gg n_{pix}$, the background term is reduced by a factor of 2 compared with the value for $n_{bck} = n_{pix}$, and the background measurement is essentially noiseless.

For an ideal CCD with measurement of the background using a large number of pixels, we can therefore write

$$\text{SNR} = \frac{n_{vS} \eta_d \Delta v t_{int}}{\left(\eta_d n_{vS} \Delta v t_{int} + \eta_d n_{pix} n_{vB} \Delta v t_{int} \right)^{1/2}}. \tag{10.8}$$

In a real CCD, there will be additional noise contributions from dark current and read noise. Dark current effectively adds to the background, contributing an uncertainty of $\left(n_{pix} \dot{q} t_{int} \right)^{1/2}$, where $\dot{q}$ is the dark current per pixel in electrons per unit time. Read noise introduces a fixed charge uncertainty, Δq_R, per pixel, regardless of the integration time, so that for n_{pix} pixels, the readout noise uncertainty is $\left(n_{pix} \right)^{1/2} \Delta q_R$.

Allowing for these noise contributions in equation (10.8), we have

$$\text{SNR} = \frac{n_{vS} \eta_d \Delta v t_{int}}{\left(\eta_d n_{vS} \Delta v t_{int} + \eta_d n_{pix} n_{vB} \Delta v t_{int} + n_{pix} \dot{q} t_{int} + n_{pix} \Delta q_R^2 \right)^{1/2}}. \tag{10.9}$$

Therefore,

$$\text{SNR} = \frac{n_{vS}\eta_d \Delta v t_{\text{int}}^{1/2}}{\left(\eta_d n_{vS}\Delta v + n_{\text{pix}}\left(\eta_d n_{vB}\Delta v + \dot{q} + \frac{\Delta q_R^2}{t_{\text{int}}}\right)\right)^{1/2}}. \tag{10.10}$$

Note that the read noise contribution to the overall noise decreases with integration time, but the dark current contribution does not.

In astronomical applications, the dark current contribution can often be made negligible by cooling the array or making the pixels sufficiently small. In various limiting cases, equation (10.10) reduces to simpler forms as summarised below.

$$\text{Signal dominates:} \quad \text{SNR} = \left(n_{vS}\eta_d \Delta v t_{\text{int}}\right)^{1/2}. \tag{10.11}$$

$$\text{Background dominates:} \quad \text{SNR} = n_{vS}\left(\frac{\eta_d \Delta v t_{\text{int}}}{n_{\text{pix}} n_{vB}}\right)^{1/2}. \tag{10.12}$$

$$\text{Read noise dominates:} \quad \text{SNR} = \frac{\eta_d n_{vS}\Delta v t_{\text{int}}}{n_{\text{pix}}^{1/2}\Delta q_R}. \tag{10.13}$$

In the signal photon noise limited regime, the SNR is, as expected, just the square root of the total number of photons registered, and there is no disadvantage in spreading the signal over a number of pixels. If the background and/or read noise are significant, then minimising n_{pix} is good for sensitivity – it is best to put as much of the signal as possible on a small number of pixels to avoid the additional background-photon number and readout uncertainties introduced by pixel averaging.

CCD operation can be read noise limited in some applications such as when signal and background are both low when the integration time needs to be kept very short or in carrying out high-resolution spectroscopy (small Δv). The SNR then increases linearly with integration time and with bandwidth because the read noise is fixed and the signal increases linearly with t_{int} or Δv. Unlike infrared arrays, which have a separate readout amplifier for each pixel, it is not possible to do sampling up the ramp with CCDs. Schemes for correlated double-sampling can be implemented however, resulting in substantial reduction in read noise.

10.3.1.5   Charge Multiplying CCDs

When a conventional CCD is used for observations of a faint source under ultra-low background conditions, or when the readout rate must be high, the sensitivity is limited by read noise. Enough electrons must be generated by incoming photons to exceed the read noise. In charge multiplying CCDs, the pixel charge packets are amplified before being read out, making the read noise contribution negligible, and allowing even single-photon detection to be achieved. The amplification is effected by adding a gain register as an extension to the output register, as shown in Figure 10.18. In the gain register, the voltages used to clock the pixels along are made much higher than necessary just to transfer the charge. As they are transferred, the stored electrons gain enough kinetic energy to cause impact ionisation of lattice atoms, so producing secondary electrons. The gain factor for one transfer is small, but cascaded over many transfers, an initially small charge packet can be boosted so that by the time it arrives are the readout stage, it is much larger than the output read noise.

10.3.1.6   CCD Performance Parameters

As an illustration of the main CCD performance parameters and typical values achieved in the case of astronomical CCDs, Table 10.2 lists the main performance parameters for the e2V CCD231-84 back-illuminated 4k × 4k CCD array:

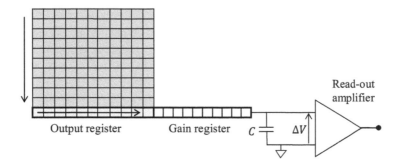

FIGURE 10.18 Use of a gain register to achieve charge amplification prior to readout in an electron-multiplying CCD.

TABLE 10.2
Key Performance Specifications of the e2V CCD231-84
Back-Illuminated 4k × 4k CCD Array

Number of pixels	4096 × 4112
Pixel size	15 μm × 15 μm
Image area	61.4 mm × 61.4 mm
Package size	63.0 mm × 69.0 mm
Amplifier sensitivity	7 μV/electron
Readout noise	5 electrons at 1 MHz
	2 electrons at 50 kHz
Max readout frequency	1000 Hz
Pixel full well capacity	350,000 electrons
Dark current	3 electrons/pixel/hour at 173 K
	0.02 electrons/pixel/hour at 153 K
Charge transfer efficiency	99.9995%
Spectral range	0.30–1.06 μm
Peak quantum efficiency	0.90

Source: From manufacturers data sheet.

10.3.1.7 Main Advantages of CCDs

The fundamental advantages of CCDs and similar electronic detector technologies, which have allowed them to become ubiquitous in astronomy, are

i. very low readout noise – as low as a few electrons;
ii. broad spectral response – arrays are available for soft X-ray to NIR wavelengths;
iii. excellent linearity – the signal scales with input power over a huge range of brightness (a factor of 10^5);
iv. convenience and reliability – low weight, low power dissipation, low voltage levels, robustness, and suitability of the output signal for digital storage and processing;
v. array size – arrays of up to 4096 × 4096 pixels are available, and multiple arrays can be used in large focal planes, providing very high observing efficiency. For instance, the Vera C. Rubin Observatory, currently under construction, will have an array of 180 4k × 4k CCDs camera, constituting a 3.2 gigapixel camera.

10.3.1.8   CCD Operation

CCDs are subject to a myriad of minor non-ideal effects which need to be corrected (but none of which detract from their excellent performance and suitability for many astronomical applications). Such effects include pixel-to-pixel variations, bad pixels or lines of pixels, blooming (as described in Section 10.3.1.1), and long-term degradation due to ionising radiation in space. Much attention and effort is devoted to the manufacture and operational procedures, and to the processing and calibration of the data, to minimise their effects and produce the highest possible data quality with detector distortions or artefacts reduced to an extremely low level.

10.3.2   THE PHOTOMULTIPLIER TUBE

Although solid-state electronic detectors such as CCDs are now used for most astronomical applications, detectors that rely on photoemission are not completely obsolete. One detector that has very high sensitivity and is suited to very low light-level application is the photomultiplier tube (PMT). This is a device, based on photoemission, which converts a low-intensity light signal to a current signal, and amplifies (multiplies) it by a large factor. It is widely used as an ancillary detector in high-energy gamma-ray detection, cosmic ray astronomy, measuring the faint light signatures created by charged particles passing through the Earth's atmosphere or generated by photons or ionising radiation in scintillation detectors. Some of these applications are described in Chapter 11.

The photoelectric effect can thus be used to count photons by producing a current proportional to the photon arrival rate. An incoming photon ejects an electron from the surface of a material (photo-emitter). The photo-emitting material is kept at a negative potential, so that the ejected electron does not fall back, and is therefore available for conduction. The negatively charged photo-emitter is called the photo-cathode. The photon energy must be at least equal to the work function of the material, W, which depends on the material, and also on its purity and surface quality. A good photo-emitter must be an efficient absorber of visible photons (so metals are no good) and have the property that the mean free path for the generated photo-electrons is greater than that for the incident photons. If this is not so, then the electrons will not be able to escape. Many different photo-cathode materials are available, optimised for different parts of the spectrum, and use materials such as alkali metals and semiconductors. A typical photo-cathode RQE is about 20%.

The essential features of the photomultiplier tube are shown schematically in Figure 10.19.

The incident photon enters the evacuated enclosure through a transparent window and ejects an electron from the photo-cathode, which is at a high negative potential. Secondary electrodes, called dynodes (of which there are typically 10), are at successively more positive potential with respect to the cathode ($V_1 < V_2 < V_3$ etc.), and are coated with a good emitter of secondary electrons. The

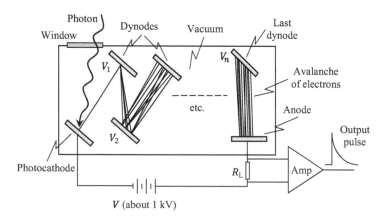

FIGURE 10.19   Essential features of a photomultiplier tube.

primary photoelectron strikes the first dynode and ejects more than one secondary electron (typically about 5). These are accelerated to dynode 2, where each causes emission of about 5 more, and so on. The result is that the single photon at the input causes an avalanche of about 10^5 electrons at the final collecting electrode (the anode). The current pulse at the anode creates a voltage pulse across the load resistor, R_L. Each detected photon gives rise to one pulse at the output, and the pulse rate is proportional to the incident photon rate.

Even with no illumination, unwanted dark current flows in the anode circuit, due to two effects: thermally excited emission of electrons from the photo-cathode, and ionising events caused by cosmic rays or radioactive decays producing ions or direct dynode hits. There are three main techniques for minimising the dark current and its effects on SNR: (i) the tube is operated at the value of anode voltage, V_{opt} which gives the maximum ratio of signal current to dark current (below V_{opt} the signal is too low, and above V_{opt} the dark current becomes prohibitively high); (ii) to reduce thermal emission, the tube is cooled (typically to –20 C), reducing the thermal emission exponentially; (iii) the output pulses are examined, and some are rejected by pulse height discrimination. All pulses with amplitudes outside a certain discrimination window are rejected because pulses from "legitimate" photon events tend to fall within a certain range of amplitudes. Figure 10.20 shows a typical plot of pulse rate vs pulse amplitude for unilluminated and illuminated conditions. There are many small noise pulses (from thermal emission) and a smaller number of very large noise pulses (from cosmic rays, etc.). Most of the signal pulses fall within quite a narrow range, and the discrimination window is set to match this range closely. This technique only works if the count rate is low enough that one can distinguish individual pulses, so the count rate must not be so large that the pulses often overlap (a typical pulse duration is 10 ns).

The overall gain of the PM tube is $G = g^N$, where g is the gain of one dynode and N is the number of dynodes. There will be random fluctuations in the number of electrons emitted per incoming electron. The probability that an incident electron will eject n_e electrons from a dynode is given by the Poisson distribution with mean g:

$$p(n_e) = \frac{g^{n_e}}{n_e!} e^{-g}. \tag{10.14}$$

Note that $n_e = 0$ is possible, with probability $p(0) = e^{-g}$, so there is a finite probability that the first dynode will not respond at all and the photon will be undetected: e.g., if $g = 4$, then $p(0) = 1.8\%$. Clearly, the gain of the first dynode should be reasonably high.

Dynode gain is proportional to the power supply voltage, V, so for the overall gain, G,

$$G \propto V^N \Rightarrow \frac{\Delta G}{G} = N\left(\frac{\Delta V}{V}\right), \tag{10.15}$$

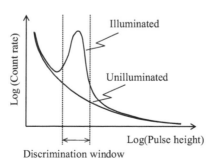

FIGURE 10.20 Pulse height discrimination.

where ΔG is a change in gain due to a change in power supply voltage of ΔV. Small fluctuations in V (for instance, due to temperature variations of the power supply) can therefore cause much larger fluctuations in the overall gain. This constitutes an extra noise contribution as it imposes a fluctuation in the relationship between input and output. Photomultiplier tube power supplies are, therefore, stabilised to a very high accuracy (typically 0.05%).

The photomultiplier tube can register the arrival of individual photons, but with the disadvantage of much lower quantum efficiency than solid-state detectors such as CCDs, and inability to register an image – it is a single-pixel detector. A device that works in a similar way and which also provides spatial information is the microchannel plate (MCP).

10.3.3 THE MICROCHANNEL PLATE

The MCP is effectively an imaging photomultiplier array and is suitable as a high-gain image intensifier for the optical, UV, and soft X-ray regions. It can be regarded as an array of tiny photomultiplier tubes. The design is illustrated in Figure 10.21.

The device consists of a thin glass plate (typical thickness ~ 1 mm), perforated with a large number of very narrow channels (typically $< 10\,\mu$m diameter). The top and bottom of the plate are coated with a metal to allow a high voltage (typically 1 kV) to be applied across the plate. An incoming photon generates a primary electron, which hits the wall of a channel producing multiple secondary electrons. These are accelerated down the channel, generating more when they strike the wall, and so on. In this way, the channel acts as a continuous dynode. A single incoming photon leads to an avalanche of 10^3–10^4 electrons at the output end. Further amplification can be achieved if desired by stacking two MCPs in sequence.

To record the image electronically, a two-dimensional grid of wires or strips is located on the output side and the centroid of the image is calculated from the ratio of the recorded signals. One type of such a readout scheme, the wedge-and-strip system, is shown in Figure 10.22. A dielectric plate has a metalised layer on which are etched four interlaced electrodes. Two of them are in the form of a pair of complementary wedges, and the other two form a complementary pair of strips. The ratios of the wedge heights and of the strip heights vary over the anode. An electron cloud emerging from an MCP pore will produce signals (X_1, X_2, Y_1, Y_2) from the four readout points, which will depend on the position of the event. The relative magnitudes of the signal pulses can be used to calculate the position of the electron cloud centroid to high accuracy so that the ultimate spatial resolution is determined by the pore size.

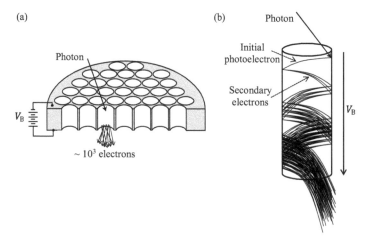

FIGURE 10.21 Schematic diagram of a microchannel plate (a); detail of electron avalanche created in one channel by a single incoming photon (b).

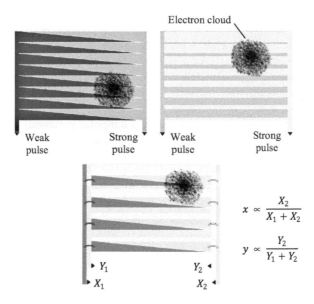

Electron cloud

Weak pulse   Strong pulse   Weak pulse   Strong pulse

$$x \propto \frac{X_2}{X_1 + X_2}$$

$$y \propto \frac{Y_2}{Y_1 + Y_2}$$

Y_1 Y_2

X_1 X_2

FIGURE 10.22 Wedge and strip microchannel plate readout system. (From Barnstedt (2019), reproduced with permission of the author.)

10.3.4 THE AVALANCHE PHOTODIODE

The avalanche photodiode (APD) is another type of optical-IR detector, which incorporates gain in the detection process. The structure of a typical APD is illustrated in Figure 10.23. A strong reverse bias (positive voltage applied to the n contact) is applied. Between the p-side of the reverse-biased junction and its heavily doped (p⁺) contact is a thick layer of intrinsic material. Most of the applied voltage appears across the depletion region of the p-n junction, resulting in a strong electric field, with a lower field in the intrinsic region. A photon enters through the p⁺ contact and is absorbed in the intrinsic region generating an electron–hole pair. The electron drifts towards the p-n junction and the hole towards the p⁺ contact. When the electron encounters the high electric field region, it is accelerated, generating enough energy to cause impact ionisation in collisions with lattice atoms. The secondary electrons are accelerated in turn causing further ionisation, resulting in an avalanche of electrons and a large current pulse. APDs have high quantum efficiency, and the arrival times of individual photons can be registered with accuracies of a few ns, making them suitable for applications where high count rate or high time resolution is required, such as observations of the optical counterparts of variable high-energy γ-ray sources (Kanbach et al. 2008).

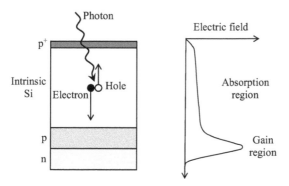

Photon

Electric field

p⁺

Intrinsic Si   Electron   Hole

Absorption region

p

n

Gain region

FIGURE 10.23 Typical avalanche photodiode structure.

10.4 ADAPTIVE OPTICS

As we discussed in Chapter 4, a telescope with input aperture diameter D produces a diffraction pattern in the focal plane with characteristic size ~ λ/D. The corresponding diffraction-limited resolution for a 1-m telescope is approximately $0.1''$ in the visible range. However, the angular resolution of a telescope that looks through the Earth's atmosphere is severely degraded by atmospheric turbulence, and the best achievable resolution at visible wavelengths is typically around $1''$ regardless of how big the telescope is. Continuous spatial and temporal variations in the refractive index of the air lead to the ideal plane wavefront that would be received from a distant target becoming distorted by the time it reaches the telescope, as shown in Figure 10.24. To make matters worse, the exact form of the wavefront distortion changes unpredictably on short timescales.

The coherence length, r_o, is the largest distance across the aperture over which the wavefront phase deviation is less than 1 radian. It depends on wavelength: $r_o \propto \lambda^{1.2}$, and is typically 10–30 cm at 500 nm wavelength. The aperture of the telescope can be regarded as being made up of a continually shifting pattern comprising a large number of "cells" of typical size r_o. The coherence time, t_o, is the timescale on which the phase distribution over the aperture can be regarded as constant. It is a measure of how rapidly the distorted wavefront is varying: the instantaneous "corrugation pattern" is continually changing randomly on a timescales ~ t_o. Typically, $t_o = 10$–100 ms.

This continually varying distortion of the wavefront degrades the image quality, as illustrated in Figure 10.25. First-order distortion (wavefront tilt) causes the image to "dance around" in the focal plane. Second-order distortion (wavefront curvature) causes the image to go in and out of focus.

Plane wavefront above the atmosphere

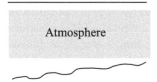

Distorted wavefront at ground level

FIGURE 10.24   Wavefront distortion caused by the atmosphere.

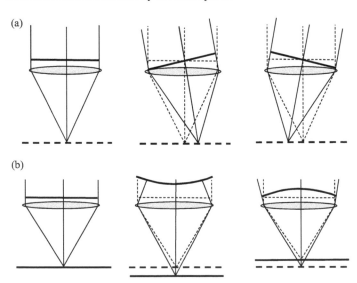

FIGURE 10.25   (a) Wavefront tilt causes the image to move in the focal plane. (b) Wavefront curvature leads to the image going in and out of focus.

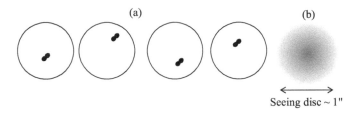

Seeing disc ~ 1"

FIGURE 10.26 Effect of wavefront tilt alone: (a) example "snapshots" of a double source taken at intervals $< t_0$; (b) the seeing disc resulting from an exposure much longer than t_0.

Higher-order distortions lead to the image becoming blurred and breaking up in more complex ways. The result is that, for an exposure longer than the turbulence coherence time, the energy from a point source is spread over a much larger area of the focal plane than the diffraction spot.

The effect of wavefront tilt alone is illustrated in Figure 10.26 for the case of a double star, which would be resolved by the telescope in the absence of distortions. For an exposure time less than t_0, an undistorted image would be recorded at some spot in the focal plane. A sequence of such exposures would produce undistorted images, but each one in a different position as the wavefront tilt changes randomly. The result of averaging such a sequence or making an exposure much longer than t_0 is that the image of the source is spread over a large area in the focal plane, the so-called seeing disc.

This atmospheric turbulence has two very unwelcome effects:

i. loss of angular resolution – for astronomical telescopes observing in good conditions, the effective angular resolution, or "seeing", might be 1"–2", which is far worse than the theoretical angular resolution of current telescopes;
ii. loss of sensitivity – the light is spread over a large area, so the amount of light received by an individual pixel is reduced. The signals from pixels can be averaged together to recover the full signal, but that introduces additional noise (proportional to the square root of the number of pixels that have to be co-added).

If the atmospheric turbulence arose just in front of the telescope, then all parts of the field of view would be subject to the same distortion, and the image quality would be the same for all angles (isoplanatic). The isoplanatic angle, θ_0, is the largest angle over which the turbulence remains correlated. Turbulence at high levels will not be shared by two sources if they are too far apart, as depicted in Figure 10.27. If the sources are farther apart than θ_0, then their images will not be affected exactly in the same way. Typically, θ_0 is quite small – a few arcseconds.

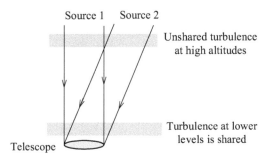

FIGURE 10.27 Low-level turbulence is shared by two nearby sources, but turbulence at higher levels in the atmosphere can be different.

10.4.1 ADAPTIVE OPTICS SYSTEMS

Advances in mirror and detector technology and computing power have made it possible to detect and correct, in real time, the distortions in wavefront flatness produced by atmospheric turbulence. This can produce major gains in spatial resolution because the size of the seeing disk is much reduced, and in sensitivity, because the source energy is concentrated onto a much smaller number of pixels in the focal plane.

A great deal can be achieved just by correcting for the gross tilting of the wavefront (first-order correction) using a "tip-tilt" secondary mirror. Further improvements can be achieved by correcting for higher-order distortions. The essential features of an adaptive optics system are shown in Figure 10.28.

A small fraction of the beam is reflected by the beam-splitter into a wavefront sensor, located at an image of the telescope primary mirror, which measures the phase distortions of the wavefront. The most common type is the Shack–Hartmann design, as shown in Figure 10.29. It uses a regular array of small lenslets (ideally of a size equivalent to the coherence length) to produce a grid of sub-images. For a flat wavefront, the sub-images have the same regular pattern as the lenslet array. Distortions in the wavefront cause shifts in the positions of the sub-images, which can be used to infer how the slope of the wavefront varies across the array.

The required corrections to the optical path are computed in real time and used to adjust the angle of the tip-tilt mirror (first-order correction) and to control a deformable mirror (higher-order correction). These corrections are made on a timescale about ten times faster than the coherence

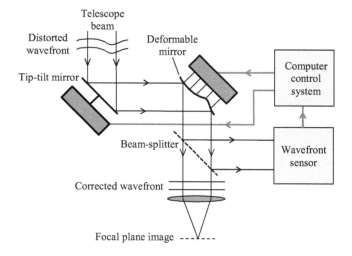

FIGURE 10.28   Essential features of an adaptive optics system.

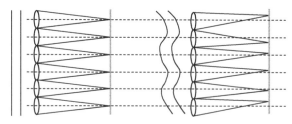

Plane wavefront incident on lenslet array
produces evenly spaces sub-images

Distorted wavefront produces
unevenly spaced sub-images

FIGURE 10.29   Principle of the Shack-Hartmann wavefront sensor.

time of the atmospheric fluctuations so that the corrected (flatter) wavefront presented to the focal plane instrument can be maintained.

Wavefront correction requires good SNR which must be achieved in a short integration time, so the source needs to be fairly bright. The science target itself is usually much too faint so that a bright object nearby (within the isoplanatic angle), termed a natural guide star, is used. Unfortunately, sufficiently bright stars are only available for a very small fraction of the sky (~1% at K-band and even less at shorter wavelengths; Beckers 1993). This problem can be overcome by the creation of a bright artificial guide star generated by directing a powerful (typically 50 W) laser beam towards a position on the sky close to the source. In most cases, the laser is designed to excite emission at 589 μm by a layer of neutral sodium atoms at an altitude of ~90 km. The artificial guide star can be used to correct higher-order distortions, but not first-order wavefront tilt because the detected laser light has gone through the atmosphere in both directions causing tip-tilt effects to cancel out. However, tip-tilt correction using a natural guide star requires only moderate brightness and slower real-time correction than for higher orders so that the probability of there being a suitable guide star near the science target is much greater.

The most advanced AO systems use multiple guide stars (typically 4–6) positioned to provide sensitivity to turbulence at different heights in the atmosphere. A guide star close to the source probes high-altitude turbulence, whereas a guide star further away on the sky probes turbulence nearer to the ground (Figure 10.27). Images of the sky at the relevant altitudes are generated within the instrument optical train and wavefront sensors, and deformable mirrors enable wavefront corrections corresponding to each level. This multi-conjugate adaptive optics (MCAO) technique creates a much larger isoplanatic area on the sky, typically an arcminute or more as opposed to a few arcseconds, and so enables imaging observations of extended sources (Rigaut & Neichel 2018). Figure 10.30 illustrates the dramatic improvements in image quality and sensitivity that are typically achieved with modern adaptive optics systems on large-aperture ground-based telescopes. It shows images of stars in the galactic centre region taken with the Keck 10-m telescope with and without adaptive optics. As well as resolving individual star images which are washed out in the AO-off image, the AO-on image also detects many faint stars which the AO-off image is not sensitive enough to register.

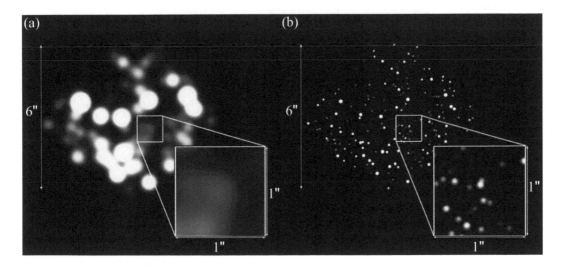

FIGURE 10.30 Images of the stars in galactic centre region taken by the W. M. Keck 10-m telescope without (a) and with an adaptive optics system (b), illustrating the gains in angular resolution and sensitivity. (Credit: UCLA Galactic Center Group - W.M. Keck Observatory Laser Team. These images were created by Prof. Andrea Ghez and her research team at UCLA.)

10.5   CASE STUDIES

10.5.1   The WFPC 3 Instrument on the Hubble Space Telescope

Since its launch in 1990, the HST has been fitted with a series of cameras operating in the UV to infrared part of the spectrum. The telescope primary mirror suffers from spherical aberration, which was not discovered until after launch. Careful measurements of the resulting distorted images allowed the effect of the aberration on the wavefront to be characterised very accurately so that it could be cancelled out by including correction plates in new instruments fitted by astronauts in Hubble servicing missions. Its third-generation Wide-Field Camera (WFC-3), installed in 2009, was designed to provide sensitive imaging over a large field of view and cover the compete UV-NIR range (0.2–1.7 μm). Figure 10.31 is a schematic diagram of the WFC-3 optical layout. There are two separate cameras, one for the UV-visible region (0.2–1 μm) and the other for the infrared (0.8–1.7 μm). The flat pick-off mirror intercepts the incoming $f/24$ beam from the telescope and directs it to the flat channel selection mirror, which allows the use of either the infrared channel (mirror in) or the UV-visible channel (mirror out).

The infrared channel optical train has a hyperbolic-ellipsoidal mirror combination to relay the focal plane onto the detector array. The ellipsoid has tip-tilt alignment and focus adjustment mechanisms to ensure that the image is correctly positioned and in focus. The spherical aberration of the HST primary is cancelled out by a refractive corrector plate, and a cold stop blocks the thermal emission from the telescope secondary and its support structure. The beam, now $f/10$, then passes through a filter wheel with 17 selectable positions (15 filters and 2 grisms) before reaching the detector focal plane. A blank position, which blocks the beam completely, is used to make dark current measurements. No mechanical shutter is needed in the infrared channel – the detector provides electronic shuttering. The detector array is a 1k × 1k HgCdTe array with a pixel size of 18 μm, corresponding to 0.13″ on the sky, and a field of view of 123″ × 136″.

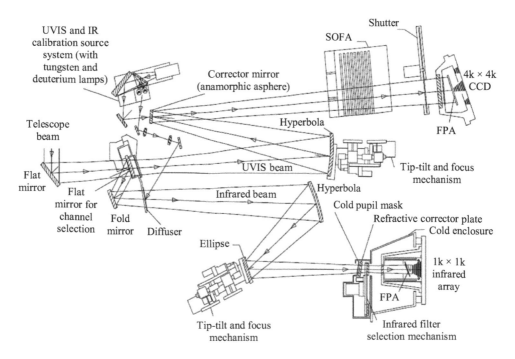

FIGURE 10.31   Optical layout of the HST Wide-Field Camera 3 (WFC 3) instrument. (From the WFC3 Instrument Handbook (Dressel 2019); credit NASA/STScI.)

The UV-visible channel has a hyperbolic re-imaging mirror with tip-tilt alignment and focus adjustment mechanisms and a refractive corrector plate to cancel the telescope spherical aberration. A selectable optical filter assembly (SOFA) contains 12 different filter wheels allowing selection of any of 48 elements: 47 filters and a UV grism. A mechanical shutter located in front of the final focal plane is used to control the exposure time for an observation. The detectors are two butted 4096×2051 e2v CCD arrays, which are thinned and back-illuminated for good UV response. The pixels are 15 μm in size, corresponding to 0.04″ on the sky with a field of view of 162″ x 162″. For optimum performance, the CCDs are operated at a temperature of −83°C, provided by a thermo-electric cooler.

Both channels make use of internal calibration sources providing a uniform flat-field illumination of the detector arrays.

10.5.2   The K-Band Multi-Object Spectrometer (KMOS) on the ESO-VLT

Large-aperture ground-based telescopes are expensive to build and to operate, and the ability to observe simultaneously over a large field of view is important to maximise their efficiency and scientific productivity. A multi-object spectrometer (MOS), as described in Chapter 7, is an instrument capable of carrying out simultaneous spectroscopy on a number of sources or regions within a larger field of view, and most large observatories are equipped with MOS instruments. An example is KMOS (Sharples et al. 2004), an infrared MOS, covering the wavelength range 0.8–2.5 μm, on the European Southern Observatory's Very Large Telescope (ESO-VLT) at the Paranal observatory in Chile.

KMOS has 24 robotically operated arms that position pick-off mirrors at selectable positions in the Nasmyth focal plane of the telescope. The overall field of view is 7.2 arcminutes in diameter, and 24 sub-fields, each 2.8×2.8 arcsecond, can be selected for simultaneous imaging spectroscopy. Each arm has internal optics that relay the sub-field to its own image slicer integral-field unit (IFU). The IFU splits the sub-field into 14 identical slices, with 14 spatial (0.2 arcsec^2) pixels along each slice. The slices are then aligned linearly onto a long spectrometer slit, the light from which is dispersed by one of three cryogenic grating spectrometers, with each spectrometer serving eight sub-fields. The light paths for a group of eight sub-fields and their spectrograph unit are illustrated schematically in Figure 10.32a. For each sub-field, a 14×14-pixel field is observed, and a spectrum with resolution $\lambda/\Delta\lambda = 2000 - 4000$ for each pixel. The three spectrometers each use a 2k × 2k H2RG HgCdTe detector array.

Like many facility instruments on the largest ground-based telescopes, KMOS is physically big. To accommodate its large mass and volume, it is mounted at the telescope-fixed Nasmyth focus. The instrument optics and detectors are contained inside a 2-m diameter cryostat and cooled by mechanical closed-cycle coolers with the optics at around 140 K and the detectors operating at less than 80 K. Figure 10.32 also shows photographs of the 24 pick-off arms and of KMOS installed at the VLT.

10.5.3   MICADO – An Imager for the Extremely Large Telescope

The European Southern Observatory's Extremely Large Telescope (ELT), which is currently under construction at Cerro Armazones in Chile, will be the largest optical telescope in the world, with a diameter of 39 m. MICADO, the Multi-AO Imaging Camera for Deep Observations, will be one of its first-generation instruments (Davies et al. 2016; 2018). It is designed to take the maximum advantage of the sensitivity and angular resolution afforded by the large aperture in combination with its advanced MCAO system.

The ELT PSF will have a diffraction-limited FWHM of between 6 and 10 mas (milliarcseconds) in the J and K bands, and MICADO will operate in the 0.8–2.4 μm range, providing diffraction-limited imaging over a field of view of up to 50 arcseconds. A key scientific requirement for MICADO is high astrometric stability so that stellar proper motions of around 10 mas can be measured over

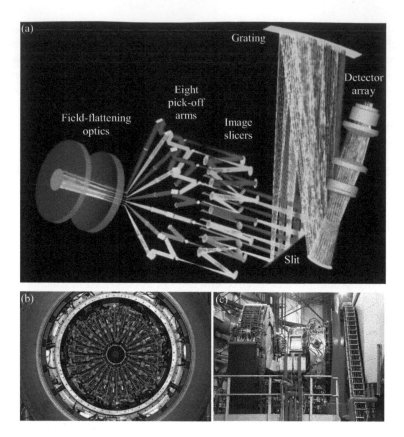

FIGURE 10.32   (a) Optical paths for a group of eight KMOS sub-fields. (Credit: UKATC/Durham University/ University of Oxford.) (b) Photograph of the 24 robotic arms that position the IFU pick-off mirrors at specified locations in the focal plane. (Credit STFC/UKATC/ESO.) (c) Photograph of KMOS installed at the Nasmyth focus of the VLT. (Credit ESO/G. Lombardi.) The instrument cryostat, in the entre of the picture is ~ 2 m in diameter.

a period of several years. MICADO is a large instrument with a complex optical layout, shown in Figure 10.33.

The cryostat maintains the optics temperature between 80 and 100 K to minimise the instrument background on the detectors. Two mask wheels at the input allow various slits for spectroscopy and field masks for imaging to be selected. The beam is collimated before passing through one or two of up to 30 filters mounted in the filter wheels. The atmospheric dispersion correction (ADC) unit compensates for the effect of the non-unit refractive index of the atmosphere, so acting as a weak prism, which would otherwise degrade the image quality. It is based on a pair of prisms that introduce an opposite dispersion to cancel out the atmospheric dispersion. The pupil wheel contains a selection of cold stops and masks for different observing modes. The main mechanism allows one of four optical trains to be inserted: one for small-field high-angular resolution (20″; 1.5 mas) imaging – shown in the beam in Figure 10.33; one for wider field imaging at lower angular resolution (50″; 4 mas); one for spectroscopy; and one for pupil imaging. Finally, the beam is imaged by the camera onto the final focal plane, containing a 3×4 array of $4k \times 4k$ H4RG arrays which are operated at 40 K to minimise the dark current.

Figure 10.34 illustrates the major advance in both sensitivity and angular resolution that MICADO will achieve. It shows simulated 2.2-μm images of the central 1×1 arcsec^2 region of the nuclear star cluster of the Milky Way as observed by an 8-m telescope, with diffraction-limited resolution of 50 mas, and by MICADO with 10 mas resolution.

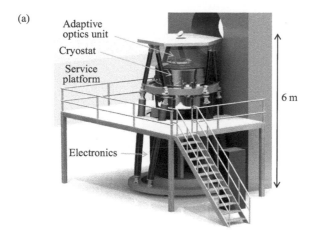

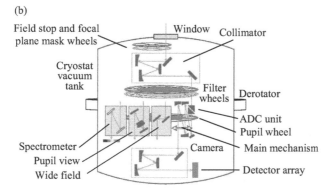

FIGURE 10.33 (a) Computer-rendered picture of the ELT MICADO instrument. (Credit: ESO/MICADO consortium.) (b) Functional diagram of the MICADO cold instrument. (Credit: MICADO Consortium.)

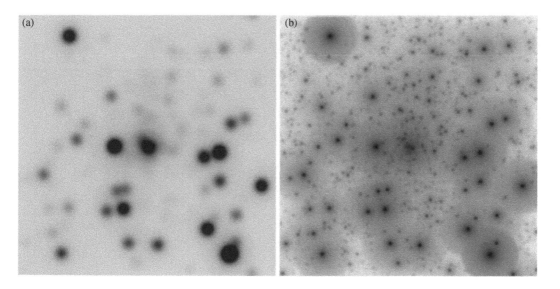

FIGURE 10.34 Simulated K-band images of $1'' \times 1''$ region near the Galactic centre with an 8-m telescope (a) and MICADO on the E-ELT (b). (From Trippe et al. (2010), reproduced with permission from Oxford University Press.)

11 X-Ray, γ-Ray, and Astro-Particle Detection

11.1 INTRODUCTION

The X-ray region of the spectrum covers wavelengths from ~10 nm down to ~10 pm, with somewhat longer wavelengths in the UV range, and shorter wavelengths in the γ-ray range. Figure 11.1 shows the (loosely defined) spectral regions in terms of wavelength and corresponding photon energy in eV. Energies of a few 100 eV to a few keV are designated as soft X-rays, and energies of a few keV to ~ 100 keV are known as hard X-rays. X- and γ-ray photons carry considerably more energy compared to those in the optical or IR, and it is more common to characterise them by referring to the photon energy rather than the wavelength. A 10-nm photon has an energy of 2×10^{-17} J or 124 eV, and 0.1 nm corresponds to 12.4 keV. For comparison, photons have has energies of 1.8–3 eV. High-energy photons are more penetrating and difficult to steer or absorb, and the characteristics of X-ray and γ-ray detection systems are largely dictated by these considerations. In this chapter, we consider the techniques and technologies used in astronomical instruments. Because the Earth's atmosphere is completely opaque to X and γ radiation, it is normally only possible to carry out observations from high-altitude balloons or better still from space. However, the signatures of very high-energy γ rays and cosmic ray particles can be observed from the ground through their interactions with the atmosphere leading to showers of particles and radiation, which can reach ground-based detectors.

11.2 X-RAY CCDs AND FANO NOISE

For X-ray energies up to around 10 keV, silicon CCD detector arrays can be used. Silicon has a sufficiently large attenuation coefficient (Figure 3.12) that single X-ray photons can be absorbed with good efficiency. A photon is absorbed in the silicon through the photoelectric effect, creating a highly energetic electron which gives rise to a cascade of ionisation resulting in the generation of a number of electron–hole pairs in proportion to its energy:

$$N_e = \frac{E}{\varepsilon},$$ (11.1)

where ε is the average amount of photon energy per created electron–hole pair.

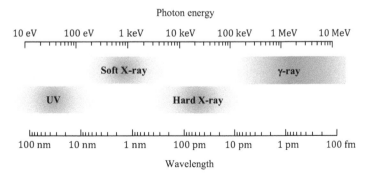

FIGURE 11.1 UV, X-ray, and γ-ray spectral ranges.

The corresponding charge packet can be collected to register the detection of the photon, and its magnitude also provides a measure of the photon energy.

Silicon is an indirect bandgap semiconductor, meaning that the minimum in the conduction band lies at a different momentum value than the maximum energy level in the valence band. This means that a transition must involve a change in the electron's momentum, and therefore not all of the energy deposited goes to promoting electrons to the conduction band – some, in fact most, of it is used up in the generation of phonons (quantised lattice vibrations) and so dissipated as heat rather than in producing signal charge. For silicon, the energy needed to create an electron–hole pair is $\varepsilon \approx 3.7\,\text{eV}$ (depending slightly on temperature), compared to the bandgap energy of $1.12\,\text{eV}$. For example, a 5-keV photon would produce on average $N_e = 1370$ pairs. If the generation of each electron–hole pair were uncorrelated with the generation of any other, then Poisson statistics would apply and we would expect, from considerations in Chapter 6, that the rms fluctuation in N_e would be $\sqrt{N_e}$. However, this is not the case as they are all generated by the same event. If all of the photon energy inevitably went to the creation of electron–hole pairs, then there would be no fluctuation in the number of charge carriers per photon, and if all of it went to generate phonons, then there would be no electron–hole pairs. In practice, the situation is intermediate between these two extremes because some of the photon energy is converted to phonon excitations. There are statistical variations in this fraction, and the resulting rms fluctuation in the number of electron–hole pairs turns out to be significantly less than $\sqrt{N_e}$, leading to better energy resolution than the Poisson prediction, as first analysed by Fano in 1947.

Let an incoming photon of energy E result in N_e electron–hole pairs and N_p phonon excitations, with E_e (the band gap energy) and E_p being the energies needed to create a single pair or phonon excitation:

$$E = N_e E_e + N_p E_p. \tag{11.2}$$

There will be fluctuations in N_e and N_p from one event to another, but for a single event, the two fluctuations must cancel out because the energy going into electron–hole pairs and that going into phonons must add up to the photon energy:

$$E_e \Delta N_e + E_p \Delta N_p = 0. \tag{11.3}$$

The rms fluctuations of N_e and N_p will be given by

$$\sigma_e = \sqrt{N_e} \quad \text{and} \quad \sigma_p = \sqrt{N_p}. \tag{11.4}$$

When averaged over a large number of photon events, the rms fluctuations in the two energy outcomes must be equal:

$$E_e \sigma_e = E_p \sigma_p, \tag{11.5}$$

so that

$$\sigma_e = \frac{E_p}{E_e} \sqrt{N_p}. \tag{11.6}$$

Substituting for N_p from equation (11.2),

$$\sigma_e = \frac{E_p}{E_e} \sqrt{\frac{E - N_e E_e}{E_p}}, \tag{11.7}$$

giving, from equation (11.1)

$$\sigma_e = \sqrt{\frac{EE_p}{E_e^2} - \frac{EE_p}{\varepsilon E_e}}. \tag{11.8}$$

Therefore,

$$\sigma_e = \sqrt{\frac{E}{\varepsilon}\left(\frac{\varepsilon E_p}{E_e^2} - \frac{E_p}{E_e}\right)} = \sqrt{N_e \frac{E_p}{E_e}\left(\frac{\varepsilon}{E_e} - 1\right)} = \sqrt{N_e F}. \tag{11.9}$$

where F, known as the Fano factor, is given by

$$F = \frac{E_p}{E_e}\left(\frac{\varepsilon}{E_e} - 1\right). \tag{11.10}$$

For silicon, $E_p=37$ meV, $E_e=1.1$ eV and $\varepsilon=3.65$ eV, giving $F=0.08$. This is close to the experimentally measured value of 0.1. The best achievable energy resolution of the detector (in the absence of readout noise) is

$$\Delta E = \sigma_e \varepsilon = \varepsilon\sqrt{N_e F} = \varepsilon\sqrt{\frac{EF}{\varepsilon}} = \sqrt{E\varepsilon F}. \tag{11.11}$$

For silicon, the ideal energy resolution for mono-energetic photons is thus smaller (better) than what would be expected from Poisson statistics by a factor of $\sim \sqrt{0.1} \approx 0.3$.

If the CCD readout noise is σ_R electrons, then the corresponding energy uncertainty of $\varepsilon\sigma_R$ must be added in quadrature giving an overall rms uncertainty in the photon energy of

$$\Delta E = \sqrt{\varepsilon^2 \sigma_R^2 + E\varepsilon F} = \varepsilon\sqrt{\sigma_R^2 + \frac{E}{\varepsilon}F}. \tag{11.12}$$

The measured spectrum for a series of mono-energetic photons will be a normal distribution with a standard deviation of ΔE, corresponding to a FWHM of $\sqrt{2\ln(2)}\Delta E = 2.35\Delta E$. Following Holland (2010), this is plotted as a function of X-ray photon energy for silicon in Figure 11.2 for different readout noise values. Ideal FWHM energy resolution for a silicon CCD thus varies between about 45 and 140 eV between 1 and 10 keV. The fixed readout noise produces a relatively greater degradation in resolution at low energies, but has only a small effect on performance above 1 keV for a typical readout noise of a few electrons.

Fano statistics also apply to other semiconductor detectors operating at X-ray and γ-ray wavelengths and also to the proportional counter detector described below.

Although it is possible to make X-ray observations from balloons or sounding rockets, most X-ray observations are made from space. The performance of CCD arrays in space can be adversely affected by damage to the crystal structure caused by ionising particle hits, especially high-energy protons. Protons can collide with atoms in the lattice, displacing them and creating traps – additional energy states in which charges can get stuck. The accumulation of trapped charge alters the effective gate bias voltage (which can be compensated by making it adjustable) and also degrades the charge transfer efficiency and increases the dark current due to charge leaking out of traps. These effects can worsen the energy resolution over time. Examples of X-ray CCDs, as used in ESA's XMM-Newton Observatory, are described in Section 11.11.1.

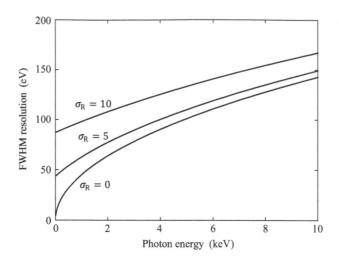

FIGURE 11.2   FWHM energy resolution for a silicon CCD as a function of X-ray energy for different values of readout noise.

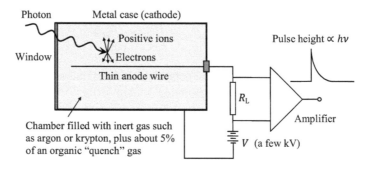

FIGURE 11.3   Essential features of a proportional counter X-ray detector.

11.3   THE PROPORTIONAL COUNTER

The proportional counter uses gas to absorb the incoming X-ray photon, producing a degree of ionisation proportional to its energy. An electric field accelerates the electrons, giving them enough energy to cause secondary ionisation, resulting in an avalanche of electrons which is detected as a current pulse. The basic features are shown in Figure 11.3.

An incident X-ray photon enters the chamber through a transparent window. It ionises an inert gas atom, and the kinetic energy of the released electron is sufficient to cause a cascade of further ionisation. Typically, one electron/ion pair is produced for every 25 eV of photon energy. The local- ised bunch of electrons liberated by the incoming photon is attracted to the anode wire, which is at a high positive voltage with respect to the case, and the ions are attracted to the case. If the electric field in the vicinity of the wire is strong enough, the electrons acquire enough energy to cause more secondary ionisation by colliding with gas atoms. This process, known as gas amplification, results in an avalanche of secondary ionisation, greatly increasing the number of electrons finally collected at the anode (typically by a factor of about 10^5). The corresponding current pulse results in a voltage pulse across the load resistor R_L, which is amplified and recorded.

For an idealised cylindrical chamber of radius R, with a long axial anode wire of radius a, at electric potential V, the magnitude of the anode electric field, E_{anode}, at perpendicular distance r from the anode axis is given by

$$E_{\text{anode}}(r) = \frac{V}{r \ln\left(\dfrac{R}{a}\right)}. \tag{11.13}$$

Therefore, a small value of r is needed to achieve the most intense field for a given voltage, which is why the anode wire is made as thin as possible.

For secondary ionisation to set in, the field must exceed a critical value, E_{crit}. The region in which secondary ionisation occurs is thus defined by a critical radius

$$r_{\text{crit}} = \frac{V}{E_{\text{crit}} \ln\left(\dfrac{R}{a}\right)}. \tag{11.14}$$

With typical values of 15 μm for the wire radius, 30 mm for the chamber radius, 1 kV for the anode voltage, and 50 kV cm^{-1} for the critical field, $r_{\text{crit}} \sim 25$ μm. Gas amplification thus occurs very close to the anode wire.

Noble gases, typically argon or krypton, are the best choice to fill the chamber as they are chemically inert, helping to avoid corrosion and contamination, and they have low critical electric field values. A small amount of a suitable "quench" gas, such as methane is added to absorb UV photons which are emitted by excited noble gas atoms. Such photons would otherwise cause further ionisation leading to a continuous discharge.

The mode of operation depends critically on the potential difference between the anode and the case, as illustrated in Figure 11.4, which shows the pulse amplitude as a function of anode voltage, V, for two different photon energies $(E_2 > E_1)$.

In region A, the anode voltage is too small, and recombination occurs before the primary electrons can reach the anode, resulting in a very small pulse. In region B, essentially all of the primary electrons are collected, but they do not gain enough kinetic energy on the way to the anode to cause secondary ionisation. The height of the output pulse is proportional to the photon energy, and the anode voltage has little effect. A device operated in this region is called an ionisation chamber. When operated in region C, the primary electrons now gain enough energy to cause secondary ionisation, resulting in gas amplification of the pulse, by a factor of typically a few thousand. The proportionality of the pulse height to the energy is still retained because the amplification factor is independent of the number of electrons. The proportional counter normally operates in this region. It gives a high SNR pulse proportional to the X-ray energy. If the anode voltage is increased further, as in region D, there is a loss of linearity because so many electrons are generated that a high space charge of electrons develops near the anode wire. This counteracts the anode electric field, reducing the gas amplification factor. Pushing the anode voltage higher still, into region E, results

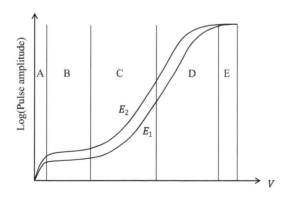

FIGURE 11.4   Pulse amplitude vs anode voltage for two different values of photon energy $(E_2 > E_1)$.

in saturation of the output – the pulse height is now very large, but completely independent of the photon energy. A Geiger counter operates in this region, and is extremely sensitive but provides no spectral (energy) information.

The useful energy range of the proportional counter is around 1–80 keV, depending on the gas used. The low-energy cut-off is mainly due to absorption by the window (typically a sheet of dielectric, such as mylar, a few tens of microns thick), and the high-energy limit is due to declining absorption efficiency of the gas. Over the optimum operating range, the quantum efficiency can be very high – up to 90%. The energy resolution is dictated by fluctuations in the degree of ionisation and is subject to Fano statistics, as for electron–hole pairs in a solid-state detector. It scales as the square root of the photon energy, with a rule of thumb being $E/\Delta E = 5E^{1/2}$ (E in keV) – about 10% energy uncertainty at 3 keV and 4% uncertainty at 20 keV. To achieve the best energy resolution, it is important to have a very stable high-voltage supply, avoiding fluctuations in the gas amplification factor. The timing accuracy of measurement of the photon arrival time is typically around 1 μs, limited by the range of transit times.

The combination of physical absorption area and quantum efficiency defines the effective area of the detector. Figure 11.5 shows the effective area of two proportional counter instruments: the Proportional Counter Array (PCA) which flew on NASA's Rossi X-Ray Timing Explorer (RXTE), launched in 1995 (Glasser et al. 1994), and the Large Area Proportional Counter (LAXPC) on board the ASTROSAT mission (O'Brien 2011), launched by the Indian Space Research Organisation in 2015. The sharp discontinuities are at 34.5 keV and are due to the K-absorption edge of the xenon atom. Photons above that energy are more efficiently absorbed as they have enough energy to liberate electrons from the innermost electron shell (the K-shell).

Background rejection is important in the use of proportional counters. Cosmic rays or charged particles from radioactive decays can pass through the chamber causing ionisation and leading to spurious pulses. Several techniques are used to overcome this. The simplest technique is to reject events outside the energy range of interest. However, unwanted events may still remain, and there are other characteristics of the signals that can be utilised for discrimination. Pulse shape analysis relies on the fact that X-rays and charged particles produce different kinds of pulses. The ionisation from an X-ray is highly localised, and although the electron bunch diffuses to some extent as it drifts towards the anode, there is a small spread in transit times and a correspondingly short pulse. A charged particle passing through the chamber creates a long ionisation trail, resulting in a large spread of transit times and a relatively longer pulse. Another technique to discriminate between X-ray and charged particles is the use of multi-anode-wire counters. The distributed ionisation of a charged particle results in pulses from more than one wire, but an X-ray will normally produce a

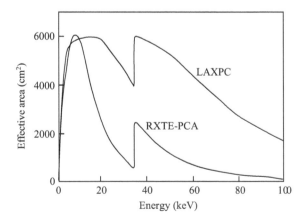

FIGURE 11.5   Effective area of proportional counter instruments: LAXPC, flown on the ASTROSAT mission, and the Proportional Counter Array (PCA) flown on the RXTE satellite. (From LAXPC instrument web site (http://astrosat.iucaa.in/?q=node/12), courtesy of Prof. D. Bhattacharya and the LAXPC team.)

detection in one wire only. A multi-wire detector can also provide positional information, with one coordinate indicated by which wire produces a signal, and the other by the relative amplitudes of the pulses measured at each end of the wire.

A variant of the proportional counter is the gas scintillation proportional counter (GSPC), which makes use of the UV emission by excited gas atoms. An X-ray photon generates an electron cloud as described above in one region of the chamber. An electric field causes it to drift into another region, the scintillation region, where the electric field is much higher. The electrons interact with gas atoms causing them to emit UV photons, which are detected by a PMT. The amount of the UV light depends on the original photon energy.

11.4   X-RAY SPECTROSCOPY

Many X-ray detectors are capable of providing a measurement of the photon energy, E, i.e., they provide spectral information. For X-ray energies up to a few keV, it is also possible to place a spectrometer in front of the detector to enhance the spectral resolution of the system. For low resolution $(E/\Delta E < 100)$, the inherent detector resolution may be enough. For medium resolution (100–500), a diffraction grating can be used, operating either in transmission or in grazing incidence reflection (the physics of grazing incidence X-ray reflection is described in Section 11.8.1). An example is the reflection grating spectrometer (RGS) flown on ESA's XMM-Newton satellite (described in Section 11.11.1). For higher resolution (500), a Bragg crystal spectrometer acts as a grazing-incidence reflection grating. This is equivalent to a diffraction grating in which the rulings are formed by the crystal lattice inter-atomic spacing. Constructive interference occurs if the path difference $(2x = 2d\sin\theta)$ in Figure 11.6 is an integral number of wavelengths:

$$2d\sin\theta = n\lambda, \tag{11.15}$$

where θ is the Bragg angle. Crystals used for Bragg spectrometers are LiF, NaCl, and quartz. To record a spectrum, the crystal must be rotated.

The Bragg crystal spectrometer has good spectral resolution, but it is very inefficient, with quantum efficiency of typically a few %. For applications that involve very low photon rates, as can be the case in X-ray astronomy, this is a serious disadvantage. A new kind of X-ray spectrometer, the X-ray calorimeter, combines high spectral resolution and high quantum efficiency.

11.5   THE X-RAY CALORIMETER

In the case of a bolometric detector used for far infrared detection (Chapter 9), the incident EM power is thermalised (converted to phonons), and the resulting temperature change leads to a change in resistance of the thermometric element. The same basic principle can be used to detect single X-ray photons and measure their energy (a process known as calorimetry). The calorimeter is very

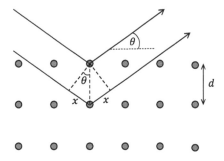

FIGURE 11.6   Principle of the Bragg crystal spectrometer.

similar to the bolometer described in Chapter 9 (Figure 9.3), but the absorber is optimised for the X-ray region instead of far infrared wavelengths to improve the X-ray absorption efficiency. The energy of the X-ray photon is quickly thermalised (in $< 1\,\mu s$), causing a rapid jump in the absorber temperature, and a corresponding step change in the resistance of the thermometer.

If the energy deposited by the photon is E, then the rise in temperature is

$$\Delta T = \frac{E}{C}, \tag{11.16}$$

where C is the calorimeter heat capacity. The change in temperature leads to an output voltage pulse with amplitude proportional to E. The pulse decays exponentially with time constant $\tau = C/G$ as the deposited energy leaks away to the heat sink:

$$\Delta V(t) = \Delta V_0 e^{-t/\tau}, \tag{11.17}$$

where ΔV_0 is the initial pulse height.

The NEP of a bolometric detector can be written as

$$\mathrm{NEP} = aT_0 (kG)^{1/2}, \tag{11.18}$$

where a is 2 for phonon noise only. Taking additional noise contributions into account, a typical value for a is 3–5. In the case of detection of an X-ray photon, the device is a detector of energy rather than power. The characteristic power dissipated in the detector during the pulse is

$$P = \frac{E}{\tau}. \tag{11.19}$$

We can define the noise equivalent energy (NEE) as the photon energy which would produce a pulse height equivalent to the detector noise. Letting $\mathrm{NEE} = (\mathrm{NEP})(\tau)$ we have

$$\mathrm{NEE} = aT_0 (kG)^{1/2} \left(\frac{C}{G}\right) = aT_0 C \left(\frac{k}{G}\right)^{1/2}. \tag{11.20}$$

By analogy with the NEP, this is the energy incident on the detector which would produce an SNR of unity in a 1-Hz post detection bandwidth (0.5-second integration time). However, in measuring the pulse response to an X-ray photon, the integration time is not 0.5 second, but is on the order of τ, the pulse duration. Since noise is inversely proportional to the square root of integration time, the rms uncertainty in the measured energy for integration time τ is

$$\Delta E = \mathrm{NEE} \left(\frac{0.5}{\tau}\right)^{1/2} = \frac{a}{\sqrt{2}} T_0 (kC)^{1/2}. \tag{11.21}$$

Note that ΔE is independent of the thermal conductance G (in contrast with NEP) so that G can be chosen for convenience (e.g. desired signal frequency range, expected count rate, etc.). Achieving the best energy resolution requires using the lowest possible operating temperature and minimising heat capacity (so as to produce the maximum temperature rise for a given photon energy). To achieve sufficiently good energy resolution, very low temperatures are needed: energy resolution of $< 2\,\mathrm{eV}$ has been achieved at $T_0 = 50\,\mathrm{mK}$ with transition-edge superconductor (TES) X-ray calorimeters (e.g. Miniussi et al. 2018; Akamatsu et al. 2020). Figure 11.7 shows an example spectrum illustrating resolution of the $^{55}\mathrm{Fe}$ doublet at around 6 keV – an important requirement for astrophysical X-ray spectroscopy.

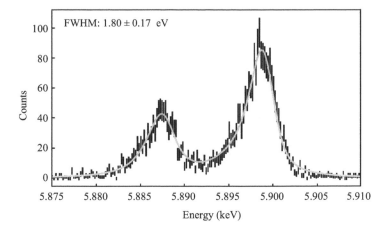

FIGURE 11.7   Measured X-ray spectrum of the ^{55}Fe 6-keV doublet using a single-pixel TES detector under development for Athena. The black bars show the data and the grey line is a model fit. (From Akamatsu et al. (2020); reproduced with permission from Springer.)

With a well-designed absorber, high quantum efficiency can also be achieved. Although low operating temperatures are difficult and costly to achieve in space, the high energy resolution and high quantum efficiency make this an attractive option. An X-ray calorimeter instrument, the X-ray Integral Field Unit (XIFU), will be flown on ESA's Advanced Telescope for High-Energy Astrophysics (Athena) mission, which is currently under development.

11.6   SCINTILLATION DETECTORS

At photon energies on the order of 100 keV and higher, gas absorption is too inefficient – more dense materials are needed. Scintillation detectors use liquid or solid material as the absorber and can be used up to energies of around 100 MeV. The photons are absorbed either by the photoelectric effect or by the Compton effect, depending on the material and the energy. When the energy of the photon is to be measured, photoelectric absorption is preferred, because all of the photon energy is absorbed, but when absorption is by the Compton effect, the scattered photon may escape from the detector.

A scintillation detector uses a material, which has luminescent centres. These have energy states which are excited by ionising radiation, and which decay with the emission of visible or UV photons. The scintillator must be optically transparent so that the burst of light can travel through the material and be detected by a PMT. Numerous different materials can be used as scintillators, and materials can be in crystal, plastic, or liquid form. Many organic hydrocarbons are natural scintillators. They have low atomic number, Z, and so are not well suited for high-energy applications (Section 3.3), but they exhibit very fast pulse decay (time constant typically 10 ns), so they can cope with high count rates. Inorganic alkali halide crystals such as NaI and CsI are also used as scintillators, normally activated by the addition of a suitable impurity such as thallium (Tl), which improves scintillation efficiency. They have higher Z, and so are better for high energy, and have a higher light output than the organic materials, but are not as fast (time constant typically 1 µs).

The scintillation process in organic materials occurs at the level of individual molecules. Most organic scintillators rely on the characteristics of carbon–carbon bonds. Energy from a γ-ray or charged particle excites electrons into various excited states, which decay very rapidly with fluorescent photon emission. The luminescence mechanism for an inorganic crystal is shown in Figure 11.8. The conduction band is normally empty, and the material is an insulator. The absorption of a high-energy photon leads to the generation of many electron–hole pairs, and the electrons migrate to the

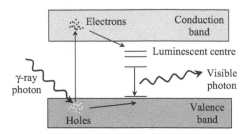

FIGURE 11.8 The scintillation process in an inorganic crystal scintillator.

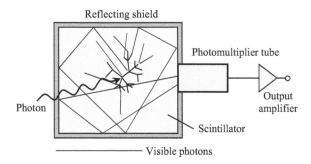

FIGURE 11.9 Essential features of a scintillation detector.

luminescent centres where they occupy excited states. As they decay from these excited states and eventually return to the valence band to neutralise the holes, visible photons are emitted.

The essential features of a scintillation detector are shown in Figure 11.9. An incoming photon is absorbed by the photoelectric or Compton processes and produces many secondary electrons. The electrons excite luminescent centres, which decay with emission of visible photons (typically, 1–5 photons are produced per 100 eV of incident photon energy). The material is surrounded by a reflecting shield everywhere except at the PMT entrance window. The photomultiplier thus detects a pulse of light every time a high-energy photon is absorbed.

The simple scintillator detector as depicted in Figure 11.9 has significant drawback for many applications: it is omnidirectional – the output conveys no information on the direction from which the photon arrives. Furthermore, ionisation can be caused by charged particles, registering a spurious output. In γ-ray astronomy, steps must therefore be taken to define a specific field of view and to discriminate between γ-ray and charged particle events. For these reasons, several scintillation detectors are often combined and operated in anti-coincidence mode. The principle is illustrated in Figure 11.10, which shows an example of a simple -ray telescope using this principle. The main detector is the NaI crystal, and the others act as anti-coincidence shields. High-energy charged particles leave ionisation trails and so will be registered either by the NaI shield or plastic scintillator, or by both. The NaI shield protects the main NaI detector from γ-rays coming from outside the desired field of view. Only when a detection is made by the main NaI detector in the absence of any response from the others will a legitimate γ-ray be registered.

A commonly used inorganic scintillator is NaI(Tl), in which around 0.1% of the Na atoms are replaced by thallium impurity atoms. It generates approximately 38,000 photons per MeV (about 25 eV per visible photon), with a mean wavelength of about 400 nm (3 eV), which is well-matched to the peak spectral response of PMTs. The overall efficiency for conversion of γ-ray photon energy to visible photon energy is thus about 11%. The pulse decay time is 0.25 μs. NaI and other alkali halide crystals are hygroscopic (they absorb moisture from the air) and so must be hermetically sealed to avoid this as it degrades transparency.

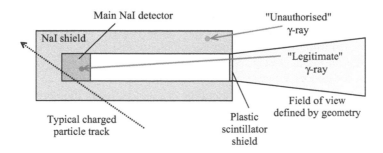

FIGURE 11.10 The use of shield detectors operated in anti-coincidence mode to reject charged particle events and define a field of view.

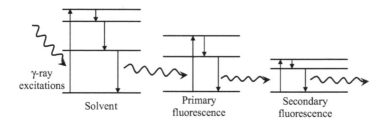

FIGURE 11.11 Operation of liquid and plastic scintillators with a solvent and one or more fluorescent materials leading to wavelength-shifting.

Typical organic scintillators are naphthalene ($C_{10}H_8$), stilbene ($C_{14}H_{12}$), and anthracene ($C_{14}H_{10}$) Anthracene has a light output peaking at around 450 nm and generates about 15,000 photons per MeV. Its light output is thus lower than that of NaI(Tl), but the pulse duration is much shorter at ~30 ns so that much higher count rates can be accommodated.

Plastic and liquid scintillators are composed of a solvent such as benzene with one or more fluorescent materials. With all three components being luminescent, wavelength shifting occurs with the final light output at the longest wavelength, as illustrated in Figure 11.11.

Inorganic scintillators have the advantages of the highest light output and high Z for better high-energy response. Their main disadvantages are that the crystals are difficult and expensive to manufacture, especially for large volumes. Organics have lower light yield (typically 3% efficiency) but produce faster pulses. They are also cheaper to manufacture and can be readily shaped.

The conversion of γ-ray energy to electrical signal in a scintillation detection system involves several stages. Consider the detection of a 1 MeV photon, which generates around 40,000 photons. Around 40% of these are lost due to light transmission inefficiency, so about 25,000 are incident on the PMT. With a typical photocathode quantum efficiency of 20%, about 5000 are incident on the first dynode, after which the electron numbers and their fluctuations are amplified. Therefore, it is the statistical fluctuations in the number of electrons at the first dynode that limit the energy resolution – in this case the SNR is $\sqrt{5000} \approx 70$, so $E/\Delta E \approx 1.4\%$. Variations in light collection efficiency within the crystal volume, and in the transmission efficiency and overall gain, also affect the energy resolution, and lead to overall values on the order of 5% in practice.

11.7 SEMICONDUCTOR DETECTORS: SILICON, GERMANIUM, AND MERCURY CADMIUM TELLURIDE

When a high-energy photon is absorbed in a semiconductor, electron–hole pairs are created in numbers proportional to the photon energy. Semiconductor detectors thus act as solid-state proportional counters. A potential difference applied across the material results in a pulse of current for each absorption event. Very high resistance devices are needed to keep the dark current low. Achieving high resistance involves using either highly pure material or "compensated" material in which residual unwanted impurities are deliberately compensated for by the controlled addition of different impurities, resulting in closely matched numbers of donor and acceptor atoms so that the carriers produced by the donors are captured by the acceptors.

For example, the basic structure of a junction-type detector is shown in Figure 11.12. The device is a reverse-biased p-n junction. The depletion region is devoid of majority carriers (holes) due to the high reverse bias. No current flows unless electron–hole pairs are generated in the depletion region. An incoming photon is absorbed in the depletion region, generating one electron–hole pair for every few eV of photon energy (depending on the material). The high reverse bias leads to a pulse of current which is converted to a voltage pulse by the load resistor R_L, and amplified. The output pulse amplitude is proportional to the number of electron–hole pairs generated and so to the photon energy.

The active volume of a semiconductor detector needs to be sufficiently large for high absorption efficiency – much larger than the size of a diode when used as a conventional electronic component. Given the strong dependence of the photoelectric interaction probability on the atomic number, (proportional to Z^5 as noted in Chapter 3), Ge ($Z = 32$), is a much more efficient absorber than Si ($Z = 14$). Cadmium and tellurium also have high atomic numbers (48 and 52, respectively), and the compound semiconductor cadmium zinc telluride (CdZnTe) is also used for high-energy photon detection.

The large absorption volume and the need to minimise the number of traps so that charge carriers can move through the crystal require very low carrier density so that a reasonable electric field ($\sim 1\,\mathrm{kV\,cm^{-1}}$) can be used to create the depletion region. In the case of germanium detectors, this was originally achieved by introducing lithium to compensate for other impurities. Modern germanium detectors are made of highly pure material (HPGe). Such detectors need to be cooled to around 100 K to achieve the lowest possible dark current.

The energy resolution of semiconductor detectors is proportional to the square root of the photon energy, as usual, but much better resolution is achieved with semiconductor detectors than with other detectors such as scintillators operating at similar energies because many more electron–hole pairs are produced per photon than visible photons in a scintillator. To illustrate this, Figure 11.13 shows the same spectrum as recorded by an HPGE detector and a NaI scintillator. The much superior energy resolution of the germanium detector is clear. Well-optimised HPGE detector systems can achieve energy resolution of ~0.3% (Smith 2010)

Semiconductor strip detectors, based on technology developed for particle physics experiments, can also be used in high-energy astronomy instruments to track the trajectories of high-energy particles produced by γ-ray photons or cosmic rays. A single strip is formed of a long thin semiconductor substrate along which many p-n diodes are formed. When an ionising particle passes through, its position along the strip is located as the ionisation trail is detected by one of the diodes. The two-dimensional position of a particle can be identified by its intersection with a layer of parallel strips,

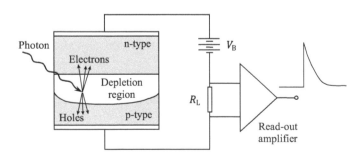

FIGURE 11.12 Structure and operation of a junction-type semiconductor for high-energy photons.

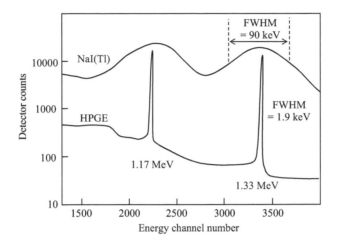

FIGURE 11.13 Measured spectra of ^{60}Co, which has sharp γ-ray lines at 1.117 and 1.33 MeV, using an NaI(Tl) scintillation detector and an ORTEC HPGE detector. (Adapted from ORTEC Experiment 7: High-Resolution Gamma-Ray Spectroscopy, available at https://www.ortec-online.com/service-and-support/library/educational-experiments. Credit: ORTEC.)

and a particle can be tracked in three dimensions by building up a stack of such layers. Examples of the use of such tracking detectors in γ-ray astronomy are given in Sections 11.8.3, 11.8.4, and 11.11.3.

CdZnTe detectors are increasingly used for X-ray energies above around 20 keV and for γ-ray astronomy. The composition is $Cd_xZn_{1-x}Te$, with x typically 0.1. The zinc increases the bandgap and decreases defect density so giving higher resistivity. With $x=0.1$, the bandgap is 1.6 eV. This gives CdZnTe detectors the attractive advantage that they can be operated at room temperature or only slightly below, simplifying cooling requirements for space operation. The higher atomic number ($Z \sim 50$) also provides better stopping power for higher photon energies and/or smaller thickness for a given stopping power. High absorption efficiencies can be achieved with a thickness of 1–2 mm. CdZnTe detectors can be made in the form of strips or multi-pixel array to provide good spatial resolution. Energy resolutions of a few keV can be achieved. Such detectors have been employed on the ESA's INTEGRAL (Winkler et al. 2003) and NASA's Swift (Gehrels et al. 2004) missions.

11.8   X- AND γ-RAY IMAGING

11.8.1   GRAZING INCIDENCE X-RAY TELESCOPES

Conventional refractive or reflective optics, with near-normal angles of incidence, are ineffective at X-ray wavelengths as the photons tend to be absorbed rather than deflected. However, photons with energies less than around 10 keV can be reflected reasonably well from some metals if the angle of incidence with respect to the surface is small (less than a few degrees).

At X-ray wavelengths, the real part of the refractive index is close to but less than unity and normally written as

$$n = 1 - \delta, \tag{11.22}$$

where δ depends on atomic number, density, and wavelength:

$$\delta \propto Z\rho\lambda^2. \tag{11.23}$$

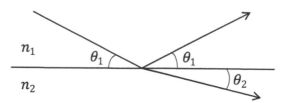

FIGURE 11.14   Grazing incidence reflection and transmission of a ray at the interface between two media of refractive indices n_1 and n_2 ($n_2 < n_1$).

Consider a beam of radiation incident on a boundary between media with refractive indices n_1 and n_2 – as shown in Figure 11.14.

Snell's law relates the angles of incidence and refraction with respect to the normal:

$$n_1 \sin\left(90° - \theta_1\right) = n_2 \sin\left(90° - \theta_2\right) \Rightarrow n_1 \cos\theta_1 = n_2 \cos\theta_2. \tag{11.24}$$

Putting $n_1 = 1$, corresponding to a vacuum, and $\cos\theta_2 = 1$ corresponding to no transmitted component, we get the critical incidence angle for total external reflection at the material interface:

$$\cos\theta_c = n_2 \tag{11.25}$$

Applying the small-angle approximation $\cos\theta_1 \approx 1 - \theta^2/2$, this gives

$$1 - n_2 = \frac{\theta_c^2}{2}, \tag{11.26}$$

so that

$$\theta_c = \left(2\delta\right)^{1/2}, \tag{11.27}$$

and

$$\theta_c \propto \left(Z\rho\right)^{1/2} \lambda \propto \frac{\left(Z\rho\right)^{1/2}}{E}. \tag{11.28}$$

The critical angle is thus larger for denser materials and lower photon energies. With ρ in kg m^{-3} and E in eV, equation (11.28) can be approximated as (Aschenbach 1985)

$$\theta_c = 37\frac{\rho^{1/2}}{E}. \tag{11.29}$$

For gold, a commonly used material for grazing incidence X-ray optics, $\rho = 1.93 \times 10^4$ kg m^{-3}, so that $\theta_c \approx 5°$ at 1 keV and 0.5° at 10 keV.

Reflectivity depends strongly on the photon energy and on the grazing incidence angle – it is necessary to have an angle of incidence considerably less than the critical angle to achieve good reflectivity. Figure 11.15 shows the reflectivity of nickel and gold as a function of energy and angle of incidence. With the low angles of incidence that need to be used, the focal lengths of X-ray telescopes must be quite long (typically ~ 10 m).

The most popular grazing incidence configuration is the Wolter type I (Wolter 1952), the principle of which is shown in Figure 11.16. The X-rays experience two grazing incidence reflections from confocal paraboloidal and hyperboloidal mirror sections. The projected area presented to the

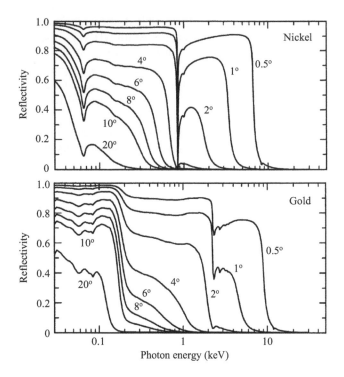

FIGURE 11.15 Reflectivity of nickel and gold vs photon energy for various angles of incidence. (From Gullikson (2009), reproduced with permission.)

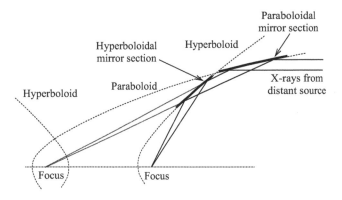

FIGURE 11.16 Wolter Type-I grazing incidence X-ray telescope design.

incoming beam by the paraboloidal mirror is small, which is a serious disadvantage if only one telescope is used, but the Wolter Type-I design is well-suited to nesting telescopes to increase the total collecting area (an example being the XMM-Newton X-ray observatory, described in a case study in Section 11.11.1). Because of the very small curvatures of the reflectors, a close approximation to ideal performance can be achieved using two cones rather than a hyperboloid and paraboloid, which simplifies the manufacturing process.

The key performance parameters for X-ray telescopes are the effective collecting area and the angular resolution. Nesting a large number of telescopes maximises collecting area but makes accurate co-alignment more difficult so that it is difficult to optimise sensitivity and angular resolution at the same time. The point spread function has a distinct energy-independent central core determined by the surface figure of the reflectors and the alignment, superimposed on a broader energy-dependent pedestal due to roughness and possible dust contamination of the surfaces. The size of the PSF is usually specified by the half-energy width (HEW) – the diameter that encloses 50% of the energy from an on-axis point source. A low value of HEW is important not just in order to resolve point sources or provide high-quality images. It also has a significant impact on sensitivity because the X-ray sky is characterised by a diffuse background from unresolved sources. This degrades sensitivity in much the same way as the diffuse background from atmospheric emission leads to photon noise at infrared wavelengths.

ESA's XMM-Newton and NASA's Chandra X-ray observatories, both launched in 1999, have such nested sets of telescopes. XMM-Newton has three telescope units each with 58 gold-coated telescopes and is optimised for the highest possible sensitivity for X-ray spectroscopy, requiring maximum photon-collecting power (effective area 0.43 m²) at the expense of angular resolution (14″ HEW). Chandra has a single telescope unit with four nested iridium-coated telescopes, with a lower effective area (0.08 m²) but better image quality (0.5″ HEW), being optimised for imaging.

The optics of ESA's Athena X-ray observatory, scheduled for launch in 2031 will be manufactured using a new technique based on silicon pore optics (SPO), designed to provide a large effective area combined with good angular resolution (Bavdaz et al. 2015; Willingale et al. 2013). The structure and fabrication process of an individual SPO module is shown in Figure 11.17a. It starts with highly flat and highly polished 300-mm diameter silicon wafers, which are commercially available as standard products of the semiconductor industry. The wafers are diced into rectangular shapes of various lengths. Wedges of SiO₂ are then deposited onto the wafers so that when they are stacked, the surfaces are arranged in a pattern that creates a common focus. Rectangular grooves are cut leaving a pore about 1 mm wide through which X-rays can propagate in vacuuo with thin membranes supporting the reflecting surfaces. The reflecting surfaces are coated with a suitable X-ray reflecting metal such as gold or iridium. Uncoated strips are left so that when the wafers are pressed together, they can cold-weld at the Si–Si interfaces to form a rigid structure. During the stacking procedure, the wafers are deliberately curved using a mandrel to match the surface profile for a Wolter Type-I system. An SPO module consists of two wafer stacks integrated together, one with a nest of surfaces corresponding to the paraboloid reflectors and the other to the hyperboloids, with a kink angle equal to the grazing incidence angle. The full telescope aperture can be filled by such modules arranged in rings. The Athena telescope will comprise 678 SPO modules arranged in 15 rows, with an overall diameter of 2.4 m.

Figure 11.18 shows the expected effective area of the Athena integral field spectrometer, XIFU, as a function of photon energy, and compared with various previous X-ray instruments.

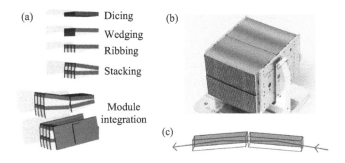

FIGURE 11.17 (a) Manufacturing steps for silicon pore optics. (Reproduced with permission from Bavdaz et al. 2010.) (b) Photograph of a prototype SPO module developed for Athena. (Courtesy of ESA and cosine measurement systems.) (c) Double grazing incidence path of a ray through two aligned pores. (Reproduced with permission from Willingale et al. 2013.)

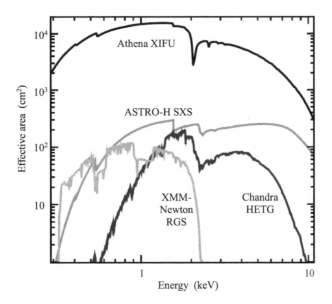

FIGURE 11.18 Effective area vs photon energy for the Athena XIFU instrument and other various X-ray instruments. (Figure courtesy of D. Barret and the XIFU team.)

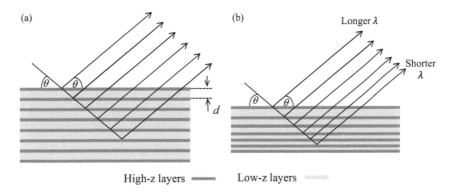

FIGURE 11.19 Uniform (a) and graded (b) multilayer X-ray reflective coatings.

Using metallic surfaces for grazing incidence reflection works well up to about 10 keV. Beyond that, as shown in Figure 11.15, the efficiency is too low for any reasonable angle of incidence. Nevertheless good performance beyond 10 keV can now be achieved using tailored multi-layer coatings.

A simple multilayer structure (Figure 11.19a) is made up of bilayers each made up of a high-Z and a low-Z film. The high-Z layers provide the reflecting surfaces and the low-Z layers act as spacers to keep the high-Z layers at the correct spacing. If a beam of wavelength λ (photon energy $E = hc/\lambda$) is incident at angle θ to the surface, the reflected components interfere constructively if the Bragg reflection condition is satisfied:

$$\lambda = \frac{2d\sin\theta}{n} \text{ or } E = \frac{nhc}{2d\sin\theta}. \tag{11.30}$$

For a given angle of incidence, such an arrangement will thus provide high reflectivity for a narrow energy range. Figure 11.20 shows the modelled reflection efficiency for a multilayer coating consisting of alternating tungsten and silicon layers. To achieve high reflectivity over a broad energy range,

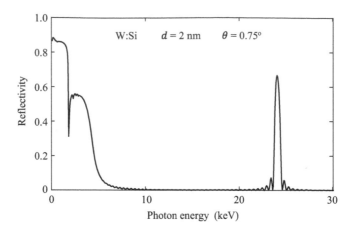

FIGURE 11.20 Reflectivity vs photon energy for a multilayer coating consisting of 50 bilayers of tungsten and silicon with d = 2 nm, with a 0.75° angle of incidence. Computed using Center for X-ray Optics on-line facility at https://henke.lbl.gov/optical_constants/multi2.html. (Based on Henke et al. 1993.)

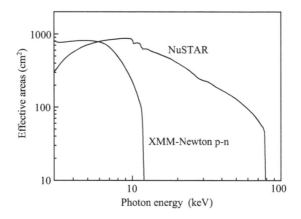

FIGURE 11.21 Effective area vs photon energy for NuSTAR compared to XMM-Newton. (Adapted with permission from NuSTAR Observatory Guide (2016); Credit: NASA.)

multilayer films can be constructed with systematically decreasing thicknesses – depth-graded multilayers – as shown in Figure 11.19b. Because of the reliance on multiple reflection, the reflection losses must be minimised to avoid attenuation of the beam, so the surfaces must also be highly polished with surface roughness less than ~ $d/10$.

NASA's Nuclear Spectroscopic Telescope Array (NuSTAR) satellite (launched in 2012) is designed to observe at X-ray wavelengths up to 70 keV and uses a nest of 133 telescopes formed from glass depth-graded multilayers with Pt:C coatings for the inner 89 shells and W:Si for the outer ones, with grazing incidence angles between 0.08 and 0.3 degrees. Figure 11.21 shows the effective area of NuSTAR compared to that of XMM-Newton. NuSTAR is sensitive to much higher photon energies, but its angular resolution is poorer (58″ HEW).

11.8.2 CODED MASK IMAGING

Photons with energy greater than ~ 100 keV are impossible to image even with grazing incidence techniques, but wide-field γ-ray imaging is possible using a coded aperture mask. The principle is based on that of the pinhole camera, which produces an image of a point source at a well-defined position but with low SNR due to the fact that only a small fraction of the incident

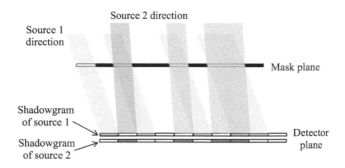

FIGURE 11.22   Principle of coded mask imaging.

radiation is detected. Sensitivity can be improved by making the pinhole aperture larger (at the expense of some blurring) and/or having a random array of apertures. A coded mask is a plate with equal-sized square regions that are either transparent or opaque, which is placed in front of a position-sensitive detector with resolution equivalent to the mask element area. As shown in the one-dimensional schematic in Figure 11.22, a point source at a certain off-axis angle will create a laterally shifted shadowgram of the mask at the detector plane with the shift encoding the position and the signal strength encoding brightness. After exposure, the accumulated detector image from the whole field of view is decoded to form an image of the sky by determining the strength of all possible shifted mask patterns. Ideally, to ensure that every sky position is uniquely encoded, the autocorrelation function of the mask should be a delta function (as for the pinhole camera). Reconstruction of the sky image is achieved by correlating the image with a decoding pattern derived from the mask. The final angular resolution is equivalent to that defined by the mask element size but with much higher SNR given the fact that 30%–50% of the collecting aperture is utilised. Commonly used masks which provide a good approximation – a sharply peaked autocorrelation function with low sidelobes – are a random pattern (as used by the Swift-BAT instrument; see Section 11.11.2) or the Uniformly Redundant Array (as used by the INTEGRAL-IBIS instrument).

11.8.3   THE COMPTON TELESCOPE

The physics of the Compton effect (Section 3.3.2) can be used to construct an imaging -ray telescope with a wide field of view and good angular resolution (a few degrees). The wavelength difference between the scattered and incident photons in a Compton scattering event is given by equation (3.35):

$$\lambda_2 - \lambda_1 = \frac{h}{m_e c}(1 - \cos\theta). \tag{11.31}$$

Converting wavelengths to photon energies and expressing the energies in terms of the electron rest mass, we have

$$\cos\theta = 1 - \frac{1}{E_1} - \frac{1}{E_2}. \tag{11.32}$$

The Compton telescope uses two arrays of detectors as shown in Figure 11.23. The upper layer is made from low-Z material and the lower one from a higher-Z material. An incident γ-ray undergoes Compton scattering in one of the upper detectors (detector A), in which the energy of the scattered

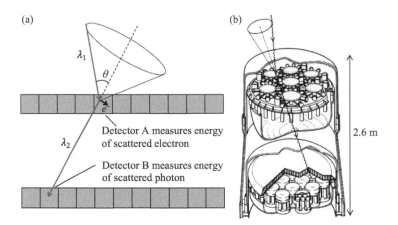

FIGURE 11.23 (a) Principle of the Compton telescope; (b) schematic diagram of the COMPTEL instrument. (Credit: NASA.)

electron $(E_1 - E_2)$ is measured. The scattered photon escapes the low-Z material and is detected and its energy (E_2) measured in one of the lower detectors (detector B). The positions of the two detectors define the direction of the scattered photon. The angle θ can be calculated from equation (11.32), specifying the direction of the incident photon as being within a conical shell (the event circle) specified by θ and the uncertainty to which it is known. The design of COMPTEL (Schönfelder et al. 1993), one of the instruments on board NASA's Compton Gamma Ray Observatory, CGRO, (launched in 1991) is shown in Figure 11.23b.

The upper array consisted of seven modules of liquid scintillator, each with eight PMTs to localise the interaction point. The lower array, 1.6 m below, had 14 modules of NaI(Tl) scintillators, each viewed by seven photomultipliers. Anticoincidence shields surrounded both arrays to allow charged particle events to be rejected, and additional background reduction was provided by discarding events that involved scattering in the lower array before the upper one. COMPTEL operated at energies between 0.8 and 30 MeV with energy resolution between 5% and 10% depending on energy. It had a field of view 1.5 sr with an angular resolution of 1°–2° depending on energy and position in the field. Instrument concepts based on Compton telescopes are currently being developed for future missions. The design of the Compton spectrometer and Imager, COSI, (Tomsick et al. 2019), uses arrays of crossed semiconductor strip detectors, which can detect γ-ray photons with high spatial and spectral resolution. The COSI design incorporates 16 layers of germanium strip detectors.

11.8.4 PAIR CREATION DETECTORS

Photons with energies greater than twice the rest-mass energy of the electron can generate electron-positron pairs, as described in Section 3.3.3. It is possible to track and measure the energies of the electron and positron, thereby inferring the direction and energy of the incident photon. This is the principle of pair creation detectors. The essential features of a modern pair creation detector are illustrated in Figure 11.24.

An incident γ-ray photon interacts with one of several layers of high-Z material (typically tungsten) producing an electron and a positron which travel through the other layers before being intercepted and detected. As they pass through the silicon strip detectors in each layer, they leave trails of ionisation which are detected, allowing their trajectories to be reconstructed. The anticoincidence shield discriminates against cosmic rays or other charged particles. By suitably configuring the geometry of the triggering detectors, a specific field of view can also be defined.

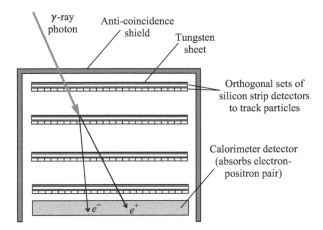

FIGURE 11.24   Principle of a pair-creation detector.

11.9   HIGH-ENERGY γ-RAY AND COSMIC RAY DETECTION

High-energy γ-rays are produced in the most energetic environments and events in the Universe. Likewise, violent processes can produce cosmic rays – atomic nuclei (from single protons to Fe nuclei) that have been stripped of their electrons and accelerated to high energies. Cosmic rays from within our galaxy have energies up to $\sim 10^{15}$ eV and are accelerated by shocks in supernova explosions. Cosmic rays of higher energies are thought to originate in the intense magnetic fields of neutron stars, from γ-ray bursts, and from the central engines of active galaxies. Unlike γ-ray photons, cosmic rays do not travel in straight lines within the Galaxy because their paths are bent by the galactic magnetic field – so the arrival direction does not indicate the origin. The arrival rate of cosmic rays at the Earth declines with particle energy, E, roughly as E^{-3}, from $\sim 1\,\mathrm{m^{-2}s^{-1}}$ at $10^{11}\,\mathrm{eV}$ to $1\,\mathrm{m^{-2}year^{-1}}$ at $10^{16}\,\mathrm{eV}$ and $1\,\mathrm{km^{-2}year^{-1}}$ at $10^{19}\,\mathrm{eV}$ – so the highest energy cosmic rays are extremely rare.

Gamma rays have been detected with energies up to 450 TeV (4.5×10^{14} eV), and cosmic rays with energies in excess of 10^{20} eV. The highest-energy cosmic ray particle yet detected was recorded as having an energy of 3.2×10^{20} eV, equivalent to ~ 50 J. This is an enormous amount of energy for a single atomic nucleus to have – comparable to the kinetic energy imparted to tennis ball served by a professional (~ 70 J).

11.9.1   EXTENSIVE AIR SHOWERS

High-energy γ-ray photons and cosmic rays interact with nuclei in the upper atmosphere, causing cascades of radiation and relativistic elementary particles. Such a cascade is known as an extensive air shower (EAS). An incoming γ-ray photon generates an electron-positron pair in an interaction with the intense electric field of a nucleus. These produce secondary γ-ray photons through bremsstrahlung, leading to a cascade of mainly photons, electrons, and positrons. Showers induced by cosmic ray particles colliding with nuclei create additional fundamental particles such as pions and muons, some of which reach the ground.

The relativistic charged particles in an air shower produce Čerenkov radiation because their speeds exceed the speed of light in air (Section 11.9.2) and atmospheric fluorescence (due to their ionisation trails). In dark-sky conditions, this radiation, which is mainly at UV and short visible wavelengths, can be detected by telescopes on the ground. Figure 11.25 shows the development of an air shower. The lateral spread of the Čerenkov light at the ground is typically 1–200 m, and the light pulse direction preserves the direction of the incoming photon or cosmic ray. The intensity at

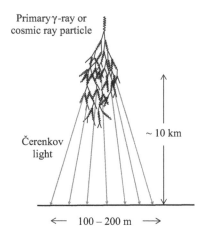

FIGURE 11.25 Characteristics of an air shower due to a primary γ-ray photon or cosmic ray particle.

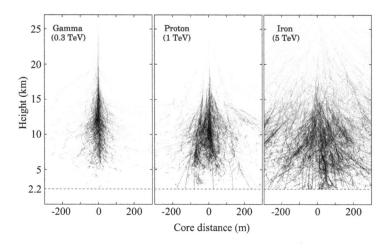

FIGURE 11.26 Simulated Čerenkov light production for different primary events for a site of altitude 2.2 km. The darkness of the particle tracks increases with increasing emission of Čerenkov emission. (From Bernlöhr (2008), reproduced with permission from Elsevier; original of figure kindly provided by K. Bernlöhr.)

the ground is proportional to the number of secondary particles in the shower, and so to the incident photon or particle energy.

Extracting reliable information about the nature and energy of the incoming photon or particle requires a detailed understanding of the generation of the air shower and of many effects in the atmosphere, such as Coulomb scattering of charged particles, geomagnetic deflections, and atmospheric absorption, together with detailed knowledge of the complete detection system. Monte Carlo models taking all of these factors into account are an essential tool in data reduction and analysis.

Figure 11.26 shows model air showers produced by a 300-GeV γ-ray photon, a 1-TeV proton, and a 5-TeV iron nucleus.

11.9.2 Čerenkov Radiation

Čerenkov radiation (Jelley 1958) arises when a charged particle passes through a medium in which the phase velocity of light is less than the speed of the particle. Like all electromagnetic radiation, it is caused by accelerating charged particles, but in this case, it is not an acceleration of the primary particle but of charges that it disturbs as it passes by. The passage of the charged particle induces dipoles in the medium (pushing like charges away and attracting unlike charges). These dipoles relax after the particle has gone. If the particle speed is less than the speed of light in the medium, then the induced dipole pattern is symmetric and no emission is observed because the radiation from the relaxing charges cancels out due to destructive interference. If the particle is moving faster than the speed of light in the medium, the induced dipole distribution is asymmetric because the region ahead of the particle has not yet been affected by its presence. After the passage of the particle, the asymmetric dipole pattern relaxes with the emission of Čerenkov radiation.

Figure 11.27 shows the geometry for Čerenkov emission. Letting the speed of the particle be

$$v = \beta c,$$
(11.33)

in time Δt it travels from A to B, with distance

$$d_1 = \beta c \Delta t.$$
(11.34)

Meanwhile, radiation emitted at A travels a distance

$$d_2 = \frac{c}{n_{air}} \Delta t,$$
(11.35)

where n is the refractive index of the air.

Huygens wavelets emitted between A and B add coherently along a line inclined to the trajectory with

$$\theta = \cos^{-1}\left(\frac{d_2}{d_1}\right) = \cos^{-1}\left(\frac{1}{\beta n_{air}}\right).$$
(11.36)

The maximum angle, for a highly relativistic particle $(\beta \rightarrow 1)$ is $\cos^{-1}(1/n_{air})$. For dry air, $n_{air} = 1.00029$ and $\theta_{max} = 1.3°$.

Čerenkov radiation has a continuous spectrum with intensity varying with wavelength as

$$I_\lambda \propto \lambda^{-2}.$$
(11.37)

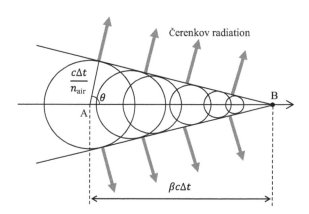

FIGURE 11.27 Čerenkov radiation from a charged particle moving through a medium with speed greater than the speed of light in the medium.

The spectrum has a cut-off in the X-ray region, when n_{air} becomes < 1, and there is a long-wavelength cut-off due to atmospheric absorption. The emitted light is most intense in the UV/visible region with a strong blue-white colour. For a highly relativistic particle in dry air, the photon emission rate is ~ 30 photons m^{-1} between 350 and 550 nm.

11.9.3 Extensive Air Shower Observatories

Čerenkov radiation from EASs provides a valuable technique for studying cosmic rays and γ rays. Although individual γ-ray events produce easily detected signals, events from cosmic rays are much more numerous – by a factor of 10^3–10^4. For γ-ray detection, these background cosmic ray events must be identified and rejected. The γ-ray events produce showers of only photons, electrons, and positrons, whereas cosmic rays induce showers containing other charged particles, especially muons, and they have distinctly different characteristics (Figure 11.26): they have a larger angular spread, a longer duration, and tend to occur lower in the atmosphere. These differences result in different observational signatures and allow the background events to be rejected by the detector triggering system. Background rejection can never be perfect, however, as cascades due to hadrons that produce an initial electron or a photon with most of the energy are difficult to distinguish, as are cosmic electron events.

The VERITAS γ-ray observatory, based on atmospheric Čerenkov radiation detection, is described as a case study in Section 11.11.4 below.

EAS observatories, especially those designed to detect high-energy cosmic rays, can also be based on arrays of ground-based detectors of various kinds to measure the signatures of particles that reach the ground. Tanks of high-purity water with reflective inner coatings and equipped with PMTs are used to detect Čerenkov radiation produced by high-energy particles passing through. Likewise, plastic scintillator detectors can register the passage of ionising particles. Charged particles in air showers also produce radio emission as a result of acceleration due to the Lorentz force as they move in the Earth's magnetic field, and an array of radio receivers can add to the capability to diagnose the nature and energy of the event. As well as producing Čerenkov radiation, cosmic ray air showers also produce light due to atmospheric fluorescence due mainly to excitation of nitrogen leading to emission at 391 nm. Detection of this light is particularly useful in the estimation of the total energy.

The largest cosmic ray facility in the world, the Pierre Auger Observatory (Pierre Auger Collaboration 2015) in Argentina, has 1660 3.6-m diameter water tank Čerenkov detectors, each equipped with a plastic scintillator on top, covering a total area of ~ 3000 km^2, and 24 atmospheric fluorescence telescopes. In addition, it has 153 radio stations operating between 30 and 80 MHz covering 17 km^2.

11.10   COSMIC NEUTRINO DETECTION

Neutrinos interact very weakly with matter, which makes them difficult to detect, but also very useful probes of many astrophysical processes that are otherwise obscured. For instance, most neutrinos emitted during nucleosynthesis in the core of the Sun escape without interaction, and most of the energy released in supernova explosions emerges immediately in the form of neutrinos. The environments of other ultra-high-energy processes in which matter undergoes phenomenally high accelerations also tend to be opaque to light, but neutrinos can emerge directly and reach the Earth without interference along the way.

Because of the low probability of a neutrino interacting with matter very large detectors are needed. Different neutrino flavours (muon, electron, tau) interact differently with nuclei. A muon neutrino interaction is shown in Figure 11.28.

An energetic muon neutrino interacts with a down quark in a neutron, mediated by a W particle, producing a muon, which escapes with about 80% of the incident neutrino energy and travels a

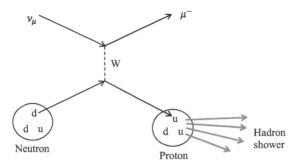

FIGURE 11.28 Interaction of a muon neutrino with a neutron, producing a muon and a hadron shower.

considerable distance. The down quark is converted to an up quark, turning the neutron into a proton which is in a highly excited and unstable state, leading to disruption of the nucleus and a shower of hadronic particles. In the equivalent reaction for an electron neutrino, an electron is produced, which travels only a very short distance, and a hadron shower. A tau neutrino creates a tau particle and a hadron shower. However, the tau particle is unstable and decays after a short time into a lower energy neutrino and a second hadron shower.

High-energy neutrino detectors use large volumes of water (e.g., the ANTARES neutrino telescope; Ageron et al. 2011) or clear Antarctic ice (the IceCube Neutrino Observatory; Gaisser and Halzen 2014, Halzen 2017), with arrays of PMTs to detect Čerenkov radiation from the high-velocity charged particles. The IceCube neutrino observatory is described below in a case study (Section 11.11.5).

The Čerenkov signatures are different for the different neutrino types. The muon from a muon neutrino interaction loses energy by bremsstrahlung and pair production as it travels, creating a long trail along which light is emitted and which preserves the direction of the incoming neutrino. The mean free path of the muon is ~ 10 km, which means that the detector array is sensitive to events occurring outside the instrumented water volume, increasing the effective volume. Čerenkov radiation from the muon track is detected along the entire length of the detector region (typically ~ 1 km). In contrast, hadron showers from electron neutrino interactions within the detector produce localised bursts of light – one in the case of the electron neutrino, and two for the tau neutrino interaction.

Detection of astrophysical neutrinos has to cope with a large background due to muons and neutrinos from cosmic rays interacting with the atmosphere. Although the detection volume is shielded by being located beneath a large depth of water or ice, there is still a large background due to downward-going atmospheric muons generated by cosmic rays – around 10^6 for every muon generated by an astrophysical neutrino. As illustrated in Figure 11.29, background muons from cosmic ray air showers in the atmosphere above the detector are downward going whereas neutrinos from cosmic ray air showers or astrophysical neutrinos arriving from the other side of the Earth produce muons with upward-going tracks, allowing them to be distinguished.

11.11 CASE STUDIES

11.11.1 XMM-NEWTON AND ITS INSTRUMENTS

XMM-Newton is an X-ray space observatory launched by ESA in 1999. Figure 11.30 is a schematic diagram of the XMM payload. There are three Wolter type-1 telescopes, each comprising 58 nested shells giving an effective area of 1550 cm^2 for each telescope at 1.5 keV. The instruments are located in the focal plane at the other end of a 7.5-m tube. One telescope is used only for imaging and has a camera based on a p-n junction array. The other two are equipped with reflection gratings and are used for both imaging and spectroscopy. Their focal planes each contain CCD arrays for imaging

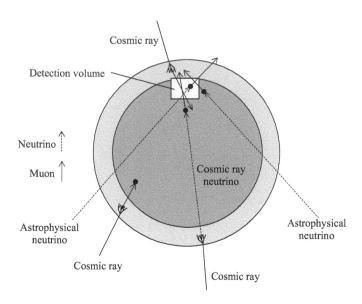

FIGURE 11.29 Muon tracks in a Čerenkov water or ice detector volume due to cosmic rays and astrophysical neutrinos. The large background due to muons from local cosmic ray air showers can be rejected as their tracks are downward in the detector while muons generated near or in the detector volume by astrophysical neutrinos and cosmic ray neutrinos from the other side of the Earth produce upward-going tracks.

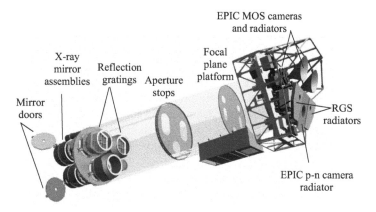

FIGURE 11.30 XMM-Newton payload. (From XMM-Newton Users Handbook (2020); credit: ESA, XMM-Newton SOC; Dornier Satellitensysteme GmbH.)

and for the reflection grating spectrometers. Radiators allow the detectors to be operated at their optimum temperatures, which are below 200 K.

The operation of the telescopes with reflection gratings is shown in Figure 11.31. Around 50% of the X-ray light passes through the stack of grazing-incidence reflection gratings without being dispersed and is focussed onto a CCD camera. Approximately 40% of the X-rays are dispersed by the stack and are incident on the spectrometer detector array.

The XMM instruments are known as EPIC (European Photon Imaging Camera) and RGS (Reflection Grating Spectrometer). There are two RGS instruments, one for each of the two telescopes with gratings, and there are three EPIC cameras – one based on a p-n sensor array (Strüder et al. 2001) at the focus of the unobscured telescope and two based on MOS CCD sensors (Turner et al. 2001) at the other two telescope focal planes. An additional optical monitor instrument provides complementary observations at visible wavelengths. So in total, XMM has six instruments (which are all capable of operating simultaneously).

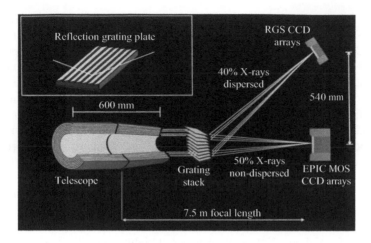

FIGURE 11.31   Optical layout of the two XMM-Newton telescopes with gratings. (From XMM-Newton Users Handbook (2020); credit: ESA; XMM-Newton SOC.)

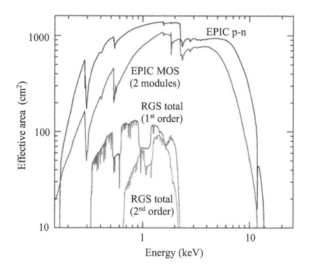

FIGURE 11.32   Overall effective area as a function of photon energy for the XMM-Newton instruments. (From XMM-Newton Users Handbook (2020); credit: ESA; XMM-Newton SOC.)

XMM-Newton was required to provide high-quality X-ray imaging, and the telescopes were carefully designed and built to provide the smallest possible point spread function (PSF) over as wide an energy range as possible. The PSF for an X-ray telescope is dictated not by diffraction but by the quality of the surfaces and the alignment between the telescopes. The achieved FWHM for the MOS and RGS instruments is around 4.4″ with half of the X-ray energy within a diameter of 17″ in the focal plane.

The X-ray energy coverage of XMM-Newton is illustrated in Figure 11.32, which shows the overall effective area of the system as a function of photon energy (including the response of the instruments). Observations with the cameras are possible up to around 15 keV, with the EPIC PN camera, which is unobscured, having the best sensitivity. The spectrometers can operate between around 0.3 and 2 keV. The discontinuities at around 2 keV are due to the Au M-edge, arising from the gold-coated mirrors. Lower energy discontinuities are X-ray absorption edges of silicon and oxygen in the detectors.

   The three EPIC cameras provide astronomers with capabilities for imaging photometry and low-resolution ($R=20$–50) spectrophotometry (using the inherent energy resolution of the detectors). The cameras use CCD detector arrays, specially developed for EPIC, operating in photon counting mode. The amount of charge generated by an individual X-ray photon is proportional to its energy, and the energy resolution is determined mainly by the statistical fluctuations in the charge generation process (Fano noise – see Section 11.2). The two types (p-n and MOS) are optimised for different requirements. The MOS cameras have smaller pixel size, and thus better spatial resolution. The p-n camera has a much faster readout system with each column having its own readout node, enabling observations with high time resolution (down to 7 μs) which can be important for rapidly varying X-ray sources and to operate in a high count-rate mode for bright sources. The p-n camera's energy response also extends to somewhat higher photon energies.

   The p-n camera uses back-illuminated buried-channel deep-depletion CCDs, similar in principle to the devices described in Section 10.3.1, except that the basic charge storage unit is based on a p-n junction rather than a MOS capacitor. The CCD is illustrated in Figure 11.33 (Strüder et al. 1998). The pixels are formed on a 270-μm thick wafer of highly pure silicon with a 12-μm epitaxial n-type layer on the top. The charge storage pixels are formed by reverse-biased p^+-n junctions. On the bottom side of the wafer is a single p^+-n junction covering the whole area, which is also reverse biased by the depletion voltage. The bulk Si substrate is hence heavily depleted of charge carriers. An incoming X-ray photon generates a population of electron–hole pairs (with about one pair produced per 3.7 eV of photon energy). The strong electric field across the substrate separates the electron–hole pairs before they can recombine, with the electrons being collected in a buried channel region, where the electric potential is highest, at the bottom of the n-type layer. The pixel size is $150\times150\,\mu$m, and the camera has 12 arrays, each with 64×198 pixels, covering an approximately 30-arcminute field of view. Each column has its own readout amplifier, making for a short readout time. A number of readout modes are available trading off count rate, field of view and sensitivity depending on the type of observation needed, with the time resolution adjustable between around 7 μs (over a small field) and 70 ms (over the full field). The arrays are cooled to $-90°$C to reduce dark current to a negligible level (< 0.1 electron/pixel/readout cycle). The energy resolution is around 100 eV below 1 keV, rising to around 180 eV at 10 keV. Compared to the MOS CCD, the p^+-n design is more radiation-hard and allows faster readout and higher full-well capacity.

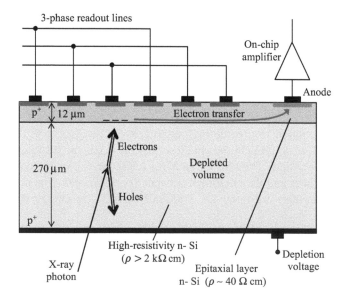

FIGURE 11.33 Structure of the EPIC p-n CCD. (Reproduced from Strüder et al. (1998) with permission from AIP Publishing.)

The MOS cameras each have seven front-illuminated three-phase frame transfer CCDs, each with 600×600 pixels and operating at a temperature of 170 K. The pixel size of 40 μm corresponds to just over 1″ on the sky, with ~15×15 pixels covering the 15″ PSF. Each CCD is 2.5×2.5 cm in size. The readout time for a full frame is 2.6 seconds.

Both cameras are equipped with filters to block optical radiation from the CCDs and with radio-active calibration sources providing spectral lines of known energy. The sensitivity of the cameras depends on the part of the sky being observed and the spectrum of the source, as well as the ionising radiation background, which varies with the orbital position (distance from the Earth) and direction (moving towards or away from perigee), and with solar activity. The background includes a contribution from detector noise, which is only significant at low energies (less than 0.3 keV) and a more significant, and variable, component due to interaction of ionising particles with structures surrounding the detectors and with the detectors themselves. Indicative plots of the overall sensitivity of the XMM cameras are shown in Figure 11.34. The 5-σ minimum detectable X-ray flux (i.e., detectable to an SNR of 5) is expressed in ergs $cm^{-2}s^{-1}$, and plotted as a function of exposure time for the nominal background rates. The EPIC MOS curves correspond to the combination of the data from the two cameras.

Two of the XMM telescope units are equipped with RGS grating stacks, each with 182 co-aligned reflection gratings. Both spectrometers cover the same field of view on the sky. Each grating is made of grooved silicon carbide face-sheets ruled with 646 lines/mm and coated with gold for high reflectivity. The stack intercepts and disperses nearly 60% of the incident light. The RGS has a linear array of nine MOS CCDs, similar to those in the EPIC MOS camera but back-illuminated to maximise the response at soft X-ray energies. Each CCD has 768×1024 pixels, each 27×27 μm in size. One side (384 ×1024 pixels) is exposed to the sky and the other side is used for frame storage and readout while the next integration is in progress. The 253-mm total length of the array covers an energy range of 0.33–2.5 keV in first order.

First- and second-order spectra overlap on the detectors, but can be separated by means of the inherent CCD energy resolution of around 160 eV. The spectral resolution depends on the combination of the grating response, telescope blur, and detector response. The measured and modelled FWHM of the line spread function is shown in Figure 11.35. For RGS-1, which has a slightly superior resolution, the wavelength resolution in first order is approximately 0.06 Å between 6 and 35 Å (corresponding to $\lambda/\Delta\lambda \approx 100 - 600$). In second order, the resolution is approximately 0.035 Å over a more limited wavelength range of 6–18 Å – nearly a factor of 2 better.

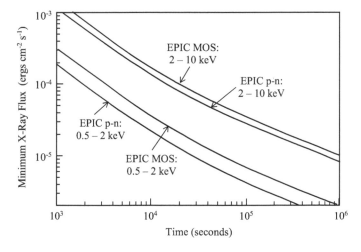

FIGURE 11.34 Sensitivity of the EPIC cameras as a function of exposure time. (From XMM-Newton Users Handbook (2020); credit: ESA; XMM-Newton SOC.)

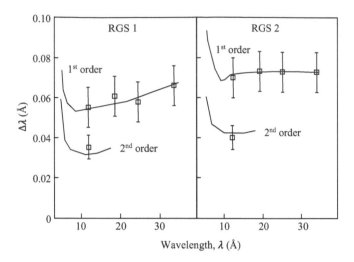

FIGURE 11.35 Measured (symbols) and modelled (lines) spectral resolution of the RGS spectrometers. (From XMM-Newton Users Handbook (2020); credit: ESA; XMM-Newton SOC.)

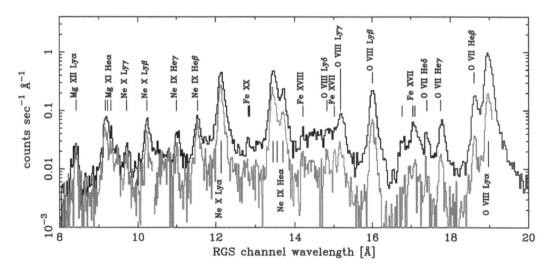

FIGURE 11.36 Portion of an RGS spectrum of supernova remnant 1E 0102.2-7219. (From Rasmussen et al. 2001.) The first-order spectrum is shown by the black line and the second-order spectrum in grey. (Reproduced with permission from Astronomy & Astrophysics; © ESO.)

As an example of an RGS spectrum, Figure 11.36 shows an observation of a supernova remnant (Rasmussen et al. 2001) showing lines of several highly ionised atoms. The plot includes both first- and second-order spectra in the 8–20 Å range, with the second-order spectrum showing higher spectral resolution.

11.11.2 THE SWIFT SATELLITE AND ITS INSTRUMENTS

The Neil Gehrels Swift satellite was launched by NASA in 2004 and is designed primarily to study γ-ray bursts (GRBs). These immensely energetic explosive events at cosmological distances are the most luminous known astronomical sources. They are associated with ultra-luminous supernovae and with mergers between neutron stars. The initial γ-ray burst can last from less than a second up to a few minutes and is followed by an afterglow at longer wavelengths. Studying GRB events and the afterglows is important to understand these events and the high-energy physics involved. They are rare events that can occur at any point on the sky at any time. Because they are unpredictable and short-lived, it is a challenge to catch them early and observe how their emission changes with time.

Swift has an instrument that views a large portion of the sky continuously and is able to detect a GRB within seconds and identify its approximate position. It then instructs the spacecraft to slew to that position immediately so that the event can be studied at X-ray and visible wavelengths. Swift detections are also immediately conveyed to a network of other space-borne and ground-based observatories for rapid follow-up.

Figure 11.37 shows a diagram of the Swift spacecraft, which is approximately 5.6 m in height. The three scientific instruments are the Burst Alert Transient instrument (BAT), responsible for initial detection and position-finding, the X-ray Telescope (XRT), and the UV-Optical Telescope (UVOT), which are then used to observe the event when the spacecraft has slewed to the target. To capture the GRB signatures as rapidly as possible, the spacecraft has a fast slew rate of up to 50° in 75 seconds. A more detailed diagram of the BAT instrument is also shown in Figure 11.37.

The BAT (Figure 11.37) is a large field-of-view coded-aperture telescope, operating between 15 and 150 keV. The 2.4 m × 1.2 m coded aperture has 52,000 absorbing tiles and is located about 1 metre above the detector array. The tiles are 5 mm square and 1.0 mm thick lead blocks arranged in a completely random pattern with a 50% open 50% closed filling factor. The geometry of the mask and the array gives the BAT a field of view of 100° by 60° (half-coded), corresponding to about 1.4 sr.

The BAT detectors are 32,768 individual crystals of the semiconductor CdZnTe (CZT), each 4 mm^2 and 2 mm thick and operating at ambient temperature with a bias voltage of around 200 V. The detectors are grouped in a hierarchical structure with 128 detector modules and 16 blocks.

The combination of the mask and detector array gives good spatial resolution, with a PSF of approximately 17 arcminutes. The CZT detectors also provide a spectral resolution of a few keV. Figure 11.38 shows a photograph of the BAT detector array unit and an example single-pixel spectrum of a 60-keV γ-ray source. The FWHM is about 3.3 keV and the threshold for response of this particular pixel is about 10 keV. The threshold varies between 10 and 18 keV from one pixel to another. The typical background count rate is approximately 0.3 count s^{-1} per pixel, varying by a factor of 2 depending on the position in Swift's low-Earth orbit which carries it through the Earth's radiation belts. The overall sensitivity for GRB flux is approximately 10^{-8} erg cm^{-2}s^{-1} integrated over the 15–150 keV range.

The BAT on-board image processing electronics continuously monitors the detector signals. When a count-rate excess is registered, the data are analysed to discover whether the characteristics are consistent with the signature of a GRB. An FFT-based algorithm correlates the pattern of detector signals with the pattern of the coded aperture, taking about 12 seconds to produce a 1024 × 512 pixel image showing the location of transient source. The position of a burst can be located within

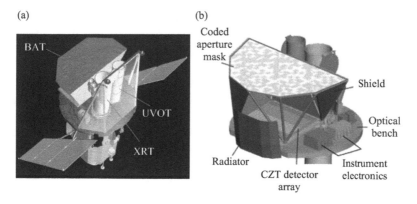

FIGURE 11.37   (a) The Swift satellite, showing the three scientific instruments; (b) cutaway diagram of the BAT instrument. (Both diagrams are from Gehrels et al. (2004). © AAS; reproduced with permission.)

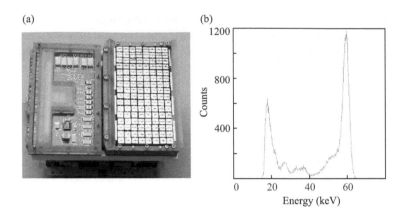

FIGURE 11.38 (a) Photograph of a 128-element BAT CZT module and its readout electronics. (Credit: NASA; SWIFT Science Center); (b) BAT spectrum of a 60-keV γ-ray line from an americium-241 calibration source. (From Swift Technical Handbook (2020), credit NASA, SWIFT Science Center.)

1–4 arcminutes. When a GRB has been detected, the spacecraft is immediately slewed to point the two narrow-FOV instruments, the XRT and the UVOT, at the source for follow-up observations.

The XRT instrument has an imaging X-ray spectrometer at the focal plane of a 3.5-m focal length, grazing incidence Wolter type I telescope. The telescope has 12 gold-coated nickel shells with an effective area of 125 cm^2 at 1.5 keV, decreasing to 20 cm^2 at 8 keV. The detector is a 600×600 pixel e2v CCD-22 (similar to those used in the XMM-Newton EPIC-MOS camera), with an energy resolution of 140 eV at 6 keV. The XRT is sensitive in the 0.2–10 keV energy range and can locate the target position to around 3″. The limiting flux for a 10-ks integration is 2×10^{-14} erg cm^{-2}.

The UVOT imager has a 30-cm aperture Ritchey-Chretien telescope and a 2048×2048 CCD camera with six filters spanning a wavelength range of 170–650 nm. It has a limiting magnitude of approximately 24 in a 1000-second integration and 0.3″ positional accuracy. The instrument also has a grism spectrometer with a resolution of 200 at 400 nm.

11.11.3 The Fermi Large Area Telescope

The Large Area Telescope (LAT) (Atwood et al. 2007; 2009) on board NASA's Fermi satellite, is a high-energy γ-ray telescope based on electron-positron pair creation (as in Section 11.8.4), and covering the energy range from about 20 MeV to 300 GeV. The LAT, shown schematically in Figure 11.39, has 18 tungsten planes and 16 silicon strip detector planes, arranged in 16 tracker modules. The first 12 tracker planes have tungsten plates 0.035 radiation lengths thick in front of the detectors, the next 4 tungsten plates are 0.18 radiation lengths thick, and the last two plates, immediately in front of the calorimeter, do not have any converters. Cosmic ray events in the tracker modules can be reconstructed from the resulting tracks, allowing the identification of the type of particle as well as its energy and incident direction.

Behind each tracker module is a calorimeter module, containing 96 CsI(Tl) scintillator crystals, in eight alternating orthogonal layers, and read out by photodiodes. A segmented anticoincidence detector (ACD) surrounds the tracker and calorimeter assembly. It comprises 89 plastic scintillator tiles. When electromagnetic cascades occur in the calorimeter, this can lead to lower-energy photons that can cause detections in the shield, leading to false vetoes. Segmenting the shield reduces the false-veto rate because only the segment nearest the incoming photon candidate needs to be considered. The ACD has better than 99.97% efficiency for detection and rejection of charged particles entering the LAT field-of-view.

The charged particle background is 10^2–10^5 times greater than the γ-ray flux. To avoid an excessively high data rate, the LAT on-board data acquisition system filters the events to reduce the number that are telemetered to the ground. The on-board analysis reduces the raw trigger rate, which

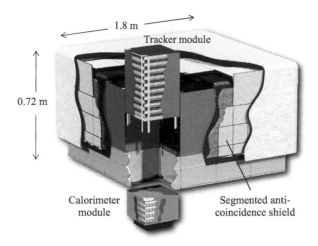

FIGURE 11.39   The Fermi Large Area Telescope (LAT) instrument. (Credit: NASA Goddard Space Flight Center.)

can approach 10 kHz, to ~400 events per second which are sent to the ground for further analysis. Of these 400, less than 5 will be due to astrophysical photons.

The LAT has a very large field of view – approximately 2.4 sr. For single photons, the angular resolution is better than 10 arcminutes above 10 GeV, decreasing to ~36 arcminutes at 1 GeV. Its energy resolving power is 10%–15% at 100 MeV, improving to < above 10 GeV. One of the LAT's most important scientific capabilities is the rapid detection of GRBs. The occurrence of a GRB can be notified to the spacecraft within 5 seconds for follow-up with other instruments.

11.11.4   THE VERITAS ČERENKOV TELESCOPE ARRAY FOR HIGH-ENERGY GAMMA-RAY ASTRONOMY

Several ground-based observatories carry out high-energy γ-ray astronomy by observing the ultra-violet Čerenkov light from EASs, as described in Section 11.9.1. One such observatory is Very Energetic Radiation Imaging Telescope Array System (VERITAS), located at Mt. Hopkins in Arizona. VERITAS can detect γ-rays with energies in the range from 85 GeV up to more than 30 TeV, achieving its best performance between 100 GeV and 10 TeV. It can operate between September and June, when the weather is dry enough, and results are best on moonless nights, when the background sky brightness is lowest.

A narrow telescope beam is not essential for Čerenkov observations, and the telescopes can be basic light-buckets, making them lighter and cheaper to manufacture. VERITAS (Holder et al. 2006; Park 2015) has four identical 12-m diameter alt-az telescopes, each consisting of 350 individual hexagonal mirrors made of aluminised glass with over 90% reflectivity at 320 nm wavelength. The individual facets are arranged to form a 12-m diameter spherical surface with a 24-m radius of curvature, corresponding to a focal length of 12 m. Each facet can be individually aligned, and the size of the well-aligned telescope point-spread function is around 3.6 arcminutes. The pointing accuracy of the telescopes is about 1.2 arcminutes. Figure 11.40 is a photograph of the VERITAS observatory showing the four telescopes and buildings for accommodation and operation.

The telescopes are each equipped with a camera comprising an array of 499 PMTs optimised for UV sensitivity. PMTs are the best detectors for this application because of their fast response time and the ability to cover the large focal plane area at low cost. The VERITAS cameras use R10560-100-20 MOD Hamamatsu PMTs, described by Otte et al. (2011), and have a quantum efficiency of over 30% at 320 nm. They have eight dynode stages providing a gain of 2×10^5 at a bias voltage of 1100 V. Fast preamplifiers are integrated in the PMT housing to boost the signals before transmission

FIGURE 11.40 The VERITAS observatory. (Credit: Stephen Criswell, Smithsonian Astrophysical Observatory.)

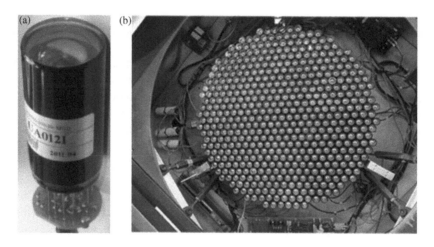

FIGURE 11.41   (a) An individual VERITAS PMT (Otte et al. 2011). (Credit: N. Otte.) (b) A complete 499-element array (Kieda et al. 2013). (Credit D. Kieda.)

to processing electronics. Each PMT is 1 inch in diameter and is equipped with a reflective light cone to concentrate the incoming light onto the PMT, to ensure no dead space between the pixels, and to restrict the pixel field of view to the telescope (reducing background light). The pixel size is 9 arcminutes on the sky, and the total field of view is 3.5° in diameter. Figure 11.41 shows photographs of a single PMT and of a complete 499-element array.

The PMT signals are digitised by fast 8-bit ADCs with 2-ns sampling intervals. A typical PMT response to a γ-ray event is shown in Figure 11.42. When triggered, 24 samples (48 ns) are stored for each detector.

Simultaneous observations by all four telescopes are used to reconstruct the properties (direction and energy) of gamma-ray photons and to reject the much more frequent background events due to sky brightness fluctuations (which dominate at low energies) and cosmic rays, especially muons (which dominate at higher energies). Muon events tend to be at lower altitudes and typically produce strong signals in only one telescope. A three-stage trigger system is adopted with cuts made at the level of single pixels, groups of pixels, and the whole four-telescope array. The nominal trigger conditions to accept an event are simultaneous detections by at least two telescopes, each with sufficiently strong signals from at least three adjacent PMTs.

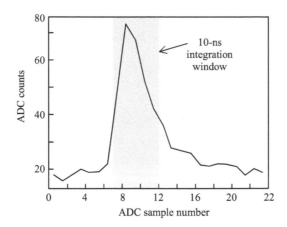

FIGURE 11.42   Typical VERITAS PMT response to a γ-ray event. (From Holder et al. 2006, reproduced with permission from Elsevier.)

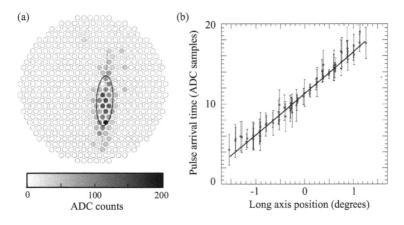

FIGURE 11.43   (a) Distribution of signals recorded by a VERITAS camera for a typical γ-ray event. (b) Typical pulse arrival time vs event position along the long axis of the event. (Both figures from Holder et al. 2006, reproduced with permission from Elsevier.)

As noted in Section 11.9, γ-ray air showers are much more compact than cosmic ray induced showers. Figure 11.43 shows typical spatial and temporal image patterns for a γ-ray event. The long axis of the ellipse points to location of the source as projected on the ground, and the intersection of the event directions recorded by the different telescopes allows the distance of that point from the system to be determined. The angular distance of the ellipse from the source position on the sky indicates the distance to the core position on the ground. The image shape and distance are the main method of identifying γ-ray events.

Single telescope (mostly muon) events are far more common than multiple telescope events and have different length:amplitude characteristics as illustrated in Figure 11.44, which allows them to be rejected very effectively.

Sources are observed in ON-OFF mode, with typically 30 minutes ON and 30 minutes OFF.

To calibrate the measured signals in terms of γ-ray energy, the overall photon conversion efficiency of the system must be known. This depends on mirror reflectivity, the PMT light collection efficiency and photocathode quantum efficiency, and the properties of the electronics (counts per photoelectron). All are carefully measured, with a typical overall conversion factor of 0.2 photo-electrons per digital count. Numerical simulations of air showers and of the detailed response of the

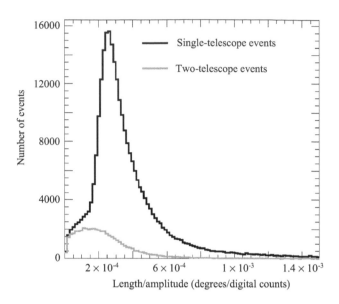

FIGURE 11.44 Black – frequency distribution of events as a function of the length:size ratio, recorded by one telescope only; grey – distribution of events common to two telescopes. (From Holder et al. 2006, reproduced with permission from Elsevier.)

system are used to recreate the characteristics and parameters of the images produced by different kinds of events with different energies. Spectral reconstruction of the event can be achieved from 150 GeV upwards, with an energy resolution of ~ 17% at 1 TeV.

The angular resolution achieved by VERITAS depends on photon energy and is 0.13° at 200 GeV, improving to around 0.08° at 1 TeV. This allows point sources to be identified unambiguously and resolution of structure in some extended sources.

The effective area of the VERITAS observatory is ~ $10^5\,\mathrm{m}^2$, much larger than the telescope collecting area because the observatory is using a large region of the Earth's atmosphere as the initial detector. The Crab Nebula, a supernova remnant, is used in γ-ray astronomy as the standard calibration source, with a spectrum given by $3.3 \times 10^{-7}\,E^{-2.4}$ photons $\mathrm{m}^{-2}\mathrm{s}^{-1}\,\mathrm{TeV}^{-1}$, with E in TeV, corresponding to 2.1×10^{-7} photons $\mathrm{m}^{-2}\mathrm{s}^{-1}$ above 1 TeV (Hillas et al. 1998). A source of 1% Crab strength (one γ-ray photon every ~ 80 minutes) can be detected in 25 hours. Integration times over 48 hours can be built up by combining observing over several nights.

Following the successful operation of VERITAS and two other Čerenkov arrays (HESS and MAGIC), a new and more sensitive facility, the Čerenkov Telescope Array (CTA) is currently being developed (Mazin 2019). There will the two CTA arrays, one in the northern hemisphere and one in the south, each array having four large (23-m) telescopes capable of detecting photons down to 20 GeV energy. The southern array will have an additional 25 medium-sized (12-m) telescopes for 0.1–10 TeV and over 70 small-sized telescopes (a few m^2 in area) for the highest energies. The northern array will have 15 medium-sized telescopes. CTA will provide an order of magnitude improvement over current facilities and extend the observable energy range beyond 100 TeV.

11.11.5 The IceCube Neutrino Observatory

The IceCube neutrino observatory is located at the South Pole, and uses Antarctic ice, which is several km thick, as the detector. It is designed to observe high-energy neutrinos (between 10 GeV and 100 TeV) produced by the interaction of cosmic rays in the emitting sources, in interstellar space, and in the Earth's atmosphere, and also to carry out research in particle physics through exploring the behaviour of neutrinos as fundamental particles. The structure of IceCube is shown in

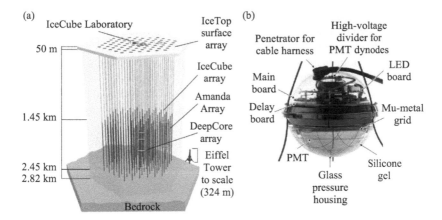

FIGURE 11.45 (a) The IceCube neutrino observatory. (b) Structure of a digital optical module (DOM). (Credit: IceCube Collaboration.)

Figure 11.45. The active detector volume is ~ 1 km³, located between approximately 1.5 and 2.5 km below the surface. At this depth scattering of light by small bubbles trapped in the ice is eliminated because the pressure at that depth becomes too great for the bubbles to exist. Čerenkov light from the particles created by neutrino interactions is detected by an array of PM tubes. Each PM tube is contained in its own Digital Optical Module (DOM), also illustrated in Figure 11.45. There are 5160 DOMs, each with a 25-cm diameter hemispherical Hamamatsu R7081-02 PM tube and associated control and readout electronics that detect and accurately time-stamp the light waveforms. All of the DOM components are enclosed in a hermetically sealed glass sphere.

The DOMs are attached to 86 strings in boreholes, most spaced about 125 m apart. The vertical spacing between most of the DOMs is 17 m. Eight of the strings in the centre have a closer spacing and also have DOMs with a smaller 7-m vertical separation (also higher sensitivity). These more densely packed strings and the surrounding standard ones constitute the DeepCore sub-array, which provides sensitivity to neutrinos down to ~ 10 GeV. After installation of the strings and DOMs immediately on completion of hot-water drilling, the water in the boreholes froze, as planned so that everything was locked in position. Because they cannot be accessed after installation, the DOMs had to be designed for long lifetime and very high reliability (better than 90% survival over 15 years). They also have low (<5 W) power dissipation (since electrical power is limited at the South Pole station) and work effectively at temperatures down to −55 C.

A layer of silicone gel between the PMT and the glass DOM enclosure ensures good light coupling. The PMTs are sensitive between 300 and 650 nm, with 25% peak quantum efficiency. They have ten dynodes with a total gain of ~ 10⁷ gain and dark noise is <500 counts s⁻¹. Large-area PMTs must have magnetic shielding to prevent signal loss due to Lorentz-force deflection of the electrons in transit. A ferromagnetic nickel–iron alloy (mu-metal) wire cage shields the PMT from the Earth's magnetic field. Each DOM is equipped with a set of light-emitting diodes (LEDs) for calibration.

Within the DOM, the PMT output is delayed by 75 ns to allow sampling electronics to receive and react to the event trigger signal. The DOMS digitise and time-stamp the event waveforms with <3 ns resolution, using a local quartz oscillator clock which is calibrated with respect to a master clock at the surface. This allows the digitised information to be transmitted over the long and variable distance to the surface using simple twisted-pair wiring. The DOM signals are processed at the surface to identify the event type and derive its parameters.

At the surface, an array of additional detectors, known as IceTop, consists of 81 ice tanks with downward-looking PMTs and acts as an anticoincidence veto to distinguish different kinds of events, especially in identifying those due to cosmic ray air showers with energy above ~ 300 TeV

The muon trajectory is determined from the arrival times of photons at the optical sensors, and the number of photons detected allows the estimation of the deposited energy. For km-length muon tracks, the source direction can be identified with an angular resolution of better than 1°. IceCube detects ~10 neutrinos per hour in a background of ~ 3000 atmospheric cosmic ray muons per second. The vast majority of detected neutrinos also originate from cosmic rays – only a few tens are astrophysical neutrinos. Above ~100 TeV, the atmospheric neutrino flux is too small to produce any event in the detection volume, so every event above that energy is likely to be of astrophysical origin.

Bibliography

Ade, P. A. R., et al., (BICEP2 Collaboration), BICEP2 II: Experiment and three-year data set, *Astrophysical Journal*, 792, 62, 2014.

Ade, P. A. R., et al., Antenna-coupled TES bolometers used in BICEP2, Keck Array, and SPIDER, *Astrophysical Journal*, 812, 176, 2015.

Audley, M. D., et al., SCUBA-2: A large-format TES array for submillimetre astronomy, *Nuclear Instruments and Methods in Physics Research A*, 520, 479, 2004.

Ageron, M., et al., ANTARES: The first undersea neutrino telescope, *Nuclear Instruments and Methods in Physics Research A*, 656, 11, 2011.

Akamatsu, H., et al., Progress in the development of frequency domain multiplexing for the X-ray Integral Field Unit on board the Athena mission, *Journal of Low Temperature Physics*, 199, 737, 2020.

Aschenbach, B., X-ray telescopes, *Reports on Progress in Physics*, 48, 579, 1985.

Atwood, W. B., et al., Design and initial tests of the tracker-converter of the gamma-ray large area space telescope, *Astroparticle Physics*, 28, 422, 2007.

Atwood, W. B., et al., The large area telescope on the Fermi gamma-ray space telescope mission, *Astrophysical Journal*, 697, 1071, 2009.

Barnstedt, J, Advanced Practical Course: Microchannel Plate Detectors, University of Tübingen, 2019.

Barret, D., et al., An Athena+ supporting paper: The X-ray Integral Field Unit (X-IFU) for Athena, arXiv1308.6784, 2013.

Baselmans, J., et al., A kilo-pixel imaging system for future space based far-infrared observatories using microwave kinetic inductance detectors, *Astronomy & Astrophysics*, 601, A89, 2017.

Bavdaz, M., et al., X-Ray pore optics technologies and their application in space telescopes, X-Ray Optics and Instrumentation, 2010, 295095, 2010.

Bavdaz, M., et al., The Athena optics, *Proceedings of SPIE*, 9603, 96030J, 2015.

Beckers, J. M., Adaptive optics for astronomy: Principles, performance, and applications, *Annual Reviews of Astronomy and Astrophysics*, 31, 13, 1993.

Berger, M. J., et al., XCOM: Photon cross section database (version 1.5), National Institute of Standards and Technology, Gaithersburg, MD, 2010, available at https://www.nist.gov/pml/xcom-photon-cross-sections-database.

Bernlöhr, K., Simulation of imaging atmospheric Cherenkov telescopes with CORSIKA and sim_telarray, *Astroparticle Physics*, 30, 149, 2008.

Bessell, M. S., Standard photometric systems, *Annual Review of Astronomy & Astrophysics*, 43, 293, 2005.

Booth, R., Goedhart, S., & Jonas, J., MeerKAT and its potential for cosmic MASER research, *Proceedings of the IAU Symposium*, 287, 483, 2012.

Booth, R., & Jonas, J., An overview of the MeerKAT project, *African Skies*, 16, 101, 2012.

Born, M., & Wolf, E., *Principles of Optics* (6th edition), Pergamon, Oxford, 1980.

Bèland, S., Boulade, O., & Davidge, T., The extinction curve at Mauna Kea in the visible range, *Canada-France-Hawaii Telescope Information Bulletin*, 19, 16, 1988.

Bracewell, R. N., *The Fourier Transform and its Applications*, McGraw-Hill, New York, 1999.

Canada-France-Hawaii Telescope, Observers Manual, Version 1.0, January, 2003, available at https://www.cfht.hawaii.edu/Instruments/ObservatoryManual/index.html.

Catalano, A., et al., The NIKA2 instrument at 30-m IRAM telescope: Performance and results, *Journal of Low Temperature Physics*, 193, 916, 2018.

Chapman, I. M., et al., BTRAM: An interactive atmospheric radiative transfer model, *30th Canadian Symposium on Remote Sensing*, 30, 22–25, 2009.

Church, S. E., et al., Performance testing of doped-germanium photoconductors for the ISO long wavelength spectrometer, *ESA SP356*, 255–260, 1992.

Condon, J. J., & Ransom, S. M., *Essential Radio Astronomy*, Princeton University Press, Princeton, NJ, 2016.

D'Addario, L. R., Synthesis imaging in radio astronomy, Lecture 4: Cross correlators, *Astronomical Society of the Pacific Conference Series*, 6, 59, 1989.

Davies, R., et al., MICADO: First light imager for the E-ELT, *Proceedings of SPIE*, 9908, 99081Z, 2016.

Davies, R., et al., The MICADO first light imager for the ELT: Overview, operation, simulation, *Proceedings of SPIE*, 10702, 107021S, 2018.

de Graauw, Th., et al., The *Herschel*-Heterodyne Instrument for the Far-Infrared (HIFI), *Astronomy & Astrophysics*, 518, L6, 2010.

Dere, K. P., et al., CHIANTI – an atomic database for emission lines, *Astronomy & Astrophysics Supplement Series*, 125, 149, 1997, available at http://www.chiantidatabase.org/.

Dicker, S, et al., MUSTANG2: A large focal plane array for the 100-meter Green Bank Telescope, *Proceedings of SPIE*, 9153, 91530J, 2014.

Dressel, L., *Wide Field Camera 3 Instrument Handbook, Version 12.0*, STScI, Baltimore, MD, 2019.

Evans, R. D., *The Atomic Nucleus*, McGraw-Hill, New York, 1955.

Formalont, E. B., & Perley, R. A., Chapter 5: Calibration and editing, synthesis imaging in radio astronomy II, *Astronomical Society of the Pacific Conference Series*, 180, 79, 1999.

Frayer, D. T., et al., The GBT 67–93.6 GHz Spectral line survey of Orion-KL, *Astronomical Journal*, 149, 162, 2015a.

Frayer, D. T., et al., Erratum: The GBT 67–93.6 GHz spectral line survey of Orion-KL, *Astronomical Journal*, 150, 39, 2015b.

Gaisser, T., & Halzen, F., IceCube, *Annual Review of Nuclear and Particle Science*, 64, 101, 2014.

Gehrels, N., et al., The Swift gamma ray burst mission, *Astrophysical Journal*, 611, 1105, 2004.

Glasser, C. A., Odell, C.E., & Seufert, S. E., The proportional counter array (PCA) instrument for the X-ray Timing Explorer Satellite (XTE), *IEEE Transactions on Nuclear Science*, 41, 1343, 1994.

Goldsmith, P. F., *Quasioptical Systems*, Wiley-IEEE Press, New York, 1998.

Grace, E. A., et al., Characterization and performance of a kilo-TES sub-array for ACTPol, *Journal of Low Temperature Physics*, 176, 705, 2014.

Grannan, S. M., Richards, P. L., & Hase, M. K., Numerical optimization of bolometric infrared detectors including optical loading, amplifier noise, and electrical nonlinearities, *International Journal of Infrared and Millimeter Waves*, 18, 319, 1997.

Green, M. A., & Keevers, M. J., Optical properties of intrinsic silicon at 300 K, *Progress in Photovoltaics Research and Applications*, 3, 189, 1995.

Griffin, M. J., Bock, J. J., & Gear, W. K., Relative performance of filled and feedhorn-coupled focal-plane architectures, *Applied Optics*, 41, 6543, 2002.

Griffin, M. J., et al., The *Herschel*-SPIRE instrument and its in-flight performance, *Astronomy & Astrophysics*, 518, L3, 2010.

Griffin, M. J., et al., Flux calibration of broad-band far-infrared and submillimetre photometric instruments: Theory and application to *Herschel*-SPIRE, *Monthly Notices of the Royal Astronomical Society*, 434, 992, 2013.

Griffiths D.J., *Introduction to Quantum Mechanics* (2nd edition), Pearson Prentice Hall, Upper Saddle River, NJ, 2005.

Gullikson, E. M., Specular reflectivities for grazing-incidence mirrors, *X-Ray Data Booklet Section 4.2*, Lawrence Berkeley Laboratory, Berkeley, CA, 2009, available at https://xdb.lbl.gov/.

Halzen, F., High-energy neutrino astrophysics, *Nature Physics*, 13, 232, 2017.

Hanel R.A., et al., *Exploration of the Solar System by Infrared Remote Sensing*, Cambridge University Press, Cambridge, 2003.

Hansen, G. L., Schmidt, J. L., & Casselman, T. N., Energy gap versus alloy composition and temperature in $Hg_{1-x}Cd_xTe$, *Journal of Applied Physics*, 53, 7099, 1982.

Harper, D. A., et al., HAWC+, the far-infrared camera and polarimeter for SOFIA, *Journal of Astronomical Instrumentation*, 7, 1840008, 2018.

Hecht E., *Optics* (4th edition), Addison Wesley, Boston, MA, 2002.

Henderson, S. W., et al., Readout of two-kilopixel transition-edge sensor arrays for Advanced ACTPol, *Proceedings of SPIE*, 9914, 99141G, 2016.

Henderson, S. W., et al., Highly-multiplexed microwave SQUID readout using the SLAC Microresonator Radio Frequency (SMuRF) electronics for future CMB and sub-millimeter surveys, *Proceedings of SPIE*, 10708, 1070819, 2018.

Henke, B. L., Gullickson, E. M., and Davis, J. C., X-ray interactions: Photoabsorption, scattering, transmission and reflection at E = 50 – 30,000 eV, Z = 1–92, *Atomic Data and Nuclear Data Tables*, 54, 181, 1993.

Heyminck, S., et al., GREAT: The SOFIA high-frequency heterodyne instrument, *Astronomy & Astrophysics*, 542, L1, 2012.

Hill, C., et al., HITRAN*online*: An online interface and the flexible representation of spectroscopic data in the HITRAN database, *Journal of Quantitative Spectroscopy and Radiative Transfer*, 177, 4, 2016, available at https://hitran.org/.

Hillas, A. M., The spectrum of TeV gamma rays from the Crab Nebula, *Astrophysical Journal*, 503, 744, 1998.

Högbom, J. A., Aperture synthesis with a non-regular distribution of interferometer baselines, *Astronomy & Astrophysics*, 15, 417, 1974.

Holder, J., et al., The first VERITAS telescope, *Astroparticle Physics*, 25, 391, 2006.

Holland, A., Chapter 24, X-ray CCDs, in *Observing Photons in Space: A Guide to Experimental Space Astronomy*, Huber, M. C. E., Pauluhn, A., Culhane, J. L., Timothy, J. G., Wilhelm, K., Zehnder, A. (Eds.), ISSI Scientific Reports Series, ESA/ISSI, New York, 443–453, 2010.

Holland, W. S. et al., SCUBA-2: The 10,000 pixel bolometer camera on the James Clerk Maxwell Telescope, *Monthly Notices of the Royal Astronomical Society*, 430, 2513, 2013.

Hollas, J. M., *Modern Spectroscopy* (4th edition), Wiley, Hoboken, NJ, 2004.

Honsberg, C. B., and Bowden, S. G., Photovoltaics Education Website, 2019, available at www.pveducation. org.

Irwin, K., SQUID multiplexers for transition-edge sensors, *Physica C Superconductivity*, 368, 203, 2002.

Irwin, K. D., & Hilton, G. C., Transition-edge sensors, in *Cryogenic Particle Detection*, Enss, C. (Ed.), Topics in Applied Physics, Vol. 99, 63, Springer-Verlag, 2005.

Jackson, N., Principles of interferometry, *Jets from Young Stars II*, Bacciotti, F., Testi, L., Whelan, E. (Eds.), Lecture Notes in Physics, Springer-Verlag, New York, Vol. 742, 193, 2008.

Jacquinet-Husson, N., et al., The 2015 edition of the GEISA spectroscopic database, *Journal of Molecular Spectroscopy*, 327, 31, 2016, available at https://geisa.aeris-data.fr/.

Jelley, J. V., *Čerenkov Radiation and Its Applications*, Pergamon Press, United Kingdom Atomic Energy Authority, London, 1958.

Kanbach, G., et al., OPTIMA: A high time resolution optical photo-polarimeter, in: Phelan, D., Ryan, O., Shearer A. (Eds.), *High Time Resolution Astrophysics, Astrophysics and Space Science Library*, 351, Springer, New York, 2008.

Kieda, D. B., and the VERITAS collaboration, The gamma ray detection sensitivity of the upgraded VERITAS Observatory, *Proceedings of 33rd International Cosmic Ray Conference*, Rio de Janeiro, arXiv:1308.4849, 2013.

Klein, B., et al., The APEX digital fast Fourier transform spectrometer, *Astronomy & Astrophysics*, 454, L29, 2006.

Kramida, A., et al., *NIST Atomic Spectra Database (version 5.7.1)*, National Institute of Standards and Technology, Gaithersburg, MD, 2019, available at https://www.nist.gov/pml/atomic-spectra-database.

Kraus, J. D., *Radio Astronomy* (2nd edition), Cygnus-Quasar Books, Powell, OH, 1986.

van der Kuur, J., et al., Multiplexed readout demonstration of a TES-based detector array in a resistance locked loop, *IEEE Transactions on Applied Superconductivity*, 25, 3, 2393716, 2015.

Landau, L. D., and Lifshitz, E. M., *Statistical Physics*, Butterworth-Heinemann, Oxford, 2013.

LeSurf, J, *Millimetre-wave Devices, Optics and Systems*, CRC Press, Boca Raton, FL, 1990.

Lord, S. D., A new software tool for computing Earth's atmospheric transmission of near- and far-infrared radiation, NASA Technical Memorandum, 103957, 1992.

Love, P., et al., 1024 x 1024 Si:As IBC detector arrays for JWST MIRI, *Proceedings of SPIE*, 5902, 590209, 2005.

Mather, J. C., Bolometer noise: Nonequilibrium theory, *Applied Optics*, 21, 1125, 1982.

Mazin, D., The Chernkov Telescope Array, *Proceedings of 36th International Cosmic Ray Conference*, ICRC2019, 741, 2019.

Meservey, R. & Tedrow, P.M., Measurements of the kinetic inductance of superconducting linear structures, *Journal of Applied Physics*, 40, 2028, 1969.

Miniussi, A. R., et al., Performance of an X-ray microcalorimeter with a 240 µm absorber and a 50 µm TES bilayer, *Journal of Low Temperature Physics*, 193, 337, 2018.

MIPS Instrument Handbook, Version 3, March 2011, Spitzer heritage archive documentation, available at https://irsa.ipac.caltech.edu/data/SPITZER/docs/mips/mipsinstrumenthandbook/.

National Radio Astronomy Observatory. GBT Observers' Guide for the 4-mm Receiver, 2014, available at http://www.gb.nrao.edu/4mm/.

National Radio Astronomy Observatory. GBT 4-mm Receiver Project Book, 2017, available at http://www.gb.nrao.edu/4mm/.

Norton, R. H., & Beer, R., New apodizing functions for Fourier spectrometry, *Journal of the Optical Society of America*, 66, 259, 1976.

NuSTAR Observatory Guide, Version 3.2, June 2016, available at https://heasarc.gsfc.nasa.gov/docs/nustar/nustar_obsguide.pdf.

O'Brien, P., ASTROSAT, *Advances in Space Research*, 47, 1451, 2011.

Otte, A. N., and the VERITAS collaboration, The upgrade of VERITAS with high efficiency photomultipliers, *Proceedings of 32nd International Cosmic Ray Conference*, Beijing, August 11–18 2011, Vol 9, 247, 2011.

Park, N., Performance of the VERITAS experiment, *Proceedings of 34th International Cosmic Ray Conference (ICRC)*, arXiv:1510.01639, 2015.

Perley, R., Fundamentals of radio interferometry, David Dunlap Summer School, 2019, available at http://www.dunlap.utoronto.ca/wp-content/uploads/2019/07/Dunlap2019.pdf.

Pierre Auger Collaboration, The Pierre Auger observatory, *Nuclear Instruments and Methods*, 798, 172, 2015.

Poglitsch, A., et al., The photodetector array camera and spectrometer (PACS) on the *Herschel* Space Observatory, *Astronomy & Astrophysics*, 518, L2, 2010.

Rasmussen, A. P., et al., The X-ray spectrum of the supernova remnant 1E 0102.2–7219, *Astronomy & Astrophysics*, 365, L231, 2001.

Rigaut, F., & Neichel, B., Multiconjugate adaptive optics for astronomy, *Annual Review of Astronomy and Astrophysics*, 56, 277, 2018.

Rieke, G. H., et al., The mid-infrared instrument for the James Webb Space Telescope, VII: The MIRI detectors, *Publications of the Astronomical Society of the Pacific*, 127, 665, 2015.

Risacher, C., et al., The upGREAT dual frequency heterodyne arrays for SOFIA, *Journal of Astronomical Instrumentation*, 7, 1840014, 2018.

Roderigo, C., & Solano, E., SVO Filter Profile Service, Version 1.0, International Virtual Observatory Alliance Note, 15 October 2012, available at http://ivoa.net/documents/Notes/SVOFPS/index.html.

Roderigo, C., & Solano, E., Filter Profile Service Access Protocol, Version 1.0, International Virtual Observatory Alliance Note, 10 May 2013, available at http://ivoa.net/documents/Notes/SVOFPSDAL/index.html.

Rogalski, A., HgCdTe photodetectors, *Mid-infrared Optoelectronics*, Tournié, E., Cerutti, L. (Eds.), Woodhead Publishing, Kidlington, UK, 235–335, 2020.

Rohlfs, K, *Tools of Radio Astronomy*, Springer-Verlag, New York, 1986.

Ruze, J., Antenna tolerance theory – a review, *Proceedings of IEEE*, 54, 633, 1966.

Schönfelder, V., et al., Instrument description and performance of the imaging gamma-ray telescope COMPTEL aboard the Compton Gamma-Ray Observatory, *Astrophysical Journal Supplement Series*, 86, 657, 1993.

Sharples, R. M., et al., KMOS: An infrared multiple object integral field spectrograph for the ESO VLT, *Proceedings of SPIE*, 5492, 1179, 2004.

Siklitsky, V., & Tolmatchev, A., Electronic Archive, New Semiconductor Materials Characteristics and Properties, Ioffe Physico-Technical Institute, 2019, available at http://www.ioffe.ru/SVA/.

Smith, D., Chapter 21, Hard X-ray and γ-ray detectors, *Observing Photons in Space: A Guide to Experimental Space Astronomy*, Huber, M. C. E., Pauluhn, A., Culhane, J. L., Timothy, J. G., Wilhelm, K., Zehnder, A. (Eds.), ISSI Scientific Reports Series, ESA/ISSI, New York, 2010.

Sprafke, T., & Beletic, J. W., High-performance infrared focal plane arrays for space applications, *Optics and Photonics News*, 19(6), 22, 2008.

Strüder, L., et al., A 36 cm² large monolithic pn-charge coupled device X-ray detector for the European XMM satellite mission, *Review of Scientific Instruments*, 68, 4271, 1998.

Strüder, L., et al., The European photon imaging camera on XMM-Newton: The pn-CCD camera, *Astronomy & Astrophysics*, 365, L18, 2001.

Sudiwala, R. V., et al., Thermal modelling and characterisation of semiconductor bolometers, *International Journal of Infrared and Millimeter Waves*, 23, 545, 2002.

Swift Technical Handbook, Version 17.0, 2020, available from the NASA Swift Science Center at https://swift.gsfc.nasa.gov/proposals/tech_appd/swiftta_v17.pdf.

Sze, S. M., *Semiconductor Devices: Physics and Technology* (2nd edition), Wiley, Hoboken, NJ, 2002.

Teledyne e2V, CCD231–84 Back Illuminated Scientific Sensor, Version 8, August 2018.

Teledyne Imaging Sensors, H2RG™ Visible & Infrared Focal Plane Array, September 2017.

Tomsick, J. A., The Compton Spectrometer and Imager, Astro2020 APC White Paper, arXiv:1908.04334v1, 2019.

Trippe, S., et al., High-precision astrometry with MICADO at the European Extremely Large Telescope, *MNRAS*, 402, 1126, 2010.

Turner, A. D., et al., Silicon nitride micromesh bolometer array for submillimeter astrophysics, *Applied Optics*, 40, 4921, 2001.

Turner, M. J. L., et al., The European Photon Imaging Camera on XMM-Newton: The MOS cameras, *Astronomy & Astrophysics*, 365, L27, 2001.

Unger H.G., *Introduction to Quantum Electronics*, Pergamon Press, London, 1970.

Weekes, T. C., The atmospheric Cherenkov imaging technique for very high energy gamma-ray astronomy, *Lectures given at the International Heraeus Summer School on "Physics with Cosmic Accelerators"*, Bad Honnef, Germany, July 5–16, 2004, arXiv:astro-ph/0508253.

Willingale, R., et al., The hot and energetic universe: The optical design of the Athena+ mirror, *Instrumentation and Methods for Astrophysics*, 2013arXiv1307, 1709W, 2013.

Winkler, C., et al., The INTEGRAL mission, *Astronomy & Astrophysics*, 411, L1, 2003.

Wolter, H., Spiegelsystem streifenden Einfalls als abbildende Optiken für Röntgenstrahlen. *Annalen der Physik*, 445, 94, 1952.

Wright, G. S., et al., The mid-infrared instrument for the James Webb Space Telescope, II: Design and build, *Publications of the Astronomical Society of the Pacific*, 127, 595, 2015.

XMM-Newton Users Handbook, Issue 2.18, July 2020, available at http://www.cosmos.esa.int/web/xmm-newton/documentation.

Index

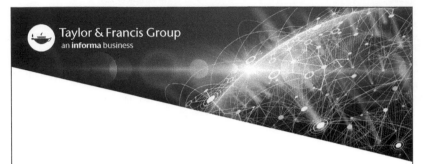

9 781032 040035